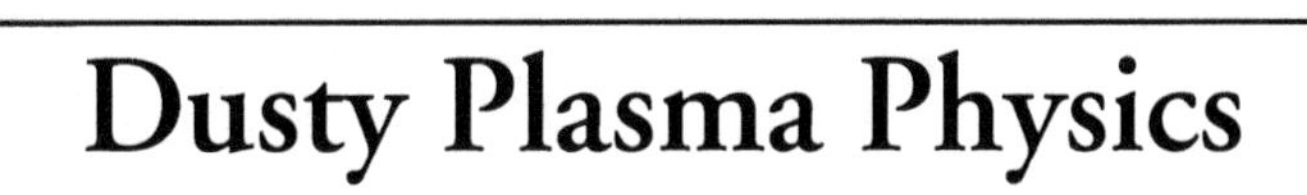
Dusty Plasma Physics

Dusty Plasma Physics

Editor: Emil Wilkinson

New York

Published by NY Research Press
118-35 Queens Blvd., Suite 400,
Forest Hills, NY 11375, USA
www.nyresearchpress.com

Dusty Plasma Physics
Edited by Emil Wilkinson

International Standard Book Number: 978-1-64725-384-4 (Hardback)

Cataloging-in-publication Data

Dusty plasma physics / edited by Emil Wilkinson.
p. cm.
Includes bibliographical references and index.
ISBN 978-1-64725-384-4
1. Dusty plasma. 2. Physics. I. Wilkinson, Emil.
QC718.5.D84 D87 2023
530.44--dc23

Contents

Preface

Every book is a source of knowledge and this one is no exception. The idea that led to the conceptualization of this book was the fact that the world is advancing rapidly; which makes it crucial to document the progress in every field. I am aware that a lot of data is already available, yet, there is a lot more to learn. Hence, I accepted the responsibility of editing this book and contributing my knowledge to the community.

A dusty plasma (or complex plasma) is plasma containing nanometer or micrometer-sized particles suspended in it. Dust particles may accumulate into larger particles resulting in grain plasmas. They exhibit the unique property to form plasma crystals, as well as liquid and crystalline states. Dusty plasmas are observed in interplanetary space dust, comets, planetary rings, dusty surfaces in space, and aerosols in the atmosphere. Currently, dusty plasmas are being researched extensively due to the presence of particles that significantly alter the charged particle equilibrium leading to different phenomena. Electrostatic coupling between the grains can differ on a broad range so that the states of dusty plasma can convert from weakly coupled (gaseous) to crystalline. Such plasmas are of interest as non-Hamiltonian systems of interacting particles. They serve as a means to study generic fundamental physics of self-organization, pattern formation, phase transitions and scaling. Different approaches, evaluations, methodologies and advanced studies on dusty plasma have been included in this book. It will help the readers in keeping pace with the rapid changes in this field.

While editing this book, I had multiple visions for it. Then I finally narrowed down to make every chapter a sole standing text explaining a particular topic, so that they can be used independently. However, the umbrella subject sinews them into a common theme. This makes the book a unique platform of knowledge.

I would like to give the major credit of this book to the experts from every corner of the world, who took the time to share their expertise with us. Also, I owe the completion of this book to the never-ending support of my family, who supported me throughout the project.

Editor

1

Tungsten Nanoparticles Produced by Magnetron Sputtering Gas Aggregation: Process Characterization and Particle Properties

Tomy Acsente, Lavinia Gabriela Carpen, Elena Matei, Bogdan Bita, Raluca Negrea, Elodie Bernard, Christian Grisolia and Gheorghe Dinescu

Abstract

Tungsten and tungsten nanoparticles are involved in a series of processes, in nanotechnology, metallurgy, and fusion technology. Apart from chemical methods, nanoparticle synthesis by plasma offers advantages as good control of size, shape, and surface chemistry. The plasma methods are also environmentally friendly. In this chapter, we present aspects related to the magnetron sputtering gas aggregation (MSGA) process applied to synthesis of tungsten nanoparticles, with size in the range of tens to hundreds of nanometers. We present the MSGA process and its peculiarities in the case of tungsten nanoparticle synthesis. The properties of the obtained particles with a focus on the influence of the process parameters over the particle production rate, their size, morphology, and structure are discussed. To the end, we emphasize the utility of such particles for assessing the environmental and biological impacts in case of using tungsten as wall material in thermonuclear fusion reactors.

Keywords: tungsten, nanoparticles, gas aggregation, nanoparticle synthesis, tungsten nanoparticle properties, fusion technology, toxicology of nanoparticles, tritium retention

1. Introduction

1.1 Tungsten nanoparticles: their applications and methods for their synthesis

Tungsten (named also as Wolfram—W) is a material presenting extreme physical and chemical properties, with applications in diverse domains, starting from common life up to high technology. First commercial applications of W started at the beginning of twentieth century when it was used for lighting bulb filaments and in steel alloys. Soon, the domain of applications of W widened, and today it is used in many fields like lighting, metallurgy, electronics, aviation, medicine, weapons, and so on. An extended description of W properties and applications may be found in literature [1, 2].

Tungsten is also considered to be used as wall material for nuclear fusion reactors, being considered "the best, if not the only, material to withstand the extraordinary operating conditions in a nuclear fusion reactor divertor" [1]. The divertor is situated at the bottom part of the reactor and it is responsible for extraction (during the reactor operation) of the heat and ash produced by the fusion reaction. The heat flux in this region is expected to be in between 10 and 20 MW/m^2 [3] in the International Thermonuclear Experimental Reactor (ITER), leading to local temperatures exceeding 1300°C [4]. Considering the high melting point of W (3422°C), it was selected as material for covering the divertor region in the current design of ITER [5, 6]. Additional properties of W supporting this application of W are: its high thermal conductivity (for removal of heat), low thermal expansion coefficient and low Young modulus (for minimizing the mechanical stresses) and low sputtering yield (for keeping low contamination of the plasma) [1]. Still, the dust generated due to plasma wall interaction may have critical effects on the functioning of the reactor, leading to: (i) cooling of the fusion plasma [7], (ii) accumulation of the fuel in the dust and thus generating possible radioactive safety issues [8] and (iii) biological safety issues regarding dust spreading in case of nuclear accidents such as loss of vacuum accidents [9]. Tests over these effects may be performed using W particles (dust, with dimension starting from nanometric and up to micron range [10, 11]). In addition, tungsten nanoparticles (W NPs) are used in different technological fields, among them being mentioned sintering metallurgy [12], biomedical applications [13] and also microelectronic [14] and spintronic applications [15]. Also, nanoparticles of W oxide (WO_3) have applications including gas sensing [16], photocatalysis (including pollutant degradation [17] and water splitting [18]).

W nanoparticles may be obtained in different ways, which are mentioned bellow. Currently in literature are mentioned various methods for producing W NPs, using different types of chemical and physical processes like: chemical [19] or solvothermal decomposition [20] of W based compounds, mechanical milling [21], vaporization of W precursors in thermal plasma [22], metallic wires explosion [23], laser ablation in liquids [24, 25], growth of W dust in sputtering discharges (i.e. complex plasmas) [26, 27], and by magnetron sputtering combined with gas aggregation (MSGA) [28, 29].

In this chapter, we will focus on the MSGA as method applied for synthesis of W nanoparticles. MSGA method is based on condensation in an inert gas flow of the supersaturated metallic vapors obtained from a magnetron discharge. It was first proposed in 1980's and it was aimed to growth *"strongly adhering thin films formed on room-temperature substrates"* by bombarding the substrate with energetic clusters; the method was first named *"energetic cluster impact"* by the authors [30]. This synthesis method was not found feasible for industrial thin films deposition, due to low efficiency of the process [31, 32]. Despite of this, it is applied to the moment by many research groups for synthesis of metallic nanoparticles (Ag, Cu, Fe, Cr, Ti, etc.) [33, 34], polymers [35] or even more complex types of structures (like bimetallic [36] or core-shell [37] nanoparticles) because, compared with other synthesis methods, MSGA has several advantages. Thus, the cluster sources based on MSGA are compatible with vacuum processes, are used in numerous technologies [38], are well controlled, and the obtained samples present a high chemical purity.

W NPs obtained by MSGA were reported for the first time by us in [28]. Further works were related to studies regarding tritium absorption and retention [39, 40] in this material and W NP cytotoxicity [41–43]. These experiments will be also presented briefly in this chapter.

1.2 Structure of the chapter

After the short Introduction presented above, the chapter continues with a short section mentioning the basic aspects of the process. Then, a typical system used for synthesis of nanoparticles, particularized in the case of Tungsten is discussed. We point out on the experimental system geometry and on the most important experimental parameters. These include the chemical nature of the working gas, the values of the electrical power applied to the discharge, and the thickness of the target. We present two characteristics, important for any material synthesis process: the deposition rate and the dimension of the deposit on the substrate. Thus, we point out on the decrease (down to near zero) of the synthesis rate in time (tens of minutes) if only Ar is used as working gas. The method for increasing and stabilizing the synthesis rate is also presented.

One of the sections is devoted to the properties of synthesized materials, respectively of the W NPs (including their morphology, size, and structure), obtained at various working parameters. Herewith, the effect of residual gases, and H_2 addition in the discharge over the nanoparticles morphology are emphasized.

Finally, we briefly describe some applications of the W NPs, namely those related to fusion technology and to cytotoxicity evaluation.

2. Formation of particles during magnetron sputtering and gas aggregation

In the field of material physics, clusters are defined as small, multiatoms objects; if their dimension is in between 0 and 100 nm, they are named as nanoparticles. One of the physical methods for atomic clusters (or nanoparticles) synthesis is gas aggregation. This is a bottom-up method and consists in condensation of a supersaturated vapor (obtained from the material of interest) in a flow of cold inert gas. The supersaturated vapor atmosphere is obtained by different methods, including: thermal evaporation, laser ablation, magnetron plasma sputtering, etc. All these techniques are feasible to be implemented for metals with relative low melting and evaporation points [32]. It is obviously that for W, possessing the highest melting (3422°C) and evaporation points (5930°C) from all metals, only laser ablation and magnetron plasma sputtering are feasible to be implemented. This work presents the results obtained using a magnetron sputtering based cluster source.

By sputtering, atoms are obtained from a solid target due to its bombardment with ions produced in plasma; the presence of the magnetic field crossed with the electrical one (typical for magnetron sputtering devices) leads to a high rate of atoms production; more details regarding magnetron sputtering technique may be found elsewhere [44, 45].

Formation of clusters/nanoparticles is described detailed in different textbooks [32, 46]; for the sake of clarity, we will present this process briefly here. Thus, synthesis of clusters can be considered to take place in two successive steps: nucleation and growth. Nucleation may be defined as the process by which embryos of a stable phase are formed in a surrounding thermodynamically metastable phase [46, 47]. In MSGA source, the metastable phase consists of a supersaturated vapor of metal particles, which is obtained due to the decrease of the temperature of the sputtered atoms when they depart from the target. Interaction between the supersaturated vapor of metal (Me) with the colder buffer gas (Ar) leads to nucleation of metal particles, this process being described in the term of a three-body process Eq. (1):

$$\mathrm{Me} + \mathrm{Me} + \mathrm{Ar} \rightarrow \mathrm{Me} - \mathrm{Me} + \mathrm{Ar} \tag{1}$$

where the inert gas atom (Ar) takes the excess of energy resulting from the Me–Me dimer formation; the dimers Me–Me are the embryos for the further growth of the nanoparticles. In MSGA process, the nucleation is usually homogeneous, appearing spontaneously and randomly, without preferential sites of nucleation. In contrast, in some situations, the Me–Me dimers are not stable in plasma and the nucleation is dependent on the presence of an external factor, like a reactive gas in a well-defined amount. For example, this situation is encountered during synthesis of Ti or Co nanoparticles by MSGA. Apparition of the nucleation sites is conditioned in this situation by the presence of small amounts of reactive gases (O_2, respective N_2) mixed with Ar [48]. Also, we observed that nucleation of W nanoparticles is dependent on the presence of H_2 mixed with Ar in the cluster source. In contrast, nucleation of Cu nanoparticles in MSGA is not dependent on the presence of any reactive gases, their nucleation appearing in sole Ar, and being practically not influenced by the presence of O_2 in discharge [49].

Briefly, the steps of nanoparticles formation and growth are the following ones [46, 47]. After the formation of the initial embryos, further growth takes place by four main processes: *attachment* of the atoms (or *condensation*), coagulation, coalescence and aggregation. During the first process, the number of the atoms from a cluster increases due to *attachment* of supplementary metal atoms; the supplementary energy gained due to bonds formation is dissipated during the collisions between the growing clusters with the buffer gas atoms. The second type of processes (*coagulation*) happens when two clusters (each containing n and respective m atoms) are unified, forming a larger cluster (containing $n + m$ atoms). This process is similar to the formation of a liquid drop following the contact of two smaller liquid drops. The third type of processes (*coalescence*), takes place when the supersaturation degree is low and it consist in growth of larger clusters by atom attachment, while the smallest ones evaporate. Finally, during *aggregation*, two clusters come in contact without modification of each one shape, forming a larger cluster.

3. Experimental

3.1 Experimental setup

A typical magnetron sputtering gas aggregation (MSGA) system for production of nanoparticles is presented schematically in **Figure 1**. It consists of a cluster source

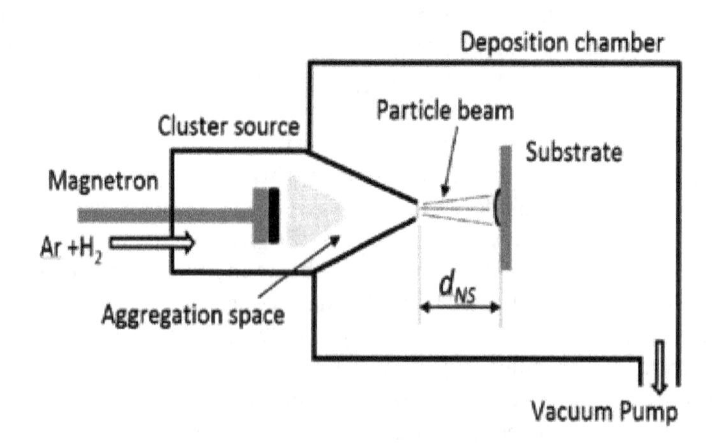

Figure 1.
Schematic of the MSGA experimental system used for nanoparticles synthesis.

ending with an aperture, attached to a high vacuum chamber similar with those used for thin film deposition. The enclosure of the cluster source is a cylindrical stainless steel tube with water cooled walls; a magnetron sputtering gun is mounted axially in it, with the target facing the exit aperture. The nanoparticles synthesis takes place in the space between them (see **Figure 1**). The working gas (an inert gas used in sputtering, usually Ar) enters in the system through the cluster source and it is evacuated by a vacuum pumping system attached to the deposition chamber.

Due to its small diameter (typically a few millimeters), the exit aperture presents a small gas flow conductance, resulting in development of a differential pressure between the cluster source and the deposition chamber. As a consequence, the nanoparticles obtained in the aggregation space of the cluster source are ejected in the deposition chamber, where they are collected on the substrate.

The properties of the nanoparticles obtained by MSGA depend on the process parameters (pressure, applied power, working gas type, etc.) and on the geometrical characteristics of the cluster source (including the distance between the target and the nozzle, and the nozzle diameter) [32]. In the present study, we considered a part of these parameters, and we report on the effects of changing the RF power, target thickness and nature of admixed gases.

3.2 Materials and characterization methods

The nanoparticles presented herein were obtained from circular W targets with 99.95% purity. Their diameter is of 2 inches while two thicknesses were used (6 and 3 mm).

The low-pressure Torus TM2 (Kurt Lesker Co.) was adapted in laboratory for use in the high pressure range (tens of Pa). The plasma discharge was sustained by a RF (13.56 MHz) power generator (model Comet CITO PLUS 1000 W) paired with a corresponding matching box (model Comet AGS 1000 W). The RF generator was operated either in continuous wave mode or in pulsed mode.

The process gas was Ar or a mixture of Ar with different other gases (H_2, O_2, water vapors), the pressure of the working gas being of 80 Pa in cluster source (0.5 Pa in the deposition chamber).

The exit aperture of the MSGA cluster source was 2.5 mm while the aggregation length was 5 cm for all studies presented here. The generated nanoparticles were collected on two types of substrates: monocrystalline Si substrates chips and microscope slides.

The deposited nanoparticles were investigated as regarding their material properties as follows.

The morphology was investigated by scanning electron microscopy (SEM), using an EVO 50XVP Zeiss device equipped with LaB6 electron gun operating at 20 kV. The nanoparticles microstructure was analyzed by high resolution transmission electron microscopy (HRTEM), using a Cs probe-corrected atomic resolution analytical electron microscope JEOL JEM-ARM 200F, operated at 200 kV.

The crystalline structure of the deposited samples was investigated by X-ray diffraction (XRD) using a PANalytical X'Pert PRO MRD diffractometer (wavelength CuK-α 0.15418 nm) equipped with a point sensitive fast detector (PIXcel) operated in Bragg–Brentano geometry.

3.3 Deposition rate: shape of the deposit on the substrate

The experiments of W NPs synthesis by MSGA were reported for first time by us [28, 29] using as working gas sole Ar. These were performed using RF power in both continuous wave (60 W [28] and 100 W [29]) and in rectangular pulsed mode [29]

(RF power was switched ON/OFF at intervals of 200 ms, duty cycle 50%, the maximum RF power being 120 W and the minimum one being zero, i.e. an average value of 60 W). The other experimental parameters were kept constant.

The deposition rate was evaluated by weighting the substrates before and after deposition. Despite of different values of electrical power applied to the discharge, we observed similar deposition rates, around 5 mg/h. On the other hand, we observed a gradual decrease of the deposition rate in time.

Figure 2a presents the image of a substrate (microscope slide), which was periodically translated vertically in front of the exit aperture, after 5 min deposition. The substrate was positioned at d_{NS} = 2.5 cm distance from the exit aperture. We can observe the decrease of the spot diameters, from around 1 cm at the process beginning (see the lower spot, at t = 0, in **Figure 2a**) down to few millimeters after 20 min (see the upper spot in the same figure). This fact proves that the deposition rate decreased down to near zero after a finite time duration. Still, after breaking the vacuum and exposing the target for few hours to the ambient atmosphere the W NPs process takes place again. A similar decrease of the deposition rate was reported in [48] during synthesis of Ti and Co nanoparticles using the MSGA technique. For these materials, it was proved that the generation of nucleation sites is conditioned by the presence of small amounts of reactive gases (O_2 for Ti, respective N_2 for Co) mixed with Ar in the aggregation chamber. Similarly, we identified by trials that mixing Ar (5 sccm) with a small amount of H_2 (0.7 sccm) revitalizes the synthesis process, leading to continuously deposition of W NPs at an average rate of 50 mg/h.

The nanoparticles are ejected from the cluster source as a conical beam, resulting that the diameter of the deposit formed on the substrate is dependent on the substrate position in respect to the exit aperture of the cluster source. For example, placing the substrate at a distance of 27 cm in respect to the nozzle, the deposited material extends over a spot area of 5 cm spot diameter (see **Figure 2b**) in case of using a mixture Ar/H_2. Thus, modifying the substrate position in the deposition chamber, the diameter of the deposited spot may be modified from 1 cm up to 5 cm, as it is described by **Figure 2c**. This aspect is important for applications where particles distributed over large areas are in demand.

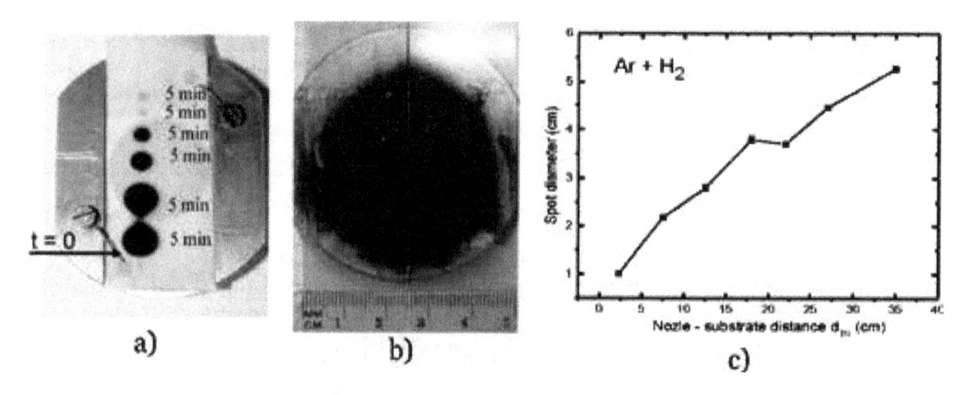

Figure 2.
(a) Time evolution of the W NPs deposit when only Ar is used as working gas; (b) example of a deposition performed in mixture Ar + H_2*, the substrate being placed at a distance of 27 cm from the nozzle; (c) dimension of the W NP deposits over the position of the substrate (using Ar +* H_2 *mixture).*

3.4 Morphology and structure of the nanoparticles: effect of the RF power applied to discharge on the particles properties

The results presented in this section were obtained using only Ar as working gas, at a mass flow rate of 5 sccm; the corresponding pressures in the aggregation and deposition chambers being 80 Pa, respective 0.5 Pa. A 6 mm thick target was used, the tangential magnetic field at the target surface being of 100 mT.

Plasma was generated either in continuous, either in pulsed power mode, as follows:

- continuous wave operation mode, applying 60 W and 80 W RF power to the discharge;
- pulsed wave operation mode, with a medium power of 60 W (P_{max} = 120 W, P_{min} = 0 W, the pulsing durations being t_{ON} = t_{OFF} = 200 ms).

No significant modification of the W NPs synthesis rate was observed when the RF power value was increased (in continuous mode) or was pulsed. Instead of this, the obtained nanoparticles present totally different material properties.

SEM images of the deposited nanoparticles are presented in **Figure 3a** (P_{RF}= 60 W), **Figure 4a** (P_{RF}= 100 W) and **Figure 5a** (pulsed mode). At first glance, it is obviously that the morphology of the nanoparticles is dependent on the power applied to the discharge. Thus, on the sample deposited at 60 W (**Figure 3a**) can be observed both individual nanoparticles (of around 70 nm in diameter) and agglomerations of such nanoparticles. These agglomerations are most probably to appear inside the aggregation chamber, being promoted by electrostatic attrac-tion of the nanoparticles [50]. Increasing the applied RF power up to 100 W the nanoparticles shape changes to concave hexapods (**Figure 4a**), with dimension in between 80 and 200 nm. In pulsed mode, the nanoparticles are round and on the sample can be observed (**Figure 5a**) two classes of nanoparticles: faceted ones, with dimension between 40 and 80 nm (higher number) and flower like ones with dimension in between 80 and 100 nm (smaller number).

The nanoparticle shapes are highlighted by the low magnification TEM images presented in **Figure 6a** (P_{RF} = 60 W), **Figure 7a** (P_{RF} = 100 W) and **Figure 8a** (pulsed mode). Supplementary, TEM images reveal that nanoparticles deposited in continuous wave mode present a dendritic growth, their branches evolving in flower like (**Figure 6a**, P_{RF} = 60 W) or star like pattern (**Figure 7a**, P_{RF} = 100 W). TEM images of faceted nanoparticles obtained in pulsed mode present projections containing 8 or 6 sides, suggesting a cube-octahedral geometry for these nanopar-ticles (see the inset image in **Figure 8a**) [51].

The XRD patterns recorded on the deposited samples are presented in **Figure 3b** (P_{RF} = 60 W), **Figure 4b** (P_{RF} = 100 W) and **Figure 5b** (pulsed mode). The peaks from these diffraction patterns show the presence in the sample of the

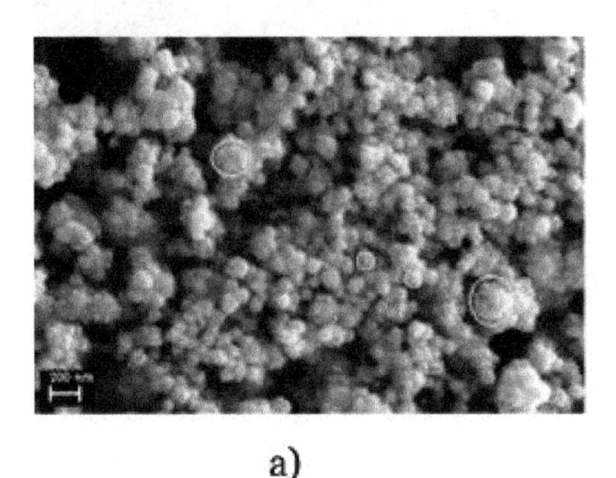

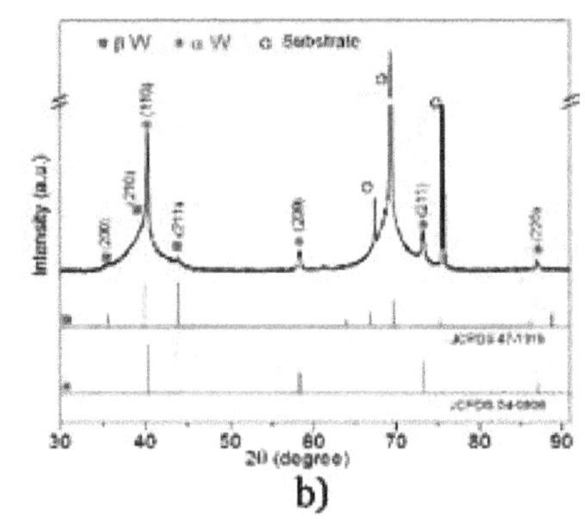

Figure 3.
SEM image (a) and the XRD pattern (b) of the W NPs deposited at 60 W applied to discharge.

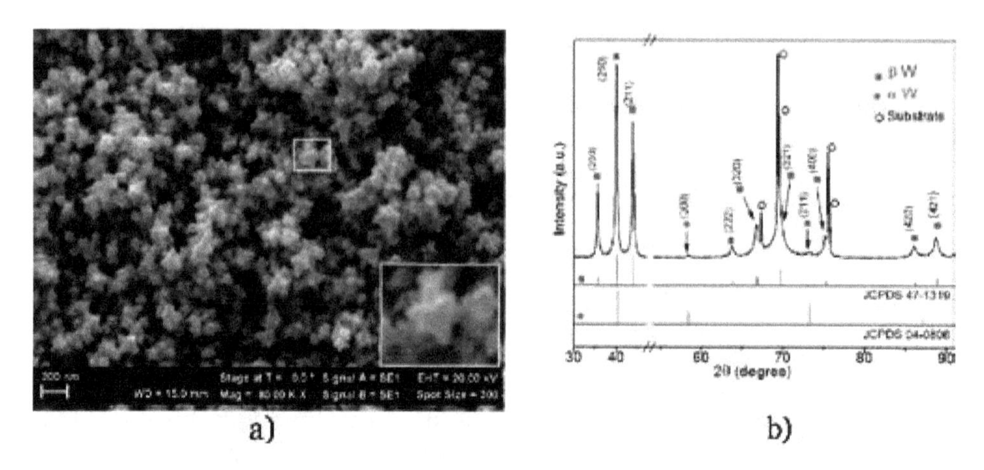

Figure 4.
SEM image (a) and the XRD pattern (b) of the W NPs deposited at 100 W applied to discharge.

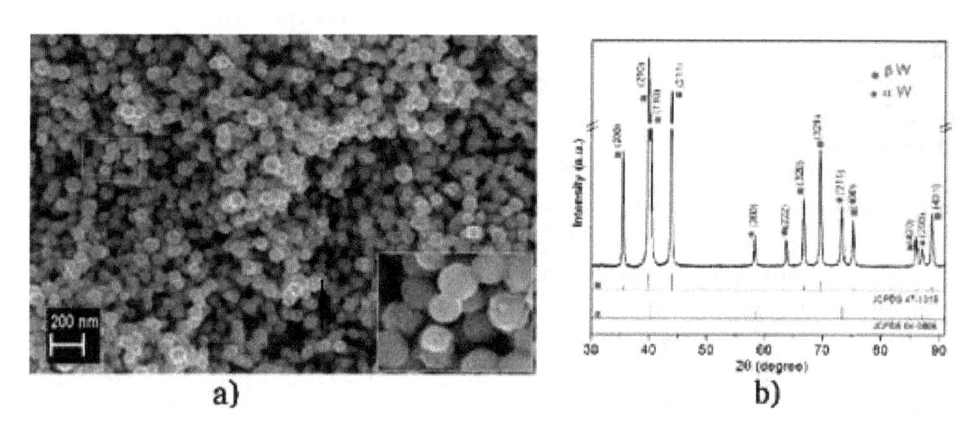

Figure 5.
SEM image (a) and the XRD pattern (b) of the W NPs deposited using pulsed RF power.

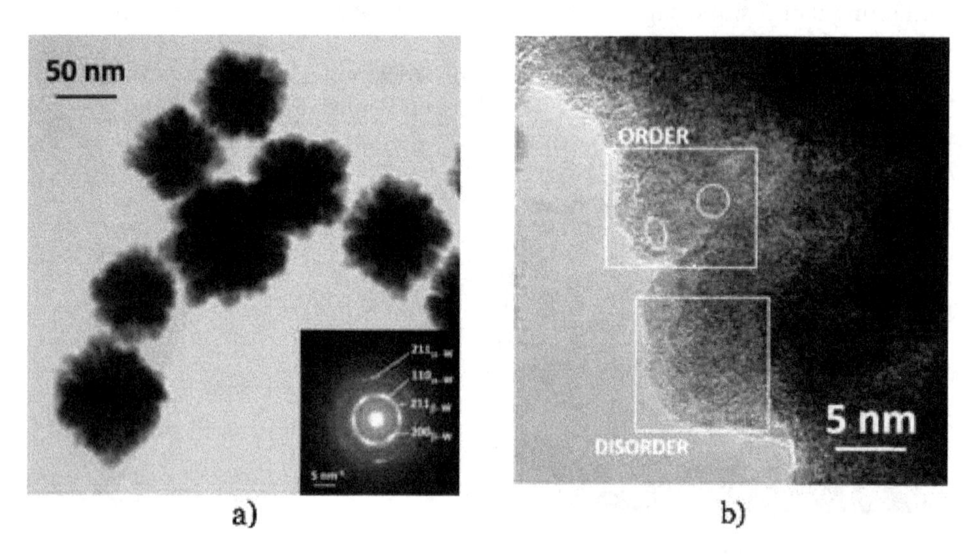

Figure 6.
TEM (a) and HRTEM (b) images recorded on W NPs deposited at 60 W applied to discharge.

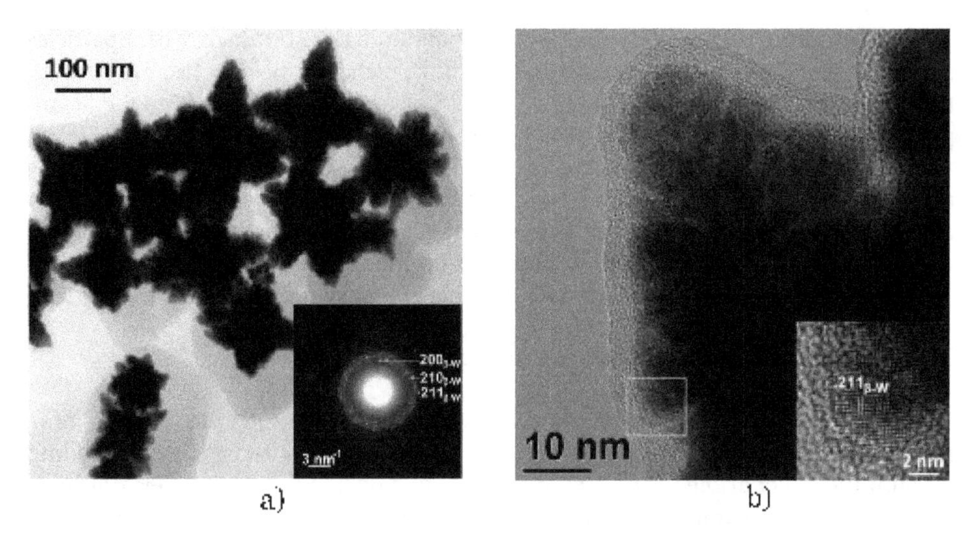

Figure 7.
TEM (a) and HRTEM (b) images recorded on W NPs deposited at 100 W applied to discharge.

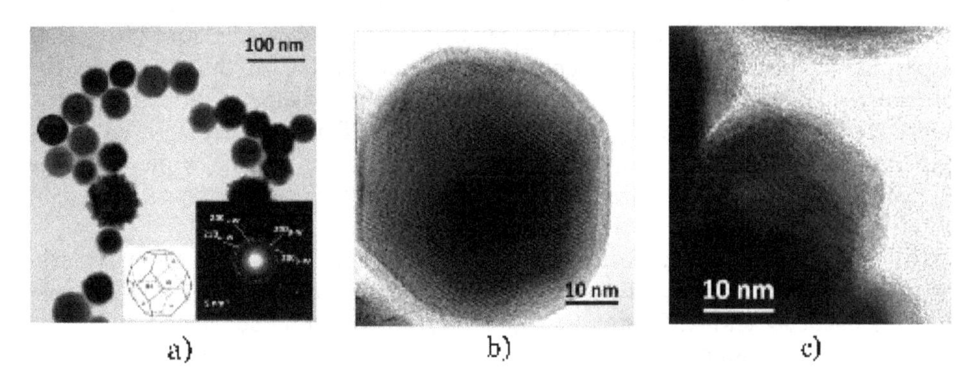

Figure 8.
TEM (a) and HRTEM images recorded on cuboctahedral W NPs (b) and on a flower like one (c), deposited using pulsed RF power.

body-centered cubic lattice (bcc) α-W (JCPDS 04-0806) and of the simple cubic A15 lattice, corresponding to metastable β-W phase (JCPDS 47-1319). It is interesting to note that the α-W phase is dominant in nanoflower samples (**Figure 3b**) while the β-W one in the nano hexapod samples (**Figure 4b**); on the other hand, the samples deposited with pulsed RF power present similar contributions from the both α-W and β-W phases. A remarkable result is the observation that the β-W (which is usually metastable, converting rapidly to α-W even at room temperature) remained unchanged even after 2 years of storage in laboratory environment (more details can be found in [28]).

The HRTEM images recorded on the samples reveal that in the branches of the nanoflowers (**Figure 6b**) coexist regions with high order or with high disorder. This observation is confirmed also by the shape of the XRD peaks (with sharp peaks and enlarged bases, (see **Figure 3b**) and by the appearance of the SAED pattern recorded (presence of both bright spots and broad diffraction rings in the diffraction pattern; see inset in **Figure 6a**). On the other hand, the hexapods and cube octahedral present a much higher degree of crystalline order in the samples.

Indeed, the HRTEM image from **Figure 7b** shows well textured regions, in the inset of the figure being presented well-defined nanocrystallites with size of around

5–6 nm, that form each branch. On the other hand, the cube octahedral nanoparticles are single nanocrystals (see **Figure 8b**). Also, the branches of the nanoflowers incidental observed in **Figure 8a** show a higher degree of order (see **Figure 8c**) when compared with those grown at 60 W (see **Figure 6b**). These observations are supported also by the SAED patterns presented as insets in **Figures 7a** and **8a**, both presenting only bright spots. Details regarding the chemical composition of the W NPs and their oxidation in ambient air can be found elsewhere [28, 52]. We briefly mention here that the amorphous layers which can be observed on the edges of the nanoparticles (see HRTEM images from **Figures 7b** and **8b**) are due to post synthesis oxidation.

3.5 The effect of target thickness on the morphology of W NPs

We already mentioned that we used W targets with thicknesses of 3 mm, respective 6 mm. In fact, modifying the thickness of the targets we took into account the effect of the magnetron magnetic field over the nanoparticles synthesis process. Thus, for 6 mm targets the measured value of the tangential magnetic field at the target surface was 100 mT, while it was 150 mT for 3 mm targets. In this section we present the morphologies of the W NPs obtained with a 3 mm thick target, following a parametric study involving the variation of the RF power applied to the discharge. The working gas is Ar only. During the study the other experimental parameters were kept constant (identical with those used in the previous section): aggregation length of 5 cm, nozzle diameter of 2.5 mm, the substrate being situated at 2.5 cm from the nozzle. Thus, similar with previous experiments, the Ar mass flow rate was kept at a value of 5 sccm, while the value RF power applied to the discharge (in continuous wave mode) was varied in between 60 and 120 W, with steps of 10 W.

All the samples were deposited on Si substrates and these were further investigated by SEM. These investigations show that in general the deposited samples contain particles with different morphologies on the same substrate. Only two samples (for the applied powers of 60 and 80 W respectively) were observed to contain particles with the same morphology; the SEM images recorded on these samples are presented in **Figure 9**.

For 60 W RF power the nanoparticles present a nanoflower like morphology, similar with those obtained in the same conditions with a thick target (see **Figure 3a**). Their dimension is also similar, being in between 50 and 100 nm. Still, we have to note that the sample of W NPs obtained with a thinner target presents a much lower degree of agglomeration (compare **Figure 3a** with **Figure 9a**).

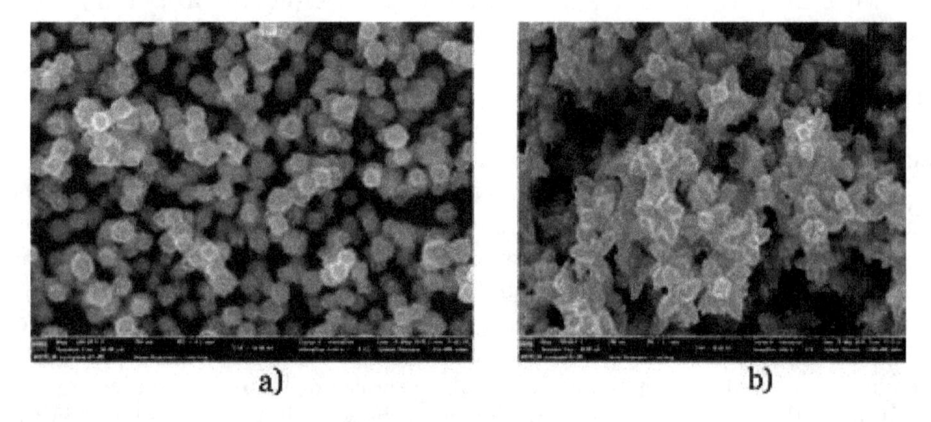

Figure 9.
SEM images recorded on substrates deposited at: (a) 60 W (nanoflowers) and (b) 80 W (multiple branches); Scale bar 100 nm.

For 80 W RF power applied to discharge, the obtained particles are in fact multi-podal nanostructures, with up to 8 emerging branches. The distance between the tips of two opposite branches ranges in between 350 and 400 nm, while the diameter of one branch measure around 80–90 nm. These structures seem to be similar with hexapod nanoparticles obtained with thick W target (see **Figure 4**).

For the remaining samples deposited with different values of the RF power, the samples are not uniform in respect with the particles morphology. Examples of SEM images recorded on two of such samples (at 70 and 110 W) are presented in **Figure 10**. Here we can observe flower like, faceted, with sharp corners, and even cubic nanoparticles.

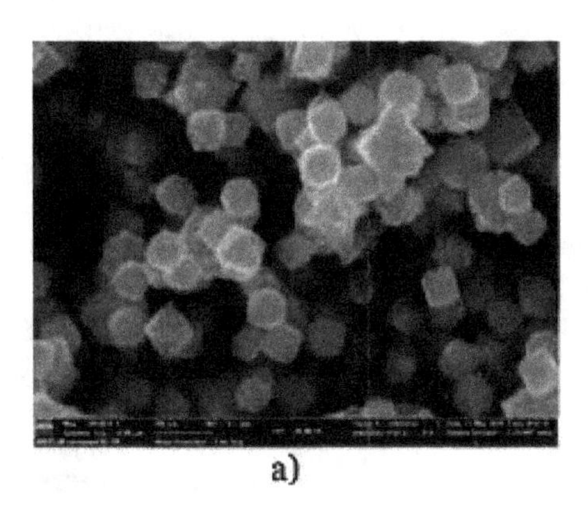
a)

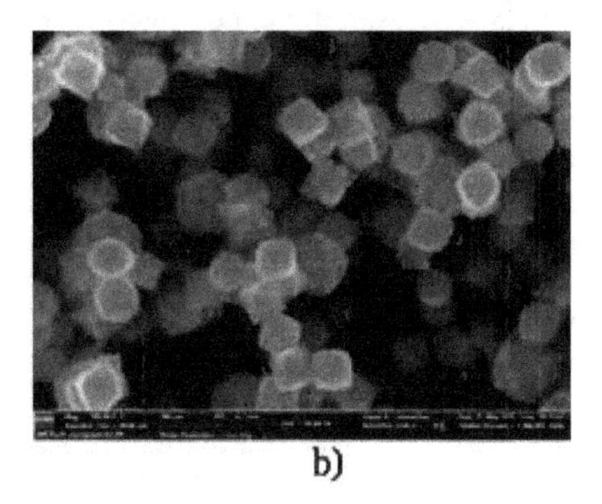
b)

Figure 10.
SEM images recorded on substrates deposited at 70 W (a) and 120 W (b); each sample contains particles with different morphologies. Scale bar 100 nm.

3.6 The effect of residual gases in discharge on the W NPs morphology and synthesis rate

As mentioned, the decrease of deposition rate down to zero in short time after the beginning of the MSGA process leads us to the hypothesis that nucleation of W NPs in MSGA may be assisted by the residual gases present in the aggregation chamber. For testing this hypothesis, we inserted deliberately small amounts of gases from the atmospheric air components in the discharge, immediately after the deposition ceasing for the process performed in Ar. We observed that mixing H_2 or water vapors with Ar in the aggregation source leads to the revitalization of the MSGA synthesis process, leading to a continuous synthesis of W NPs, at higher deposition rate; it seems that O_2 does not have such effect over the deposition rate. However, a side effect of mixing of other gases with Ar in the aggregation chamber is the modification of the nanoparticles morphologies.

We note that these experiments were performed in continuous wave, with a 6 mm thick W target. **Figure 11** presents SEM images recorded with different gases added in the discharge.

By comparing **Figure 11a** with **Figure 3a** it results that addition of H_2O vapors in the discharge at an applied power of 60 W leads to less agglomerated particles. Apart from nanoflower morphology, also appear particles with sharp corners, while the dimension of particles increases (from up to 50–100 nm in dry Ar, up to 100–250 nm in humid Ar). Like we already mentioned, the synthesis rate increases significantly.

Addition of O_2 in the discharge does not increase the synthesis rate, but modifies significantly the particles shape. This can be observed by comparing **Figure 11b** (round nanoparticles) with **Figure 4a** (hexapod nanoparticles). These samples

Figure 11.
SEM images recorded on substrates deposited with W NPs in Ar (5 sccm) mixed with different other gases: (a) H_2O vapors (0.5%), at 60 W; (b) O_2 (0.03 sccm) at 100 W and (c) O_2 (0.03 sccm) + H_2 (0.1 sccm) at 100 W.

were deposited in similar conditions, at 100 W RF, the single difference being addition of a very small amount of O_2 (0.03 sccm) in the main process gas (Ar, 5 sccm). Enriching this Ar/O_2 mixture with H_2 (0.1 sccm) leads to further change in morphology, from round nanoparticles (**Figure 11b**) to particles presenting corners (**Figure 11c**). H_2 seems to favor anisotropic growth (particles with sharp corners), while O_2 seems to favor isotropic growth (round NPs). Therefore, addition of H_2 sustains the synthesis of W NPs and leads to the increase of deposition rate.

4. Use of tungsten nanoparticles for toxicology studies and assessment of tritium retention in nuclear fusion technology

Expected power and plasma duration of future fusion devices, such as ITER, requires the divertor plasma-facing components (PFC) to withstand considerable plasma fluxes (up to 10 MW/m^2 and 10^{23} H m^{-2} s^{-1}). In this area, tungsten has been chosen for its good resistance to high temperature and its low plasma sputtering yield. However, experiments have shown that the combination of different phenomena (e.g. melting of edges of PFC, material fatigue, intense particle fluxes, material erosion followed by accretion in the plasma edge, ...), particularly during off normal events such as Edge Localized Modes (ELMs) or disruptions, can trigger the formation of particles. They will experience variable size, from tens of nanometers to hundreds of micrometers. Compared to the current tokamaks, the expected energy outflows are much higher in ITER. Larger quantities of dust will be produced. Besides, impurities in the Scrape-Off Layer (SOL), such as oxygen, metals and especially gas radiators are expected to enhance physical sputtering of the PFC materials. Generated particles will be hence tritiated as formed in a tritium environment.

Safety limits have been set for the inventory of tritium in the vessel (for which dust particles can contribute), but also for the quantity of dust to avoid explosive hazards. Even if dust production in present tokamaks is less important, it is of a major importance, in order to prepare a safe ITER operation, to investigate how relevant dust can store tritium. Indeed, such tritiated particles could be released in case of a Loss Of Vacuum Accident (LOVA) and be accidentally inhaled by ITER workers. Harmful consequences will depend on the dust physicochemical characteristics and the inhaled dose. It is the reason why we have decided to investigate their potential biological toxicity. To this end, several actions are to be undertaken: understanding and characterizing the interaction of particles with tritium and studying the biological impact of these tritiated dusts on human lung cells.

As said above, the current dust tokamak production is scarce and dust relevant particles have to be produced. Different production methods have been approached, among them being MSGA, milling, and laser ablation. The properties of the nanoparticles produced by magnetron sputtering gas aggregation are of major interest. Indeed, such particles size have the size ranging from 100 to 200 nm, therefore present the maximum probability to escape the High Efficiency Particulate Air (HEPA) filters, normally used to prevent as much as possible the release of particles out of the vacuum chamber.

4.1 Tritium retention in tungsten dust

To study this topic, after particle production, tritiation has been undertaken using the usual gas loading procedure with tritium followed at the Saclay tritium laboratory [53]. The procedure involved various steps, like native oxide layer removed by reduction under hydrogen atmosphere, loading of tritium, cooling the sample at liquid nitrogen temperature to freeze the tritium detrapping processes, thus leading to tritium saturation in particles. For comparison during toxicity studies, particles are also hydrogenated in the same way than for the tritium loading. Particle tritium inventory is obtained by desorption and by full dissolution of the samples in water peroxide and liquid scintillation counting. In case of MSGA type of particle, it is about 5 Gbq/g. These values are two or three orders of magnitude higher than in massive samples, where 11.5 MBq/g were measured [39].

4.2 Cytotoxicity of tungsten nanoparticles

The study of consequences of inhaling small W particles that can reach the human inner lung or can come in contact with the skin, as first organ barriers, is of high importance. Different topics are thus addressed such as the behavior of particles and embedded tritium in biologic media or the in vitro toxicity of these tritiated particles [54]. It is worth to note that the particles are rapidly dissolved (less than 1 week) in different biologic media as saline solution, TRIS buffer, and cell culture medium lung solution. TRIS solution is usually used to prepare particle stock solution for in vitro studies. The consequence of the dissolution of the particles is the release of tritium into the solution. Tritium, once solubilized, will be removed quickly as it has been shown that tritium in tritiated water form is eliminated very quickly from the human body (half-life of 10 days).

The toxicity of W particles produced by milling was evaluated on the MucilAir® model [55], a 3D in vitro cell model of the human airway epithelium grown on air-liquid-interface [42]. Epithelia were exposed to tungsten nanoparticles (tritiated or not) for 24 h. Thanks to the long shelf life of this model, cytotoxicity was studied immediately after treatment and in a kinetic mode up to 1 month after cell exposure to assess the reversibility of toxic effects. Acute and long-term toxicities were monitored by several endpoints: (1) epithelial integrity, (2) cellular metabolic activity, (3) pro-inflammatory response, (4) mucociliary clearance, and (5) morphological modifications. Transmission Electronic Microscopy (TEM) observations, inductively coupled plasma mass spectrometry (ICP-MS) measurements and liquid scintillation were performed to determine tungsten and tritium lung absorption as well as intracellular accumulation. On this human lung epithelium model, no significant toxicity was observed after exposure to ITER-LIKE particles produced by milling followed by 28-day kinetic analysis. Moreover, tritium transfer through the epithelial barrier was found to be limited in comparison with the complete transfer of tritium from tritiated water. These data provide preliminary information to biokinetic studies aimed at defining biokinetic lung models to establish new safety rules and radiation protection approaches.

In respect with usage of MSGA and laser ablation particles in toxicology studies, the behavior of cells from bronchial human-derived BEAS-2B cell line with such particles in their pristine, hydrogenated, and tritiated forms is described in [43]. Cell viability, cytostasis and DNA damage were evaluated by specific biological assays. The study concludes that long exposures (24 h) induced significant cytotoxicity, the effect being enhanced in case of the hydrogenated particles. Epigenotoxic alterations were observed for both MSGA and laser ablation types of particles. Due to the observed oxidative dissolution of W nanoparticles in liquid media, such effects might be related to the presence of W^{6+} species in the medium.

The toxicity of MSGA tungsten nanoparticles on human skin fibroblast cells (BJ ATCC CRL 2522), was evaluated in [56]. Different concentrations of tungsten nanoparticles (1–2000 μg/mL) added in liquid medium were used. MTS colorimetric tests and Scanning Electron Microscopy in secondary electrons (SE) and backscattering (BSE) operating modes were used to observe the effects. At low concentrations of nanoparticles in suspension, no toxic effects were observed by the MTS colorimetric test. However, when the nanoparticle concentration added to the cellular medium increases above 100 μg/mL, reaching even 1 mg/mL and 2 mg/mL, the toxicity of tungsten nanoparticles is high. The cell morphology changed, in comparison with the control sample. Cells tend to become round, a sign that precedes their death. At a higher magnification, in BSE mode, it is noticeable that the nanoparticles may be internalized under the cell membrane.

5. Conclusions

This chapter describes a physical method of fabrication of nanoparticles, namely Magnetron Sputtering Gas Aggregation (MSGA). The advantages come from a clean plasma-based process which do not make use of wet chemistry, and is well controlled, and reproducible. The synthesis of tungsten particles was selected to illustrate the potential of the method. Not only those particles are promising materials for nanotechnological devices, photocatalysis, and energy storage, but environmental and toxicological questions raised by the occurrence of tungsten dust in fusion experiments request answer of high importance for the scientific community working in the fusion energy research. The produced particles, analyzed in this chapter, have size in the range 100–200 nm, and have, depending of the applied plasma power, flower like, hexapod, or cube-octahedral morphologies. Mixture of these morphologies in the same sample is also possible at other parameters. The particles are single crystals in case of cube-octahedral morphology (pulsed applied power), or consists of small nanocrystallites (3–5 nm) which are assembled in a disordered manner (in nanoflowers, at P_{RF} = 60 W), or in a much-ordered manner (in nanohexapods, at P_{RF} = 100 W). The W dust consisting of such particles incorporates tritium amounts of two to three orders of magnitude higher than the massive tungsten material. Although limited in respect with the conditions of exposure and types of cells, the biological studies mentioned in the chapter indicate that the toxic effects should be considered in the activities involving W dust, especially at large concentrations of particles.

Acknowledgements

Part of this work has been carried out within the framework of the Eurofusion consortium and has received funding from the Euratom research and training program 2014–2018 and 2019–2020 under grant agreement no. 633053. The views and opinions expressed herein do not necessarily reflect those of the European Commission. In addition, we acknowledge the support in the frame of the Romanian PN-III-P1-1.2-PCCDI-2017-0637/33 PCCDI (MultiMonD2) and INFLPR Nucleus projects, and the support of the A*MIDEX project (no. ANR-11-IDEX-0001-02) funded by the "Investissements d'Avenir" French Government program, managed by the French Research Agency (ANR).

Author details

Tomy Acsente[1], Lavinia Gabriela Carpen[1,2], Elena Matei[3], Bogdan Bita[1,2], Raluca Negrea[3], Elodie Bernard[4], Christian Grisolia[4] and Gheorghe Dinescu[1,2*]

1 National Institute for Lasers, Plasma and Radiation Physics, Bucharest, Romania

2 Faculty of Physics, University of Bucharest, Bucharest, Romania

3 National Institute of Materials Physics, Bucharest, Romania

4 CEA, IRFM, Saint Paul lez Durance, France

*Address all correspondence to: dinescug@infim.ro

References

[1] Lassner E, Schubert W-D. Tungsten, Properties, Chemistry, Technology of the Element, Alloys, and Chemical Compounds. USA: Springer; 1999. p. 447. DOI: 10.1007/978-1-4615-4907-9

[2] 20 Interesting Facts About Tungsten [Internet]. 2019. Available from: https://briandcolwell.com/20-interesting-factsabout-tungsten/ [Accessed: 31 January 2020]

[3] ITER [Internet]. Available from: https://www.iter.org/mach/Divertor [Accessed: 31 January 2020]

[4] Gunn JP et al. Surface heat loads on the ITER divertor vertical targets. Nuclear Fusion. 2017;**57**(4):046025. DOI: 10.1088/1741-4326/aa5e2a

[5] Pitts RA, Carpentier S, Escourbiac F, Hirai T, Komarov V, Lisgo S, et al. A full tungsten divertor for ITER: Physics issues and design status. Journal of Nuclear Materials. 2013;**438**:S48-S56. DOI: 10.1016/j.jnucmat.2013.01.008

[6] Philipps V. Tungsten as material for plasma-facing components in fusion devices. Journal of Nuclear Materials. 2011;**415**:S2-S9. DOI: 10.1016/j.jnucmat.2011.01.110

[7] Winter J, Gebauer G. Dust in magnetic confinement fusion devices and its impact on plasma operation. Journal of Nuclear Materials. 1999;**266-269**:228-233. DOI: 10.1016/S0022-3115(98)00526-1

[8] Rosanvallon S, Grisolia C, Delaporte P, Worms J, Onofri F, Hong SH, et al. Dust in ITER: Diagnostics and removal techniques. Journal of Nuclear Materials. 2009;**386-388**:882-883. DOI: 10.1016/j. jnucmat.2008.12.195

[9] Malizia A, Poggi LA, Ciparisse J-F, Rossi R, Bellecci C, Gaudi P. A review of dangerous dust in fusion reactors: From its creation to its resuspension in case of LOCA and LOVA. Energies. 2016;**9**(8):578. DOI: 10.3390/en9080578

[10] Baron-Wiechec A, Fortuna-Zaleśna E, Grzonka J, Rubel M, Widdowson A, Ayres C, et al. First dust study in JET with the ITER-like wall: Sampling, analysis and classification. Nuclear Fusion. 2015;**55**(11):113033 (7 pp). DOI: 10.1088/0029-5515/55/11/113033

[11] Ratynskaia S, Tolias P, De Angeli M, Weinzettl V, Matejicek J, Bykov I, et al. Tungsten dust remobilization under steady-state and transient plasma conditions. Nuclear Materials and Energy. 2017;**12**:569-574. DOI: 10.1016/j. nme.2016.10.021

[12] Wang H, Fang ZZ, Hwang KS, Zhang H, Siddle D. Sinter-ability of nanocrystalline tungsten powder. International Journal of Refractory Metals & Hard Materials. 2010;**28**: 312-316. DOI: 10.1016/j.ijrmhm.2009. 11.003

[13] Syed MA, Manzoor U, Shah I, Bukhari SH. Antibacterial effects of tungsten nanoparticles on the Escherichia coli strains isolated from catheterized urinary tract infection (UTI) cases and Staphylococcus aureus. New Microbiologica. 2010;**33**:329- 335. Available from: http://www.newmicrobiologica.org/PUB/allegati_pdf/2010/4/329.pdf

[14] Lita AE, Rosenberg D, Nam S, Miller AJ, Balzar D, Kaatz LM, et al. Tuning of tungsten thin film superconducting transition temperature for fabrication of photon number resolving detectors. IEEE Transactions on Applied Superconductivity. 2005;**15**:3528-3531. DOI: 10.1109/TASC.2005.849033

[15] Pai CF, Liu L, Li Y, Tseng HW, Ralph DC, Buhrman RA. Spin transfer torque devices utilizing the giant spin Hall effect of tungsten. Applied Physics Letters. 2012;**101**(12):122404. DOI: 10.1063/1.4753947

[16] Long HW, Zeng W, Zhang H. Synthesis of WO_3 and its gas sensing: A review. Journal of Materials Science Materials in Electronics. 2015;**26**(7):4698-4707. DOI: 10.1007/s10854-015-2896-4

[17] Adhikari S, Chandra KS, Kim DH, Madras G, Sarkar D. Understanding the morphological effects of WO_3 photocatalysts for the degradation of organic pollutants. Advanced Powder Technology. 2018;**29**(7):1591-1600. DOI: 10.1016/j.apt.2018.03.024

[18] Zhao Y. WO_3 for photoelectrochemical water splitting: from plain films to 3D architectures [thesis]. Eindhoven: Technische Universiteit Eindhoven; 2019. p. 135

[19] Schottle C, Bockstaller P, Gerthsenb D, Feldmann C. Tungsten nanoparticles from liquid-ammonia-based synthesis. Chemical Communications. 2014;**50**(35):4547-4550. DOI: 10.1039/c3cc49854a

[20] Sahoo PK, Kamal SSK, Premkumar M, Kumar TJ, Sreedhar B, Singh AK, et al. Synthesis of tungsten nanoparticles by solvothermal decomposition of tungsten hexacarbonyl. International Journal of Refractory Metals & Hard Materials. 2009;**27**:784-791. DOI: 10.1016/j.ijrmhm.2009.01.005

[21] Oda E, Ameyama K, Yamaguchi S. Fabrication of nano grain tungsten compact by mechanical milling process and its high temperature properties. Materials Science Forum. 2006;**503-504**: 573-578. DOI: 10.4028/www.scientific. net/MSF.503-504.573

[22] Antony LVM, O'Dell JS, McKechnie TN, Power C, Hemker K, Mendis B. US pioneers 'high speed' tiny tungsten. Metal Powder Report. 2006;**61**(11):16-19. DOI: 10.1016/S0026-0657(06)70761-5

[23] Sarathi R, Sindhu TK, Chakravarthy SR, Sharma A, Nagesh KV. Generation and characterization of nano-tungsten particles formed by wire explosion process. Journal of Alloys and Compounds. 2009;**475**(1-2):658-663. DOI: 10.1016/j.jallcom.2008.07.092

[24] Xiao J, Liu P, Liang Y, Li HB, Yang GW. Super-stable ultrafine beta-tungsten nanocrystals with metastable phase and related magnetism. Nanoscale. 2013;**5**(3):899-903. DOI: 10.1039/c2nr33484d

[25] Stokker-Cheregi F, Acsente T, Enculescu I, Grisolia C, Dinescu G. Tungsten and aluminium nanoparticles synthesized by laser ablation in liquids. Digest Journal of Nanomaterials and Biostructures. 2012;**7**(4):1569-1576. Available from: http://www.chalcogen.ro/1569_Stokker.pdf

[26] Samsonov D, Goree J. Particle growth in a sputtering discharge. Journal of Vacuum Science & Technology A. 1999;**17**:2835. DOI: 10.1116/1.581951

[27] Kumar KK, Couedel L, Arnas C. Growth of tungsten nanoparticles in direct-current argon glow discharges. Physics of Plasmas. 2013;**20**:043707. DOI: 10.1063/1.4802809

[28] Acsente T, Negrea RF, Nistor LC, Logofatu C, Matei E, Birjega R, et al. Synthesis of flower-like tungsten nanoparticles by magnetron sputtering combined with gas aggregation. European Physical Journal. 2015;**69**(6):161. DOI: 10.1140/epjd/e2015-60097-4

[29] Acsente T, Negrea RF, Nistor LC, Matei E, Grisolia C, Birjega R, et al. Tungsten nanoparticles with controlled shape and crystallinity obtained by magnetron sputtering and gas aggregation. Materials Letters. 2017;**200**:121-124. DOI: 10.1016/j.matlet.2017.04.105

[30] Haberland H, Karrais M, Mall M, Turner Y. Thin films from energetic cluster impact: A feasibility study. Journal of Vacuum Science & Technology A. 1992;**10**:3266. DOI: 10.1116/1.577853

[31] Smirnov BM, Shyjumon I, Hippler R. Flow of nanosize cluster-containing plasma in a magnetron discharge. Physical Review E. 2007;**75**(6):066402. DOI: 10.1103/PhysRevE.75.066402

[32] Huttel Y. Gas-Phase Synthesis of Nanoparticles. 1st ed. Weinheim: Wiley-VCH; 2017. p. 402. DOI: 10.1002/9783527698417

[33] Haberland H, Karrais M, Mall M. A new type of cluster and cluster ion-source. Zeitschrift Fur Physik D-Atoms Molecules and Clusters. 1991;**20**:413-415. DOI: 10.1007/BF01544025

[34] Majumdar A, Ganeva M, Kopp D, Datta D, Mishra P, Bhattacharayya S, et al. Surface morphology and composition of films grown by size-selected Cu nanoclusters. Vacuum. 2008;**83**(4):719-723. DOI: 10.1016/j.vacuum.2008.05.022

[35] Polonskyi O, Kylian O, Solar P, Artemenko A, Kousal J, Slavınska D, et al. Nylon-sputtered nanoparticles: Fabrication and basic properties. Journal of Physics D-Applied Physics. 2012;**45**(49):495301. DOI: 10.1088/0022-3727/45/49/495301

[36] Krishnan G, Negrea RF, Ghica C, ten Brink GH, Kooi BJ, Palasantzas G. Synthesis and exceptional thermal stability of Mg-based bimetallic nanoparticles during hydrogenation. Nanoscale. 2014;**6**:11963-11970. DOI: 10.1039/c4nr03885a

[37] Llamosa D, Ruano M, Martinez L, Mayoral A, Roman E, Garcia-Hernandez M, et al. The ultimate step towards a tailored engineering of core–shell and core–shell–shell nanoparticles. Nanoscale. 2014;**6**:13483. DOI: 10.1039/c4nr02913e

[38] Wegner K, Piseri P, Vahedi Tafreshi H, Milani P. Cluster beam deposition: A tool for nanoscale science and technology. Journal of Physics D-Applied Physics. 2006;**39**:R439-R459. DOI: 10.1088/0022-3727/39/22/R02

[39] Grisolia C, Hodille E, Chene J, Garcia-Argote S, Pieters G, El-Kharbachi A, et al. Tritium absorption and desorption in ITER relevant materials: Comparative study of tungsten dust and massive samples. Journal of Nuclear Materials. 2015;**463**:885-888. DOI: 10.1016/j.jnucmat.2014.10.089

[40] El-Kharbachi A, Chêne J, Garcia-Argote S, Marchetti L, Martin F, Miserque F, et al. Tritium absorption/desorption in ITER-like tungsten particles. International Journal of Hydrogen Energy. 2014;**39**(20):10525-10536. DOI: 10.1016/j.ijhydene.2014.05.023

[41] Bernard E, Jambon F, Georges I, Sanles Sobrido M, Rose J, Herlin-Boime N, et al. Design of model tokamak particles for future toxicity studies: Morphology and physical characterization. Fusion Engineering and Design. 2019;**145**:60-65. DOI: 10.1016/j.fusengdes.2019.05.037

[42] George I, Uboldi C, Bernard E, et al. Toxicological assessment of ITER-like

tungsten nanoparticles using an In vitro 3D human airway epithelium model. Nanomaterials (Basel). 2019; **9**(10):1374. DOI: 10.3390/nano9101374

[43] Uboldi C, Sanles Sobrido M, Bernard E, Tassistro V, Herlin-Boime N, Vrel D, et al. In vitro analysis of the effects of ITER-like tungsten nanoparticles: Cytotoxicity and epigenotoxicity in BEAS-2B cells. Nanomaterials (Basel). 2019;**9**(9):1233. DOI: 10.3390/nano9091233

[44] Shi F. Introductory Chapter: Basic Theory of Magnetron Sputtering [Online First]. Rijeka: IntechOpen; 2018. DOI: 10.5772/intechopen.80550. Available from: https://www.intechopen.com/online-first/introductory-chapter-basic-theory-of-magnetron-sputtering

[45] Kelly PJ, Arnell RD. Magnetron sputtering: A review of recent developments and applications. Vacuum. 2000;**56**(3):159-172. DOI: 10.1016/S0042-207X(99)00189-X

[46] Smirnov BM. Nanoclusters and Microparticles in Gases and Vapors. Berlin: De Gruyter; 2012. p. 264. DOI: 10.1515/9783110273991

[47] Ganeva M. Formation of metal nano-size clusters with a DC magnetron-based gas aggregation source [thesis]. Greifswald: Mathematisch-Naturwissenschaftliche Fakultät/Institut für Physik; 2013

[48] Peter T, Polonskyi O, Gojdka B, Ahadi AM, Strunskus T, Zaporojtchenko V, et al. Influence of reactive gas admixture on transition metal cluster nucleation in a gas aggregation cluster source. Journal of Applied Physics. 2012;**112**:114321. DOI: 10.1063/1.4768528

[49] Marek A, Valter J, Kadlec S, Vyskocil J. Gas aggregation nanocluster source-reactive sputter deposition of copper and titanium nanoclusters. Surface and Coating Technology. 2011;**205**(S2):S573-S576. DOI: 10.1016/j.surfcoat.2010.12.027

[50] Vollath D. Plasma synthesis of nanopowders. Journal of Nanoparticle Research. 2008;**10**:39-57. DOI: 10.1007/s11051-008-9427-7

[51] Kimoto K, Nishida I. The crystal habits of small particles of aluminium and silver prepared by evaporation in clean atmosphere of argon. Japanese Journal of Applied Physics. 1977;**16**(6):941-948. DOI: 10.1143/JJAP.16.941

[52] Arnas C, Chami A, Couedel L, Acsente T, Cabie M, Neisius T. Thermal balance of tungsten monocrystalline nanoparticles in high pressure magnetron discharges. Physics of Plasmas. 2019;**26**:053706. DOI: 10.1063/1.5095932

[53] EL KA, Chene J, Garcia-Argote S, Marchetti L, Martin F, Miserque F, et al. Tritium absorption/desorption in ITER-like tungsten particles. International Journal of Hydrogen Energy. 2014;**39**(20):10525-10536. DOI: 10.1016/j.ijhydene.2014.05.023

[54] Grisolia C, Gensdarmes F, Peillon S, Dougniaux G, Bernard E, Autricque A, et al. Current investigations on tritiated dust and its impact on tokamak safety. Nuclear Fusion. 2019;**59**:086061. DOI: 10.1088/1741-4326/ab1a76

[55] Constant S, Huang J, Derouette JP, Wisniewski L. MucilAir: A novel in vitrohuman 3D airway epithelium model for assessing the potential hazard of nanoparticles and chemical compounds. Toxicology Letters. 2008;**180S**:S32-S246. DOI: 10.1016/j.toxlet.2008.06.042

[56] Carpen LG, Acsente T, Acasandrei MA, Matei E, Chilom CG, Savu DI, et al. The Interaction of Tungsten Dust with Human Skin Cells [Online First]. Rijeka: IntechOpen; 2019. DOI: 10.5772/intechopen.86632. Available from: https://www. intechopen.com/online-first/the-interaction-of-tungsten-dust-with-human-skin-cells

2

Comparison of Concentration Transport Approach and MP-PIC Method for Simulating Proppant Transport Process

Junsheng Zeng and Heng Li

Abstract

In this work, proppant transport process is studied based on two popular numerical methods: multiphase particle-in-cell method (MP-PIC) and concentration transport method. Derivations of governing equations in these two frameworks are reviewed, and then similarities and differences between these two methods are fully discussed. Several cases are designed to study the particle settling and conveying processes at different fluid Reynolds number. Simulation results indicate that two physical mechanisms become significant in the high Reynolds number cases, which leads to big differences between the simulation results of the two methods. One is the gravity convection effect in the early stage and the other is the particle packing, which determines the shape of sandbank. Above all, the MP-PIC method performs better than the concentration transport approach because more physical mechanisms are considered in the former framework. Besides, assumptions of ignoring unsteady terms and transient terms for the fluid governing equations in the concentration transport approach are only reasonable when Reynolds number is smaller than 100.

Keywords: proppant transport, two-phase flow, gravity convection, multiphase particle-in-cell method, concentration transport approach

1. Introduction

In unconventional oil and gas industry, there exists a significant granular flow process, which is known as the proppant transport [1]. It is necessary to pump high-strength granular materials such as ceramic particles and sand into the stimulated fracture networks with carrying fluid. Eventually after the flow-back of fluid, the granular materials remain in the fractures and fracture networks are efficiently propped, which contributes to a high conductivity for gas/oil exploitation. There-fore, it is important to reveal the physical mechanisms in the proppant transport process.

Essentially, proppant transport process is a two-phase flow problem constrained in a channel with various widths. In previous works, concentration transport approach was very popular for simulating the proppant transport. In the approach, proppant is considered as a continuum, and is quantitatively described using concentration. The motion of the particle phase is solved based on a concentration transport equation. This method is firstly established by Mobbs and Hammond [2]. They derived the governing equations for the fluid-particle mixture (i.e., slurry) by combining mass conservation laws of two phases and convection models. Then a Poisson equation for the fluid pressure is obtained. They proposed an important dimensionless number, that is, the Buoyancy number, to quantitatively describe the relative intensity of horizontal convection effect and the vertical settling effect. Their pioneering work is then adopted and extended in many later works. For example, Gadde and Sharma [3] and Gu and Mohanty [4] extended this framework by considering the effects of fracture propagation. Wang et al. [5] introduced a blocking function in order to consider the proppant bridging effect. Dontsov and Peirce [6] utilized a more accurate velocity retardation model based on their theoretical analysis. Roostaei et al. [7] applied the WENO (weighted essentially non-oscillation) scheme to solve the concentration transport equation, which greatly reduces the numerical diffusion.

The MP-PIC method [8, 9] is another numerical method for simulating large-scale fluid-particle coupling system, which is popular in chemical engineering. In the MP-PIC method, fluid motion is governed by the volume-averaged Navier-Stokes equation, and particle motion is solved using the Newton's second law under the Lagrangian framework, which is different from those in the concentration approach. Due to its Lagrangian feature and high fidelity, the MP-PIC method is also shown to be a powerful tool for simulating proppant transport process.

In this work, the two above numerical methods are both applied in simulating the proppant transport process. Though the two numerical methods are built under different frameworks, there exist both similarities and differences between them. The hidden facts are revealed based on the analysis of the governing equation sets, as well as the numerical results. The remaining contents of this work are organized as follows. In Section 2, basic governing equation sets of the two methods are demonstrated, and relationship between the two methods are discussed. In Section 3, several numerical cases are designed to illustrate the performance of the two methods, and the numerical results are then discussed. Finally, conclusions are drawn in the Section 4.

2. Methodology

2.1 Concentration transport approach

Assume that there exists only one kind of proppant (same density and size, or mono-disperse) in the fracture and the particle phase is well distributed so that in the large scale we can take the derivative of the particle concentration in most regions (except discontinuity), and the particle and fluid phase are both incompressible. Then we have following unknown variables: fluid velocity $\vec{u}_f$, particle velocity $\vec{u}_p$, particle concentration C, and fluid pressure P.

Let us start from the continuity equation. **Figure 1** shows the fluxes and accumulation in a control volume. It is clear that for the particle phase we have the mass balance equation:

$$\frac{C^{n+1}w^{n+1} - C^n w^n}{\Delta t} = \frac{\left(Cwu_{p,x}\right)_W \cdot dy - \left(Cwu_{p,x}\right)_E \cdot dy}{dx} + \frac{\left(Cwu_{p,y}\right)_N \cdot dx - \left(Cwu_{p,y}\right)_S \cdot dx}{dy}, \tag{1}$$

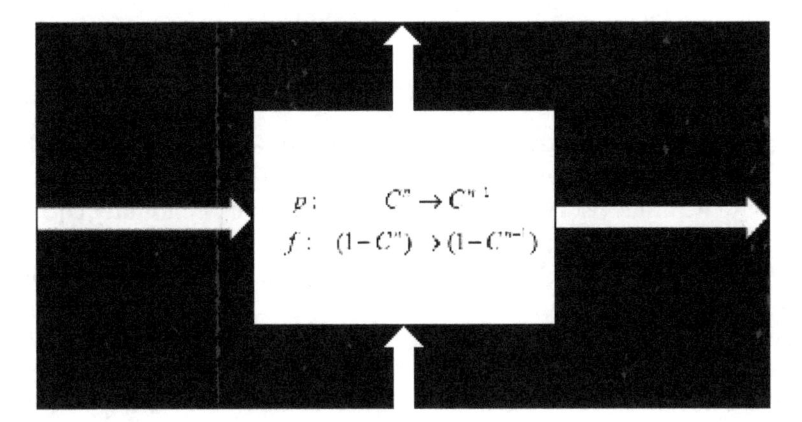

Figure 1.
Control volume in concentration transport approach.

where w is the fracture width. Then it is trivial to obtain the differential form [2]:

$$\frac{\partial(Cw)}{\partial t} + \nabla \cdot \left(Cw\vec{u}_p \right) = 0. \tag{2}$$

Similarly, we have the mass balance equation for fluid phase:

$$\frac{\partial((1-C)w)}{\partial t} + \nabla \cdot \left((1-C)w\vec{u}_f \right) = 0. \tag{3}$$

Considering fluid leak-off, it is trivial to add a source term in the RHS:

$$\frac{\partial((1-C)w)}{\partial t} + \nabla \cdot \left((1-C)w\vec{u}_f \right) = wu_{leak}, \tag{4}$$

where u_{leak} is the leak velocity, which has the dimension of 1/[TIME].

In large-scale cases, we assume that in the horizontal direction, the velocities of the two phases are the same (homogeneous slurry flow), and in the vertical direction, the particle phase velocity differs from the fluid phase due to particle settling. Then we have the following formula:

$$u_{p,x} = u_{f,x}, \quad u_{p,y} = u_{f,y} + u_{settling}, \tag{5}$$

where $u_{settling}$ is the particle settling velocity. There are also other modifications indicating that in the horizontal direction particle velocity does not equal the fluid velocity, which is called proppant retardation [3, 4].

Using Eqs. (2) and (4), we can obtain the fluid/particle mixture (slurry) continuity equation.

$$\frac{\partial(w)}{\partial t} + \nabla \cdot \left(w\vec{u}_s \right) = wu_{leak}, \tag{6}$$

where $\vec{u}_s = (1-C)\vec{u}_f + C\vec{u}_p = \vec{u}_f + Cu_{settling}\vec{j}$ is the slurry velocity.

Eq. (6) is necessary for solving fluid pressure and it is illustrated below. As we know, for pure Newtonian fluid, we have the constitutive laws in which viscosity is a significant parameter. In viscous case (low Re number), based on the Poiseuille's Law, we can derive the relationship between the pressure gradient and average velocity in a channel/fracture:

$$\vec{u}_f = -\frac{w^2}{12\mu_f}\nabla\left(P + \rho_f g z\right), \tag{7}$$

where μ_f is the fluid viscosity and w is the fracture width. The term $w^2/12$ is also considered as the effective permeability of a fracture. This relationship connects the fluid pressure with the velocity, and if we substitute it into continuity equation, we will obtain a Poisson equation for pressure and it can be easily solved.

In the case of fluid/particle mixture, we also expect there exists a similar relationship between fluid pressure and slurry velocity. There are many previous literatures including experimental and numerical works revealing this relationship. It is well known that the apparent viscosity of the mixture is higher than that of the pure fluid and a formula similar to Eq. (7) can be obtained introducing the effective viscosity of the mixture [7]:

$$\vec{u}_s = -\frac{w^2}{12\mu(C)}\nabla(P + \rho_s g z), \tag{8}$$

where $\mu(C)$ is the effective viscosity, which depends on particle concentration C and obviously $\mu(0) = \mu_f$, $\rho_s = \rho_f(1 - C) + \rho_p C$ is the slurry density.

By substituting Eq. (8) into Eq. (6), we get the following pressure Eq. (7):

$$\frac{\partial w}{\partial t} - \frac{\partial}{\partial x}\left(\frac{w^3}{12\mu(C)}\frac{\partial P}{\partial x}\right) - \frac{\partial}{\partial z}\left(\frac{w^3}{12\mu(C)}\frac{\partial(P + \rho_s g z)}{\partial z}\right) = w u_{leak} \tag{9}$$

If the fracture width does not vary with time, then the time derivative vanishes, with the pressure Poisson equation remaining, and the sand concentration C can be solved explicitly or implicitly using Eq. (2) with high-order accuracy schemes.

2.2 MP-PIC method

The MP-PIC method is an Eulerian-Lagrangian method, in which fluid motion is solved in the Eulerian grids and particle motion is solved under the Lagrangian framework. The governing equation of fluid motion reads [10]:

$$\frac{\partial(\alpha_f w)}{\partial t} + \nabla\cdot\left(\alpha_f w \vec{u}_f\right) = w u_{leak}, \tag{10}$$

$$\frac{1}{w}\frac{\partial\left(\alpha_f w \vec{u}_f\right)}{\partial t} + \frac{1}{w}\nabla\cdot\left(\alpha_f w \vec{u}_f \otimes \vec{u}_f\right) = -\frac{1}{\rho_f}\nabla P + \vec{f}_w + \vec{f}_p, \tag{11}$$

where w is the fracture width, α_f is the volume fraction of fluid, $\vec{f}_w$ is the wall friction term and $\vec{f}_p$ is the fluid-particle coupling source term, $\vec{u}_f$ is the fluid velocity, symbol "$\otimes$" indicates the union product.

In the MP-PIC method, the particle phase is discretized into parcels, and every parcel represents several physical particles, which possess the same properties such as density and size, and also similar kinetic behavior such as the velocity and acceleration. Parcel motion is governed by the Newton's second law listed as follows:

$$\frac{d\vec{u}_p}{dt} = D_p\left(\vec{u}_f - \vec{u}_p\right) + \left(1 - \frac{\rho_f}{\rho_p}\right)\vec{g} - \frac{1}{\rho_p}\nabla\cdot\left(P\bar{\bar{I}} - \bar{\bar{\tau}}_f\right) - \frac{\nabla\tau_p}{\alpha_p\rho_p} - \frac{1}{\tau_w}\vec{u}_p \tag{12}$$

where $\vec{u}_p$ is the parcel velocity, D_p is the drag coefficient, α_p is the volume fraction of the particle phase, τ_p is the particle normal stress, and τ_w is the damping relaxation time due to particle-wall interaction, P is the local average fluid pressure, and $\overline{\overline{\tau}}_f$ is the local average fluid viscous stress tensor. Here we consider the Archimedes buoyancy force in the term $\left(1-\frac{\rho_f}{\rho_p}\right)\vec{g}$ and the fluid pressure is the net pressure excluding the hydrostatic pressure.

2.3 Comparison between the two methods

First let us summarize the governing equations of the two approaches.

Concentration transport approach:

$$\begin{cases} \frac{\partial w}{\partial t} - \frac{\partial}{\partial x}\left(\frac{w^3}{12\mu(C)}\frac{\partial P}{\partial x}\right) - \frac{\partial}{\partial z}\left(\frac{w^3}{12\mu(C)}\frac{\partial(P+\rho_s gz)}{\partial z}\right) = wu_{leak} \\ \frac{\partial(Cw)}{\partial t} + \nabla\cdot\left(Cw\vec{u}_p\right) = 0 \\ u_{p,x} = u_{f,x},\ u_{p,y} = u_{f,y} + u_{settling} \\ \vec{u}_s = (1-C)\vec{u}_f + C\vec{u}_p = \vec{u}_f + Cu_{settling}\vec{j} = -\frac{w^2}{12\mu(C)}\nabla(P+\rho_s gz) \end{cases} \tag{13}$$

MP-PIC approach:

$$\begin{cases} \frac{\partial(\alpha_f w)}{\partial t} + \nabla\cdot\left(\alpha_f w\vec{u}_f\right) = wu_{leak} \\ \frac{1}{w}\frac{\partial\left(\alpha_f w\vec{u}_f\right)}{\partial t} + \frac{1}{w}\nabla\cdot\left(\alpha_f w\vec{u}_f\otimes\vec{u}_f\right) = -\frac{1}{\rho_f}\nabla P - f_{wall} + f_{particle} \\ \frac{d\vec{u}_p}{dt} = D_p\left(\vec{u}_f - \vec{u}_p\right) + \left(1-\frac{\rho_f}{\rho_p}\right)\vec{g} - \frac{1}{\rho_p}\nabla\cdot\left(P\overline{\overline{I}} - \overline{\overline{T}}_f\right) - \frac{1}{\theta_p\rho_p}\nabla\tau_p - \frac{1}{\tau_w}\vec{u}_p \end{cases} . \tag{14}$$

In Eq. (13), we need to introduce models for settling velocity $u_{settling}$ and effective viscosity $\mu(C)$. In Eq. (14), we need to model the two terms: wall friction f_{wall} and fluid-particle interaction force $f_{particle}$.

Next we will discuss the similarity and difference between these two equation sets, which are denoted as set I and set II, respectively:

a. Fluid part: it is clear that in set II unsteady and convection/inertial terms are considered. Set I is suitable for homogeneous slurry flow, which indicates that particles settle pretty slowly and Reynolds number is low. In slick water cases, set II is preferred. Actually set II shall converge into set I in low Re number cases. In steady cases, the unsteady terms vanish and the fluid-particle interaction term converges into additional gravity force of particle phase, then the second equation in set II is simplified as:

$$\nabla\langle P\rangle = -f_{wall} - \alpha_p\left(\rho_p - \rho_f\right)\vec{g} \tag{15}$$

If the term f_{wall} converges into the formula of $\frac{12\mu(\alpha_p)}{\rho_f w^2}\vec{u}_s$ we can recover the Eq. (8) without hydrostatic pressure. Note that here $\left\langle\vec{u}_s\right\rangle = \alpha_f\left\langle\vec{u}_f\right\rangle + \frac{1}{V_f}\sum V_p\vec{u}_p$

according to definition or we can just consider it as the average fluid velocity for simplicity. However, how the wall friction term changes in high Re case is still unrevealed.

b. Particle part: obviously we solve particle phase motion under Eulerian framework in set I while under Lagrangian framework in set II. Density/size distribution is easier to be considered in set II. In set I we directly assign a settling velocity for the particle phase while in set II we can resolve the settling history with the drag relaxation term if the time step is set to be small enough. The terminal velocity in set II is determined by which drag force model is utilized. Both of the settling velocity model in set I and drag force model in set II are modifications of Stokes settling theoretical results. Also unsteady terms including fluid pressure and viscous stress effects are considered in set II.

In both of the two 2D large-scale equation sets, models are necessary for closure issues and they cannot exactly describe the full-scale fluid/particle behavior. "Large scale" has two meanings: large time scale and large spatial scale. Different physical mechanisms play significant roles in different scales. For the time scales, p-p collision occurs in the time scale of "μs," f-p drag occurs in the time scale of "ms," fluid leak-off and fracture width change have significant effects in the time scale of "s" or "min." For the spatial scales, fluid viscous force dominates in the Kolmogorov length scale (mm), gravity convection dominates in the length scale of "m." We expect that in our 2D large-scale models, we can capture the large-scale fluid/particle behavior. However, small-scale fluid/particle interactions shall affect the large-scale phenomena.

Using 3D DNS (direct numerical simulation) we can obtain the full-scale details; however, due to computational limits simulation time at most reaches several minutes and simulation length at most reaches 1 dm. Though it is not possible to perform 3D DNS even for an experimental-scale case, useful models can be extracted from the DNS data, for example effective viscosity, settling velocity, retardation factor etc.

In this work, Barree and Conway's [11] effective viscosity model is utilized to calculate slurry viscosity for the concentration transport approach and calculate wall friction force for the MP-PIC method, which is expressed as follows:

$$\mu_f(\alpha_p) = \frac{\mu_0}{\left(1 - \frac{\alpha}{\alpha_{cp}}\right)^{1.82}}, \tag{16}$$

where α_p is the particle volume fraction, or proppant concentration, α_{cp} is the close packing particle volume fraction, which is set as 0.6 in this work.

Besides, Wen and Yu's drag force model [12] is utilized for calculating the fluid-particle coupling drag force for the MP-PIC method and determining particle settling velocity for the concentration transport approach, which is expressed as follows:

$$D_p = \frac{3\rho_f}{8\rho_p} C_d \frac{\left|\vec{u}_f - \vec{u}_p\right|}{r_p}, \tag{17}$$

$$C_d = \begin{cases} \dfrac{24}{\mathrm{Re}_p}\left(1 + 0.15\,\mathrm{Re}_p{}^{0.687}\right)\alpha_f^{-2.65} & \mathrm{Re}_p < 1,000 \\ 0.44\alpha_f^{-2.65} & \mathrm{Re}_p \geq 1,000 \end{cases}, \tag{18}$$

where $\mathrm{Re}_p = \frac{2\rho_f |\vec{u}_f - \vec{u}_p| r_p}{\mu_f}$ is the particle Reynolds number, and r_p is the physical radius of proppant particle.

3. Case study

3.1 Parameter setting

Numerical simulation is performed in a rectangle domain as shown in **Figure 2**. The left side of the domain is set as inlet. Proppant is uniformly injected from the whole inlet and proppant concentration is set as 20%. Here the proppant concentration means the volume ratio of particles to total volume of fluid/particle mixture. The right side of the domain is set as outlet. The upper and bottom sides of the rectangle domain are set as the non-flow boundary. Particle deposition mechanisms are different in these two methods. In the concentration transport approach, if proppant concentration reaches the close packing limit, particle settling velocity is set to be approaching zero, and it is easy to implement the non-flow condition for a Eulerian approach. In the MP-PIC method, if proppant concentration reaches the close packing limit, additional forces due to particle stress gradient are exerted on parcels to make sure these parcels shall move away from the high-concentration region. When parcels move across the non-flow boundary, the displacements of these parcels are modified following a bounce-back way.

Parameters used in the simulation are listed in **Table 1**. Cases with different fluid viscosities, that is, 1, 10, and 100 cP are considered in this section. Because characteristic length is 0.005 m (fracture width) and characteristic velocity is 0.2 m/s (inlet velocity), the Reynolds numbers in these cases are 1000, 100, and 10 respectively.

3.2 Simulation results and discussions

Figures 3–5 show the numerical results of three different Reynolds number cases of two methods. The results are contoured by the volume fraction of particle (denoted as "vfp"), or the proppant concentration. Here we denote the high Reynolds number case as case I, the moderate Reynolds number case as case II, and the low Reynolds number case as case III respectively.

In case I the viscosity of carrying fluid is very low, that is, 1 cp, which indicates a poor capability of proppant transport. From **Figure 3**, it is clear that both of the two numerical methods illustrate the packing process during the transport process. In the figure, blue, white, and red regions indicate the pure fluid, suspending slurry, and sandbank regions respectively. In case II, settling behavior of proppant is weaker than that of case I, and instead gravity convection of the slurry front in the

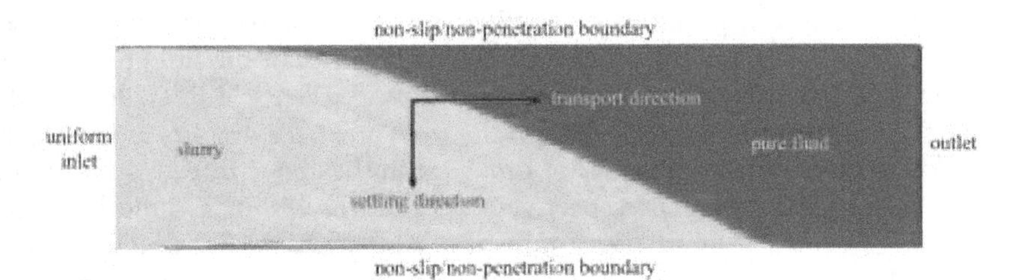

Figure 2.
A sketch for the numerical simulation domain.

Parameter	Value	Parameter	Value
Fluid density	1000 kg/m^3	Particle density	2500 kg/m^3
Fluid viscosity	1 and 10 and 100 cP	Particle diameter	0.6 mm
Time-step	5×10^{-3} s	Simulation time	4 s
Inlet velocity	0.2 m/s	Inlet concentration	20%
Domain size	$1 \times 0.25 \times 0.005$ m^3	Mesh size	100×25

Table 1.
Parameters for simulating proppant transport.

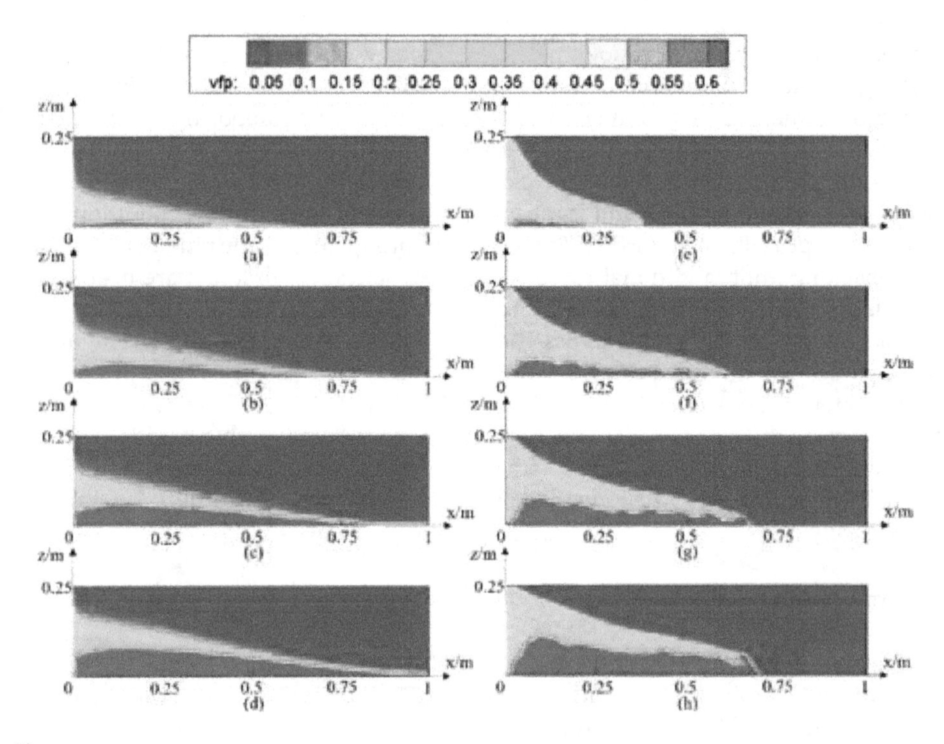

Figure 3.
Simulation results of high Reynolds number case (Re = 1000). (a)–(d) are the concentration transport results at time = 1, 2, 3, and 4 s respectively, and (e)–(h) are the MP-PIC results at time = 1, 2, 3, and 4 s respectively.

vertical direction is obvious as shown in **Figure 4**. In case III, it is obvious from **Figure 5** that proppant settling behavior is hard to recognize and gravity convection is much weaker than that of case II.

By comparing the results at different Reynolds numbers, it can be summarized that as the slurry is injected into the domain, there are several significant mechanisms that determine the eventual proppant distribution listed as follows.

The first mechanism is proppant settling. Proppant settling is due to the net gravity force of the proppant if particle density is larger than the fluid density, and the terminal velocity of proppant is determined by the particle Reynolds number and particle volume fraction. According to the Wen and Yu's drag force model, it is trivial to obtain the settling velocities of proppant in the above three tests, and they are 53.2, 11.8, and 1.28 mm/s, respectively. Therefore, within a same horizontal transport distance, the level of slurry-pure water interface declines more dramatically in case I compared to the other two cases with lower Reynolds numbers.

The second one is gravity convection. In case II and case III, proppant settling can be ignored compared to the inlet velocity, however the slurry fronts in these

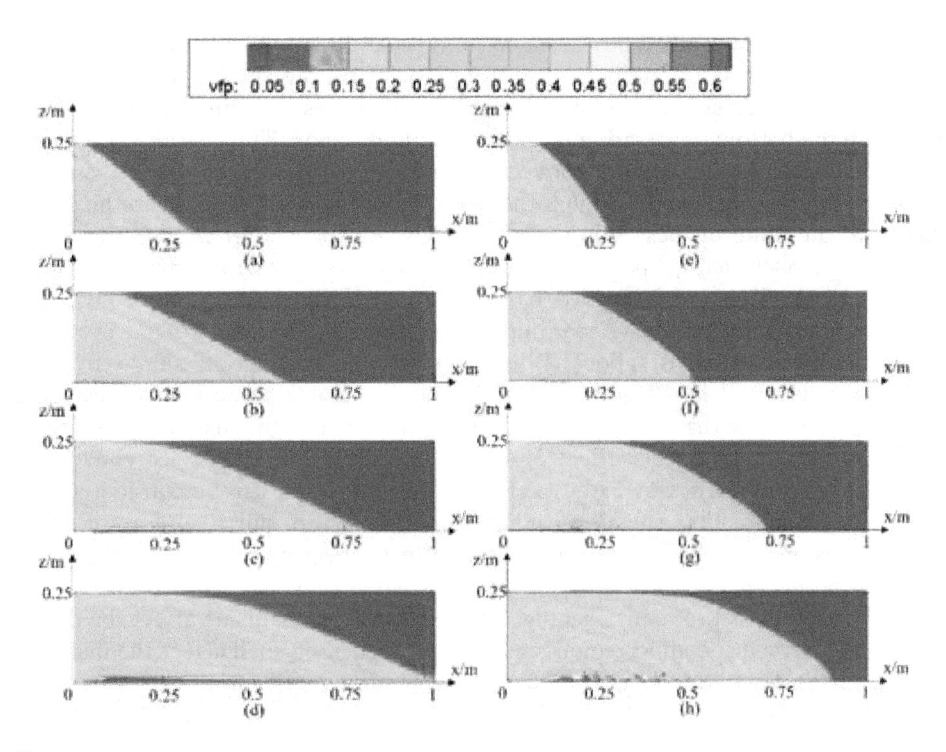

Figure 4.
Simulation results of moderate Reynolds number case (Re = 100). (a)–(d) are the concentration transport results at time = 1, 2, 3, and 4 s respectively, and (e)–(h) are the MP-PIC results at time = 1, 2, 3, and 4 s respectively.

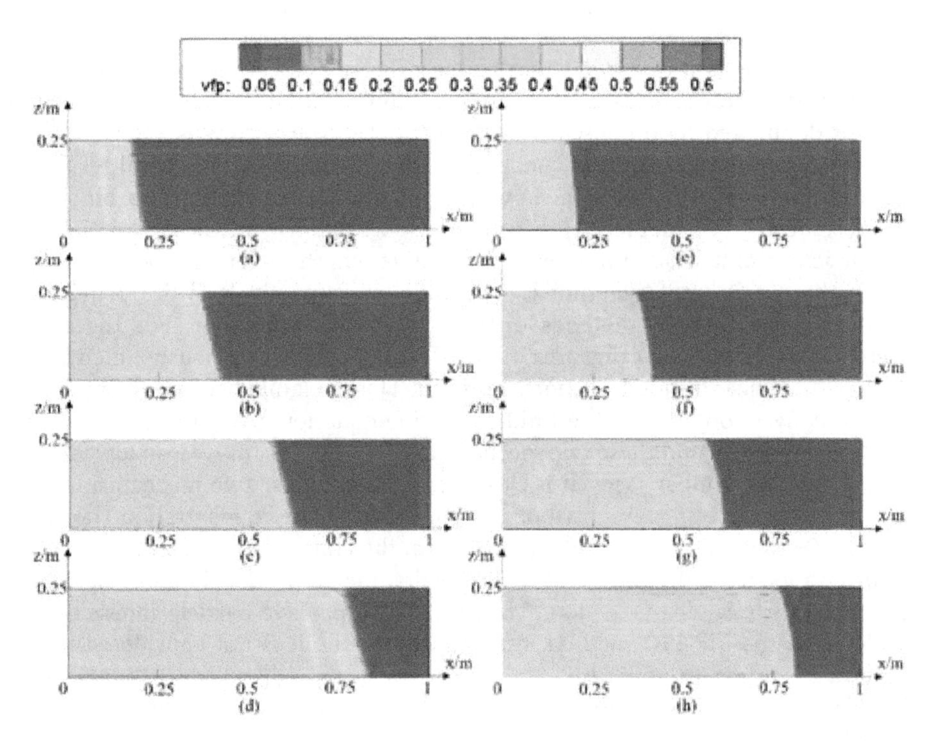

Figure 5.
Simulation results of low Reynolds number case (Re = 10). (a) – (d) are the concentration transport results at time = 1, 2, 3, and 4 s respectively, and (e)–(h) are the MP-PIC results at time = 1, 2, 3, and 4 s respectively.

two cases still evolve and both trend from vertical direction to inclined direction. This is because of the horizontal pressure gradient on the slurry front due to the difference between slurry and pure fluid, which provides a driving force and keeps slurry on the bottom penetrating into the pure fluid region. This mechanism is then intuitively described as gravity convection. Gravity convection in case III is much weaker than that in case II. Though the horizontal pressure gradients on the slurry front are the same in these two cases, fluid viscosity is greater in case III, which leads to a smaller channel permeability for slurry flows. Therefore, the penetration velocity in case III is much smaller than that in case II. In case I, gravity convection also plays a significant role. According to the ratio of inlet velocity and proppant settling velocity, that is, about 4, without gravity convection slurry front is expected to be exact the diagonal line of the domain. However, numerical results of both methods show that the suspending region is far below the diagonal line, which indicates that fluid velocity field is greatly changed due to gravity convection compared to a uniform flow. Above all, gravity convection can be considered as a global effect reflecting the density difference between slurry and pure fluid, whereas proppant settling represents a local effect reflecting the density difference between proppant particle and pure fluid.

The third one is proppant packing. The two prior mechanisms affect the distribution of slurry suspending region, and proppant packing shall affect the distribution of sandbank. In case I, as time evolves proppant concentration increases at the bottom of the simulation domain. When proppant concentration reaches a maximum value, that is, close to the packing state, particle-particle interactions and particle-wall interactions become more frequent. Then, the early formed sandbank stays unmovable, and the fluid velocity also dramatically decreases in this region due to the fluid-particle coupling and particle-particle/wall damping effects.

By comparing the numerical results of two different methods based on the above mechanisms, similarities and differences between the two methods discussed in Section 2.3 are verified.

Firstly, in case III it is obvious that numerical results of methods are almost the same with each other. In case III, Reynolds number is pretty low and the fluid motion is dominated by the viscous mechanism. The slurry flow is approximately plug flow. Secondly, in case II of moderate Reynolds number, numerical results of concentration transport approach show that the gravity convection is a bit severe than those of MP-PIC method. However, transport patterns of these two methods are in general similar. Quantitatively, the relative length of top and bottom slurry fronts at the end of simulation time is 0.5 and 0.4 m, respectively. Thirdly, in case I, numerical results of two methods differ a lot from each other. For the slurry suspending region, the covering area of the MP-PIC results is obviously much larger than those of concentration transport approach. This is mainly because the transient term and convection term in the fluid governing equation are ignored in the concentration approach, and these two mechanisms play significant roles in low viscous or high Reynolds number cases. It is clear that losses of these two mechanisms shall under-estimate the transport capability when utilizing the concentration transport approach. Besides, sandbank packing process in the concentration transport results also differ a lot from that in the MP-PIC results, including the slopes of upstream and downstream sandbank region. This is because particle-particle interaction is considered in the MP-PIC method in some way, while it is not considered in the concentration transport approach.

It is worth noting here that in this work the fifth order WENO (weighted essentially non-oscillation) scheme is adopted for solving the concentration transport equation, thus the effect of numerical diffusion of concentration transport approach can be considered insignificant, and differences between the numerical results of two methods can exactly reflect the effects of different modeling strategy on proppant transport. Above all, both methods can precisely capture the proppant settling mechanism when the same drag force model is adopted in both methods, and concentration transport approach can capture the gravity convection mechanisms precisely when Reynolds number is smaller than 100. However, proppant packing mechanism is not captured very well in the concentration approach.

4. Conclusion

In this work, two numerical methods are adopted to simulate proppant transport: the concentration transport approach and the MP-PIC method. The first one is a typical Eulerian-Eulerian method and the second one is a typical Eulerian-Lagrangian method. With full discussions on their frameworks and comparisons between the numerical results, the following conclusions are then obtained:

1. From the view of pure horizontal proppant advection, numerical diffusion can be insignificant in the concentration transport approach if high-order scheme like WENO scheme is adopted, and the interfaces between slurry front and pure fluid can be clearly captured.

2. When fluid Reynolds number is smaller than 100, assumptions of ignoring velocity convection term and transient term for fluid governing equation adopted in the concentration transport approach are reasonable, and numerical results of both methods show similar transport patterns.

3. When fluid Reynolds number reaches 1000, that is, in the low viscous cases, numerical experiments prove that the concentration transport approach shows a low fidelity for capturing both the slurry suspending region and the sandbank packing process.

4. Generally, the more physical mechanisms, including particle-particle/wall interaction and fluid-particle interaction, accurately considered in a numerical framework, the better a simulator performs on capturing proppant transport behaviors.

Acknowledgements

This work is funded by the National Natural Science Foundation of China (nos. U1730111, 11972088, 91852207), the State Key Laboratory of Explosive Science and Technology (no. QNKT19-05), and the National Science and Technology Major Project of China (Grant no. 2017ZX05039-005).

Author details

Junsheng Zeng[1*] and Heng Li[2]

1 College of Engineering, Peking University, Beijing, China

2 School of Earth Resources, China University of Geosciences, Wuhan, China

*Address all correspondence to: zengjs@pku.edu.cn

References

[1] Novotny EJ. Proppant transport. In: Paper SPE 6813 Presented at the 52nd Annual Fall Technical Conference and Exhibition of the Society of Petroleum Engineers of AIME, Held in Denver, Colorado, 9-12 October 1977

[2] Mobbs AT, Hammond PS. Computer simulations of proppant transport in a hydraulic fracture. In: Paper SPE 29649 Prepared for Presentation at the 1995 SPE Western Regional Meeting Held in Bakersfield, California, 8-10 March 2001

[3] Gadde PB, Sharma MM. The impact of proppant retardation on propped fracture lengths. In: Paper SPE 97106 Prepared for Presentation at the 2005 SPE Annual Technical Conference and Exhibition Held in Dallas, Texas, USA, 9-12 October 2005

[4] Gu M, Mohanty KK. Effect of foam quality on effectiveness of hydraulic fracturing in shales. International Journal of Rock Mechanics and Mining Sciences. 2014;**70**:273-285

[5] Wang J, Elsworth D, Denison MK. Propagation, proppant transport and the evolution of transport properties of hydraulic fractures. Journal of Fluid Mechanics. 2018;**855**:503-534

[6] Dontsov EV, Peirce AP. A lagrangian approach to modelling proppant transport with tip screen-out in KGD hydraulic fractures. Rock Mechanics and Rock Engineering. 2015;**48**:2541-2550

[7] Roostaei M, Nouri A, Fattahpour V, Chan D. Numerical simulation of proppant transport in hydraulic fractures. Journal of Petroleum Science and Engineering. 2018;**163**: 119-138

[8] Patankar NA, Joseph DD. Lagrangian numerical simulation of particulate flows. International Journal of Multiphase Flow. 2001;**27**:1685-1706

[9] O'Rourke PJ, Zhao P, Snider DM. A model for collisional exchange in gas/liquid/solid fluidized beds. Chemical Engineering Science. 2009;**64**:1784-1797

[10] Zeng J, Li H, Zhang D. Numerical simulation of proppant transport in propagating fractures with the multi-phase particle-in-cell method. Fuel. 2019;**245**:316-335

[11] Barree RD, Conway MW. Experimental and numerical modeling of convective proppant transport. In: Paper SPE 28564 Prepared for Presentation at the SPE 69th Annual Technical Conference and Exhibition, Held in New Orleans, LA, USA, 25-28 September 1994

[12] Wen C, Yu Y. Mechanics of fluidization. The Chemical Engineering Progress Symposium Series. 1966;**162**: 100-111

3

Integral Photography Technique for Three-Dimensional Imaging of Dusty Plasmas

Akio Sanpei

Abstract

The integral photography technique has an advantage in which instantaneous three-dimensional (3D) information of objects can be estimated from a single-exposure picture obtained from a single viewing port. Recently, the technique has come into use for scientific research in diverse fields and has been applied to observe fine particles floating in plasma. The principle of integral photography technique and a design of a light-field camera for dusty plasma experiments are reported. The important parameters of the system, dependences of the size of the imaging area, and the spatial resolution on the number of lenses, pitch, and focal length of the lens array are calculated. Designed recording and reconstruction system is tested with target particles located on known positions and found that it works well in the range of dusty plasma experiment. By applying the integral photography technique to the obtained experimental image array, the 3D positions of dust particles floating in an RF plasma are identified.

Keywords: dusty plasma, integral photography, three-dimensional reconstruction, particle measurements, light-field, plenoptic camera

1. Introduction

Fine particles immersed in plasma are charged up negatively, show three-dimensional (3D) motion, and form 3D-ordered state, i.e., Coulomb crystal [1–5]. Diagnostic methods for 3D information about the positions of fine particles in a plasma have therefore been widely researched. Among the various dusty plasma experiments, 90° separated two CCD cameras with helping 3D computed tomographic reconstruction [6] and stereoscopic [7, 8] are widely used to determine the 3D position of each fine particle [9]. They require two or more detectors; however, the locations and numbers of observation ports are considerably restricted in many plasma experiment devices. Planar laser scanning technique can obtain the 3D information of particles with one CCD camera [10, 11], but it requires a little while to scan across the wide field of view. In-line holographic techniques [12] and two-color gradient methods [13, 14] can obtain 3D position of dust particles from a single-exposed photograph taken from a direction; however, these methods require a 12 bit or higher dynamic range sensors. It is required that a technique can acquire the 3D information of a dusty plasma with a single-exposed photograph taken from one viewing port with a conventional dynamic range sensor.

The "integral photography technique" [15] is known as a principle used in naked eye 3D display and in commercial refocus cameras. Such refocus camera is also called as "plenoptic camera" or "light-field camera." It provides 3D imaging technologies based on a small lens array or a pinhole array to capture light rays from slightly different directions. This technique has an advantage in which instantaneous 3D information of objects can be estimated from a single-exposure picture obtained from a single viewing port.

Recently, the integral photography technique has come into use for scientific research in diverse fields. Example applications are particle tracking for velocimetry [16–19], microscopy measurement [20, 21], spray imaging [22], etc. In the research filed of plasma, 3D reconstructions of positions of particles levitating in a plasma have been demonstrated using commercial light-field Lytro cameras [23], and the time evolution of dusty plasmas has been measured using a commercial light-field Raytrix camera [24]. An open-ended plenoptic camera, which is constructed with a lens array and a typical reflex CMOS camera, obtained the 3D positions of dust particles in a radio-frequency (RF) plasma [25, 26]. Dual-filter plenoptic imaging system has been applied to observe lithium pellets in a high-temperature plasma [27].

In this chapter, the principle, design, and experimental results of the integral photography technique for 3D imaging of dusty plasmas will be presented.

2. Principle of the integral photography analysis

Figure 1 shows a schematic of recording and 3D reconstruction system with integral photography for dusty plasma experiment. A small lens array is placed in front of the particles levitating in a plasma to obtain an array of projected image. The rays emerging from 3D objects, i.e., scattering light rays from dusts pass through the small lens array and are captured on a sensor device. A 3D spatial point of object that is a position of a dust particle should be projected to two-dimensional (2D) image points $(X^{(i,j)}, Y^{(i,j)})$ through each (i,j) th lens of the array on a detector. The 3D reconstruction is then carried out computationally by creating inverse propagating rays within a virtual system similar to the recorded one.

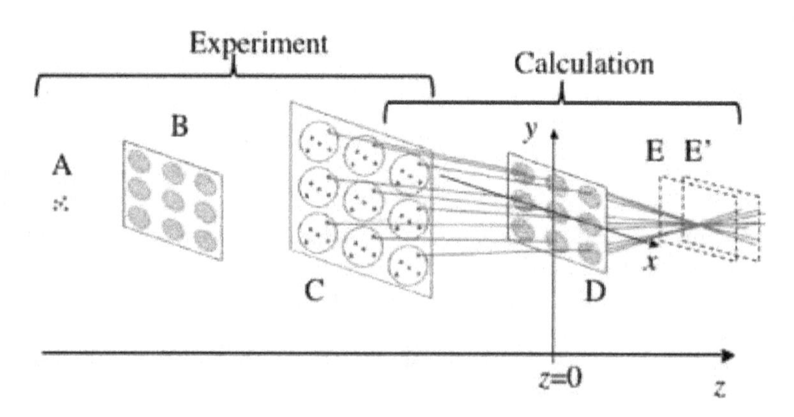

Figure 1.
Schematic of recording and 3D reconstruction system with the integral photography technique: (A) object, (B) lens array, (C) array of projected images on the detector, (D) virtual lens array, and (E and E') virtual observation planes [25].

2.1 Recording system for dusty plasma experiments

In this section, it is shown how to design a recording system for dusty plasma experiments. **Figure 2** shows a schematic of relationship among lens array, detector, and imaging area on the recording system. The Cartesian (x, y, z) coordinate axes are indicated in the lower right corner of the figure. Rays are projected directly onto the sensor after passing through the lens array, which is placed at $z = 0$. Therefore, 3D spatial point of light source, i.e., the position of a dust particle, is projected through each lens of the array to 2D image points $(X^{(i,j)}, Y^{(i,j)})$ on the detector. The array of projected images on the sensor is called as "elemental image array."

The considerable parameters of the lens array are the number of lenses, lens pitch, and focal length. In the following discussion, we deal only with convex lens array. The focal length F of a convex lens array is calculated from the lens law as follows:

$$F = \frac{Ll}{L + l}, \tag{1}$$

where L is the distance between the object and the lens array and l is the distance between the lens array and the sensor. **Figure 3** shows F dependence on L with three values of l. The most suitable value of F can be determined by changing the experimental configuration. Particles located at misaligned positions from focused z-plane create blurred spots on the sensor.

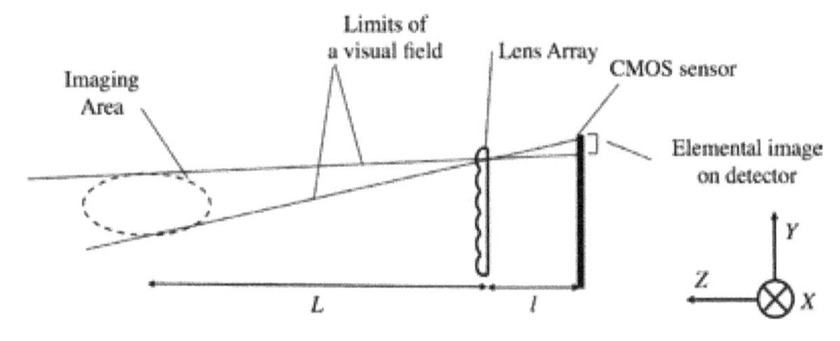

Figure 2.
The schematic of relationship among the small lens array, the detector, and the imaging area of the recording system. Rays are projected directly onto the sensor after passing through the lens array to form an elemental image array [26].

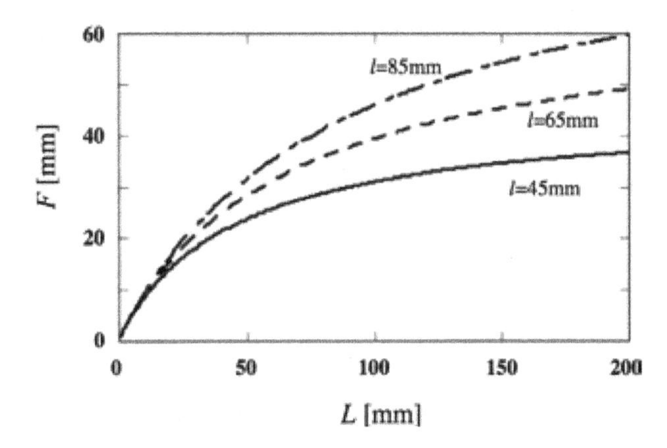

Figure 3.
Focal length F of a convex lens array plotted as a function of the distance L between the object and the lens array for various l values between the lens and a sensor [26].

In order to make an efficient recording system, the configuration of the CMOS sensor must be considered. The number of lenses in the array should be a multiple of the aspect ratio of the CMOS sensor. In addition, the lens pitch must take into account the size of the imaging area. After passing through the lens, rays are projected onto the sensor directly. If we assume that rays from a given dust particle are projected onto all elemental images, the limit of the imaging area is calculated using straight lines passing through the center of the outermost lens

$$(x, y, z) = (x, p(n_y - 1)/2, 0) \tag{2}$$

and through pixels on the edges of the sensor area corresponding to the outermost lens

$$(x, y, z) = (x, H/2, -l), \; (x, H(n_y - 2)/2, -l). \tag{3}$$

Here, n is the number of lenses along a given axis, the subscript "y" denotes the direction of the y-axis, H is the length of the CMOS sensor in the y-axis, and p is the lens pitch. Furthermore, the limits for the plane perpendicular to the z-axis are expressed as

$$y_1 = p\frac{n_y - 1}{2} - L\frac{H - (n_y - 1)p}{2l}, \tag{4}$$

$$y_2 = p\frac{n_y - 1}{2} + \frac{L}{l}\left\{p\frac{n_y - 1}{2} - \frac{H}{n_y}\left(\frac{n_y - 2}{2}\right)\right\}. \tag{5}$$

In the same manner, the limits for the z-axis of the imaging area are expressed as

$$z_1 = \frac{n_y - 1}{2}p \times \frac{2n_y l}{(n_y - 2)H - n_y(n_y - 1)p}, \tag{6}$$

$$z_2 = \frac{n_y - 1}{2}p \times \frac{2l}{H - (n_y - 1)p}. \tag{7}$$

The number n_x of lenses along the x-direction is calculated from n_y using the aspect ratio of the CMOS sensor. Then we can design p and n from experimental requirement of L and the size of the imaging area using above equations.

An uncertainty in the reconstructed image will be attributed to the spatial resolution of an elemental image on CMOS sensor [24]. The length h of a side of a pixel on the CMOS sensor produces uncertainties in the plane perpendicular to the z-axis as

$$\frac{hL}{l} \tag{8}$$

and along the z-axis itself as

$$\frac{hL}{2l} \times \frac{(n_y - 1)(L + l)}{H}. \tag{9}$$

The above equation indicates that the large ratio of distances L/l and finite size of a pixel on the CMOS sensor cause a relatively large uncertainty along the z-axis.

2.2 Reconstruction of 3D position of light source

To extract the positions of projected particles $(X^{(i,j)}, Y^{(i,j)})$ from the elemental image array, a subtraction technique is applied together with the color and profile thresholds. Background image is subtracted from obtained experimental image to reduce background noises such as the reflection of the laser light from the wall of the vacuum vessel. Then, the color space of the subtracted image is converted from RGB to "Lab" scale [28]. To obtain bright pixels colored with irradiating laser from the image, the threshold values of "L" and "a," which depend on the experimental configurations, were considered. For example, by analyzing the dust experiment shown in the Section 3.3 (see **Figure 9**), pixels with "L" values greater than 110 out of 255 and "a" values less than 110 out of 255 are used. Filtering with the geometrical features of the luminance distribution patch on pixels was also applied to detect the positions of dust particles from noise. The luminance centroid of each luminance distribution patch was estimated, and it was treated as the position of the particle.

Using the observed elemental image array, the 3D image of particle distribution is reconstructed in the computer. The light path arriving at the point (p_z, q_z) on a virtual observation plane located on z from a point $(X^{(i,j)}, Y^{(i,j)})$ corresponding to (i, j)th lens on the elemental image array is calculated with ray tracing according to geometrical information, such as distance from lens array to detector and focal length of lenses. If the thin lens approximation can be applied, (p_z, q_z) on giving z is easily calculated as

$$p_z = -\frac{\left(X^{(i,j)} - X^{(i,j)}_{center}\right) \times z}{l} + X^{(i,j)}_{center} \tag{10}$$

$$q_z = -\frac{\left(Y^{(i,j)} - Y^{(i,j)}_{center}\right) \times z}{l} + Y^{(i,j)}_{center} \tag{11}$$

where $\left(X^{(i,j)}_{center}, Y^{(i,j)}_{center}\right)$ is the center of the (i, j)th lens.

Luminosity $I(x, y, z)$ of 3D light-field is estimated as a summation of light intensity $I'(X^{(i,j)}, Y^{(i,j)})$ over all (i, j)th lenses [25, 29, 30] as

$$I(x,y,z) = \frac{\sum_i \sum_j I'\left(X^{(i,j)}, Y^{(i,j)}\right) G_{(i,j)}\left(X^{(i,j)}, Y^{(i,j)}\right) \cos^2\alpha / r^2}{\sum_i \sum_j G_{(i,j)}\left(X^{(i,j)}, Y^{(i,j)}\right)}, \tag{12}$$

where r is the distance between $(X^{(i,j)}, Y^{(i,j)})$ and the position (x, y, z) and α is the angle between the optical axis and the incident ray. $G_{(i,j)}$ is a function indicating whether (i, j)th lenslet exists in a field of view or not defined as

$$G_{(i,j)} = \begin{cases} 1, & \textit{if } (X, Y) \in \text{lenslet} \\ 0, & \text{otherwise} \end{cases} \tag{13}$$

A target light source should locate on where $I(x, y, z)$ shows the extremal value with respect to z, i.e., at convergent points of the rays. To detect the positions of dust particles, therefore, the virtual observation plane, indicated as E and E′ in **Figure 1**, is scanned along z-axis.

3. Experimental setup and results

3.1 Estimation of measurement parameters for a dusty plasma experiment

In order to determine the parameters of multi-convex lens array for a dusty plasma experiment, we adjusted the side of the imaging area and L to be approximately 5 and 80 mm, respectively. The commercial, standard reflex camera (D810, Nikon) was applied as the CMOS sensor. The sensor has dimensions of 36×24 mm^2 and $h = 1/205$ mm. Due to the geometrical limit of the camera D810, the distance l must be more than 65 mm. Rays are projected onto the CMOS sensor directly after passing through the lens array. Subsequently, F is estimated as 35 mm from Eq. (1). The number n of lenses is inversely proportional to p and to the size of a lens. Because of the increased number of tracing rays, an increase in n increases the resolution; however, the number of pixels illuminated by each lens decreases. For a lens array including $(n_x, n_y) = 9 \times 6$ lenses, p is led as 2.2 mm and the D810 camera has 818×818 square pixels for each lens. The imaging area for the plane perpendicular to the z-axis lies between $y_1 = -2.5$ mm and $y_2 = 2.423$ mm, as estimated from Eqs. (4) and (5). The area of the measurable plane is a function of the position z of virtual plane, and it is maximized to be approximately 24 mm^2 at $z = 80$ mm. Regarding the z-axis, the imaging area lies between $z_1 = 143$ mm and $z_2 = 55$ mm, as estimated from Eqs. (6) and (7). **Table 1** shows the parameters of the designed multi-convex lens array and those of the recording system.

A picture of the designed lens array using acrylic plastic is shown in **Figure 4**. The rim around the periphery facilitates the holding of the array. **Figure 5** shows a sample image obtained with the designed lens array system. The object appears as a single green circle with ~1 mm of diameter which is located on $z = 80$ mm.

3.2 Reconstruction of known target light sources

Developed recording and reconstruction system has been tested using target light sources located on known positions. In this test experiment, the typical exposure time of the camera is 1/400 s. The elemental image array obtained with the system was stored in a computer as 10 bits of data. In addition to the position $(X^{(i,j)}, Y^{(i,j)})$ of the particles identified in each elemental images, the intensity $I'(X^{(i,j)}, Y^{(i,j)})$ for each particle is recorded as well. Then $I(x, y, z)$ on a virtual observation plane located on z is calculated according to Eq. (12). The position of the maxima of the brightness of the light-field $I(x, y, z)$ determines the z, which represents the depth of particles. After calibrating the apparent pixel sizes on the images to the real

Parameter	Value
F	35 mm (convex)
$n_x \times n_y$	9×6
Size of a lens	2.2×2.2 mm^2
Pixels per lens	818×818 pixels
l	$\geq$65 mm
Working area on x–y plane at $z = 80$ mm	~24 mm^2
Working range of z from lens array	55–143 mm

Table 1.
Parameters of multi-convex lens array and the recording system for dusty plasma experiment with the CMOS camera D810.

Figure 4.
Photograph of the designed lens array. It includes 9 × 6 convex spherical lenses with rectangular boundaries. All lenses have the same F value of 35 mm [26].

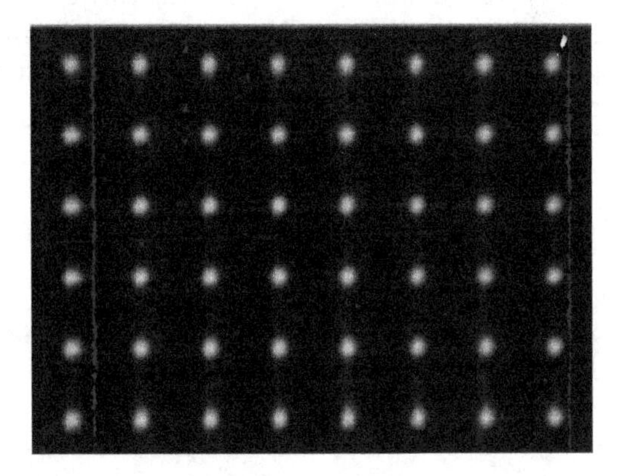

Figure 5.
Sample image obtained with the designed lens array system. Single green circles are recorded as 9 × 6 circles on the CMOS sensor.

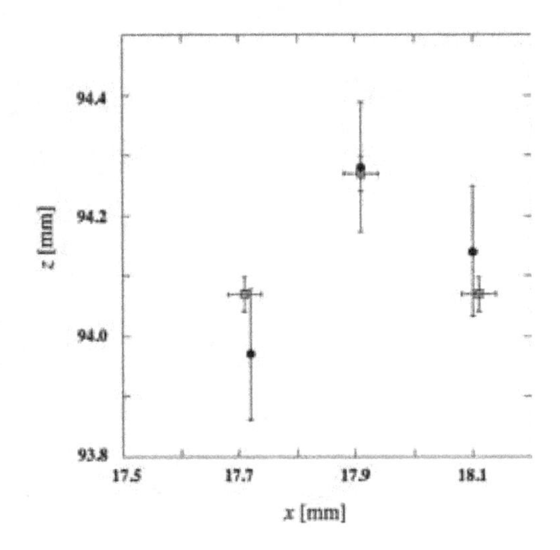

Figure 6.
Top view of reconstruction result for the test data. The target and reconstructed data are indicated by open squares and closed circles, respectively [26].

scales of the dust particle cloud, the absolute (x, y, z) coordinates can be determined.

In **Figure 6**, the 3D positions of the known target and reconstructed particles are marked by open squares and closed circles, respectively. In this test experiment, the optical axis is set along the z-axis, the multi-convex lens array is located at $z = 0$, and l is set to be 77 mm. Mechanical setting errors cause 30 μm of the error bars for the target data points. The error bars of the reconstructed data points are determined according to the pixel dimensions of the recording system. From Eqs. (8) and (9), the uncertainties for the x- (y-) and z-directions are estimated as ~6 and 108 μm, respectively. The relative error between the positions of the target and reconstructed data fits into the known range. Therefore it is concluded that the developed recording and reconstruction system works well in the range of dusty plasma experiment.

3.3 Apply to dusty plasma experiment

Finally, the developed system is applied to a dusty plasma comprising monodiverse polymer spheres (diameter = 6.5 μm) floating in a horizontal, parallel-plate RF plasma. **Figures 7** and **8** show a photograph and schematic of the experimental setup. A piezoelectric vibrator is contained in an RF electrode as the injector of fine particles into a plasma. A grounded counter electrode is positioned at the upper side of the 13.56-MHz-powered electrode at the distance of 14 mm. Fine particles levitate in the plasma generated between the electrodes. A solid-state laser, which radiates light of 532 nm in wavelength, 4 mm in diameter, and ~10 mW in radiation power, was used in our experiment to observe fine particles in the plasma using scattered laser light.

The lens array and the CMOS sensor were located at a side port of the chamber with a distance of l = 65 mm to obtain the elemental image array. An enlarged

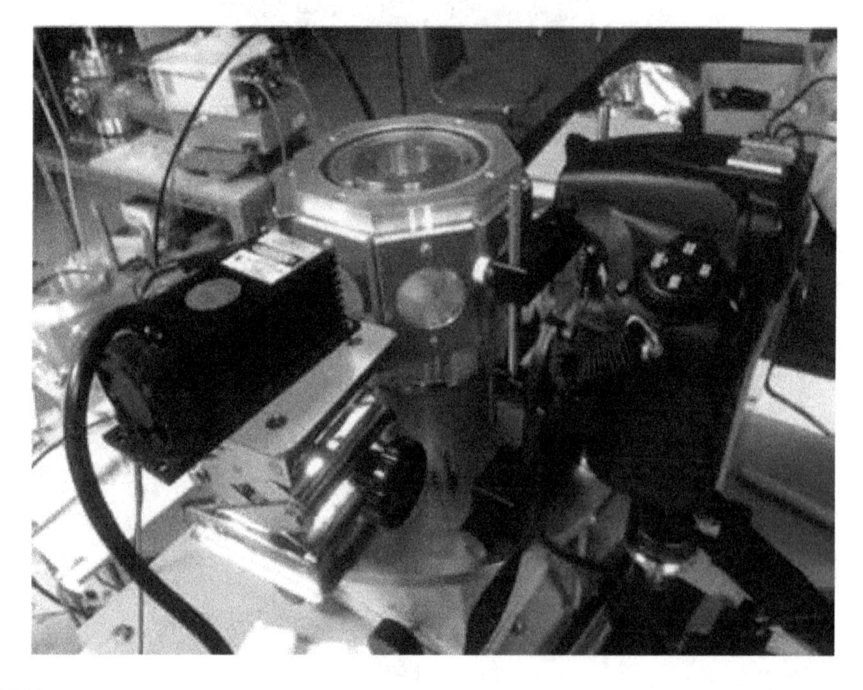

Figure 7.
A photograph of experimental setup. Plasma is generated inside of an octagonal pillar shape chamber shown in the center of the figure. Designed lens array is located between the chamber, and the camera is shown in right-hand side of the figure.

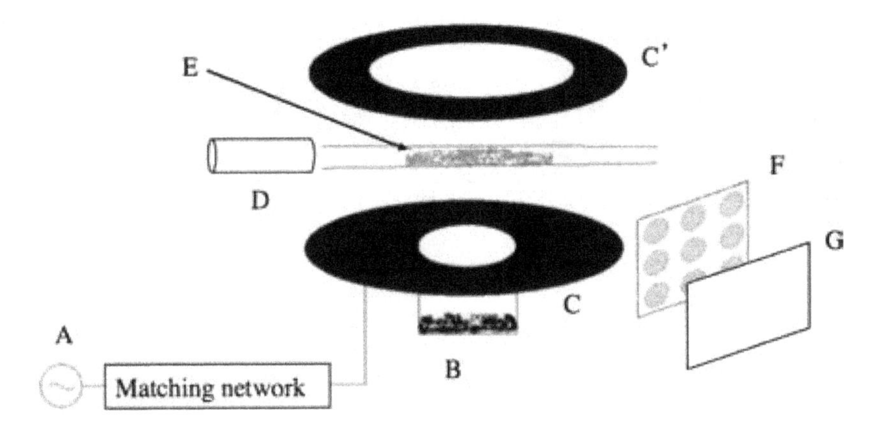

Figure 8.
Schematic of experimental setup (not to scale). (A) 13.56 MHz RF source, (B) dust source, (C and C′) powered and grounded electrode respectively, (D) illuminating laser (10 mW at 532 nm), (E) dust particles in plasma, (F) lens array, (G) CMOS sensor.

Figure 9.
Roughly 3 × 2 elements are enlarged from experimentally obtained array of 54 elemental images after subtraction of a background image. The green dots indicate the presence of dust particles. Each image element is recorded with approximately 818 × 818 square pixels [26].

experimentally obtained elemental image array from above experimental device is shown in **Figure 9**. This figure shows roughly 3×2 elements out of the 9×6 elements captured by the CMOS camera. Each image element is recorded with approximately 818×818 square pixels on the sensor. Scattered light from the dust particles levitating in the RF plasma appears as green dots, and slightly different elemental images result from parallax differences. As shown in **Figure 9**, only five dust particles floated in the plasma in this experiment. They oscillated vertically in the field of view and did not form any ordered array. The 3D positions of particles determined from **Figure 9** are shown in **Table 2** and **Figure 10**. Note that the z-axis is along the optical axis and value of z is the distance from the lens array. **Figure 10 (a)** shows a bird's-eye view of the reconstructed distribution. Green dots indicate the 3D positions of the levitating dusts, and cross symbols indicate the projected position on the x–y plane with z = 90 mm. The x–y distribution of dust particles in **Figure 10(c)** agrees with the observed configuration of dust particles in **Figure 9**. From the reconstructed image, we can recognize that particles are randomly

x	y	z
−0.4226	0.7535	88.542
−0.2322	0.2773	88.950
0.1896	0.4043	89.676
0.3075	0.3272	88.860
0.6975	0.1367	89.540

Table 2.
x, y, *and* z *coordinates in mm of each particle, as determined from* ***Figure 9***.

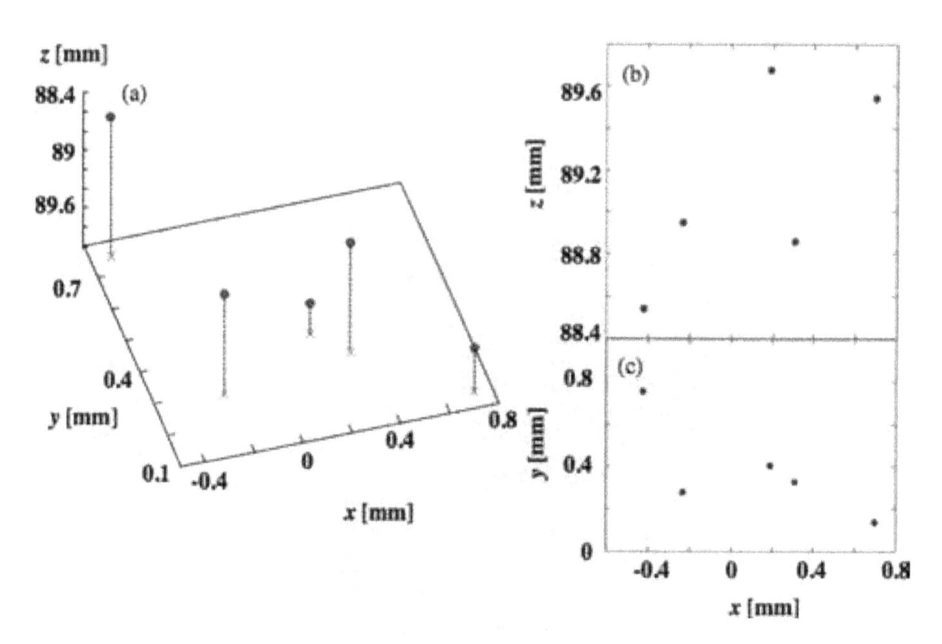

FIgure 10.
(a) Bird's-eye view. Cross symbols indicate projected position of the x−y plane with z = 90 mm. (b) Projection of the x−z plane. (c) Projection of the x−y plane. Each panel shows the reconstructed positions of the levitating particles, as obtained from ***Figure 9***. *Dust are randomly distributed between z = 88.4 and 89.8 mm, which are the distance from the lens array [26].*

distributed between z = 88.4 and 89.8 mm, which are the distance from the lens array. The mean distance among particles is estimated approximately 780 μm.

4. Expected future of the integral photography technique for plasma measurement

The integral photography technique has great potential of versatile applications for plasma measurement. With the help of Mie-scattering ellipsometry technique [31, 32], it would bring information about the size of particles in addition to six-dimensional information about position and velocity. Combined with intrinsic fluorescence spectroscopy [33], specification of dust's materials will be available not only for standard polymer but also for unusual target such as microorganisms [34, 35]. Moreover, deconvolution techniques [36, 37] will extend the integral photography to determine 3D distribution of spatially continuous light sources [38]. 3D information of bremsstrahlung emissivity distribution should be obtained with pinhole ultraviolet or soft X-ray detector [39–41], instead of lenslet array for visible light.

It is still unclear how many dust particles can be counted by the system using the integral photography technique because many parameters trade off against each other. Well-designed optical and recording systems are required to identify the 3D positions for a large number of particles. The defocusing effect of objects is another considerable problem. The effect makes the positions of objects on sensor difficult to identify, and uncertainties in the reconstructed image may increase. In order to avoid such a problem, some commercial light-field cameras mount different F lens arrays simultaneously. Moreover, subpixel analyses, modern particle detection, and interpolation algorithms would enable the achievement of enhanced accuracy [42, 43].

5. Conclusions

The integral photography technique is useful for 3D observation of dusty plasmas. This technique has an advantage in which instantaneous 3D information of objects can be estimated from a single-exposure picture obtained from a single viewing port. The principle of integral photography technique and its analytical method has been explained in detail. A design of a light-field camera for dusty plasma experiments has been reported. The important parameters of the system, dependences of the size of the imaging area, and the spatial resolution on the number of lenses, pitch, and focal length of the lens array are calculated. Then, the recording and reconstruction system has been tested with target particles located on known positions and found that it works well in the range of dusty plasma experiment. By applying the integral photography technique to the obtained experimental image array, the 3D positions of dust particles floating in an RF plasma are identified.

Acknowledgements

The author appreciates Prof. Y. Hayashi and Prof. Y. Awatsuji of Kyoto Institute of Technology for fruitful suggestions from the perspective of researches for the dusty plasma and the integral photography technique, respectively. The author also thanks Prof. S. Masamune of Chubu University and Prof. H. Himura of Kyoto Institute of Technology for the comments on this study. Finally, the author would like to thank Mr. K. Tokunaga of Kyoto Institute of Technology for experimental assistance. This research is partly supported by JSPS KAKENHI Grant Numbers 15K05364, 18K18750, and 24244094.

Author details

Akio Sanpei
Kyoto Institute of Technology, Kyoto, Japan

*Address all correspondence to: sanpei@kit.ac.jp

References

[1] Hayashi Y, Tachibana K. Observation of Coulomb-crystal formation from carbon particles grown in a methane plasma. Japanese Journal of Applied Physics. 1994;**33**:L804. DOI: 10.1143/JJAP.33.L804

[2] Chu JH, I L. Direct observation of Coulomb crystals and liquids in strongly coupled rf dusty plasmas. Physical Review Letters. 1994;**72**:4009. DOI: 10.1103/PhysRevLett. 72.4009

[3] Thomas H, Morfill GE, Demmel V. Plasma crystal: Coulomb crystallization in a dusty plasma. Physical Review Letters. 1994;**73**:652. DOI: 10.1103/ PhysRevLett.73.652

[4] Piel A. Plasma crystals—Structure and dynamics. Plasma Fusion Reserch. 2009;**4**:013. DOI: 10.1585/pfr.4.013

[5] Hayashi Y. Structural transitions of 3-dimensional Coulomb crystals in a dusty plasma. Physica Scripta. 2001; **T89**:112. DOI: 10.1238/Physica. Topical.089a00112/meta

[6] Kak AC, Slaney M. Principles of Computerized Tomographic Imaging: Classics in Applied Mathematics. Vol. 33. Philadelphia, PA: Society for Industrial and Applied Mathematics; 2001. p. 49

[7] Thomas E, Williams JD, Silver J. Application of stereoscopic particle image velocimetry to studies of transport in a dusty (complex) plasma. Physics of Plasmas. 2004;**11**:L37. DOI: 10.1063/1.1755705

[8] Käding S, Melzer A. Three-dimensional stereoscopy of Yukawa (Coulomb) balls in dusty plasmas. Physics of Plasmas. 2006;**13**:090701. DOI: 10.1063/1.2354149

[9] Hayashi Y. Radial ordering of fine particles in plasma under microgravity condition. Japanese Journal of Applied Physics. 2005;**44**:1436. DOI: 10.1143/ JJAP.44.1436

[10] Thomas HM, Morfill GE, Fortov VE, Ivlev AV, Molotkov VI, Lipaev AM, et al. Complex plasma laboratory PK-3 plus on the International Space Station. New Journal of Physics. 2008;**10**:033036. DOI: 10.1088/1367-2630/10/3/033036

[11] Samsonov D, Elsaesser A, Edwards A, Thomas HM, Morfill GE. High speed laser tomography system. The Review of Scientific Instruments. 2008;**79**:035102. DOI: 10.1063/ 1.2885683

[12] Kroll M, Harms S, Block D, Piel A. Digital in-line holography of dusty plasmas. Physics of Plasmas. 2008;**15**: 063703. DOI: 10.1063/1.2932109

[13] Goldbeck DD. Analyse dynamischer Volumenprozesse in komplexen Plasmen Dr. Thesis. [in German]. Fakultät für Physik, LMU München; 2003. Available from: https://edoc.ub .uni-muenchen.de/1111/

[14] Annaratone BM, Antonova T, Goldbeck DD, Thomas HM, Morfill GE. Complex-plasma manipulation by radiofrequency biasing. Plasma Physics and Controlled Fusion. 2004;**46**:B495. Available from: https://iopscience.iop. org/article/10.1088/0741-3335/46/12B/ 041/meta

[15] Lippmann G. PHOTOGRAPHY—Reversible prints. Integral photographs. Comptes Rendus de l'Académie des Sciences. 1908;**146**:446. Available from: https://people.csail.mit.edu/fredo/ PUBLI/Lippmann.pdf

[16] Lynch K, Fahringer T, Thurow B. Three-dimensional particle image velocimetry using a plenoptic camera. In: Proceeding of 50th AIAA Aerospace

Sciences Meeting Including the New Horizons Forum and Aerospace Exposition; 2012. p. 1056. Available from: https://arc.aiaa.org/doi/pdf/10.2514/6.2012-1056

[17] Cenedese A, Cenedese C, Furia F, Marchetti M, Moroni M, Shindler L. 3D particle reconstruction using light field imaging. In: Proceeding of 16th Int Symp on Applications of Laser Techniques to Fluid Mechanics; 2012. Available from: http://ltces.dem.ist.utl.pt/lxlaser/lxlaser2012/upload/253_paper_qoisoj.pdf

[18] Shi S, Wang J, Ding J, Zhao Z, New TH. Parametric study on light field volumetric particle image velocimetry. Flow Measurement and Instrumentation. 2016;**49**:70-88. DOI: 10.1016/j.flowmeasinst.2016.05.006

[19] Hall EM, Thurow BS, Guildenbecher DR. Comparison of three-dimensional particle tracking and sizing using plenoptic imaging and digital in-line holography. Applied Optics. 2016;**55**:6410-6420. DOI: 10.1364/AO.55.006410

[20] Pégard NC, Liu H-Y, Antipa N, Gerlock M, Adesnik H, Waller L. Compressive light-field microscopy for 3D neural activity recording. Optica. 2016;**3**:517-524. DOI: 10.1364/OPTICA.3.000517

[21] Kim J, Jeong Y, Kim H, Lee C-K, Lee B, Hong J, et al. F-number matching method in light field microscopy using an elastic micro lens array. Optics Letters. 2016;**41**: 2751-2754. DOI: 10.1364/OL.41.002751

[22] Nonn T, Jaunet V, Hellman S. Spray droplet size and velocity measurement using light-field velocimetry. In: Proceeding of ICLASS 2012, 12th Triennial International Conference on Liquid Atomization and Spray Systems; 2012. Available from: http://www.ilasseurope.org/ICLASS/iclass2012_Heidelberg/Contributions/Paper-pdfs/Contribution1219_b.pdf

[23] Hartmann P, Donkó I, Donkó Z. Single exposure three-dimensional imaging of dusty plasma clusters. The Review of Scientific Instruments. 2013; **84**:023501. DOI: 10.1063/1.4789770

[24] Jambor M, Nosenko V, Zhdanov SK, Thomas HM. Plasma crystal dynamics measured with a 3D plenoptic camera. The Review of Scientific Instruments. 2016;**87**:033505. DOI: 10.1063/1.4943269

[25] Sanpei A, Takao N, Kato Y, Hayashi Y. Initial result of three-dimensional reconstruction of dusty plasma through integral photography technique. IEEE Transactions on Plasma Science. 2016;**44**:508. Available from: https://ieeexplore.ieee.org/abstract/document/7336542

[26] Sanpei A, Tokunaga K, Hayashi Y. Design of an open-ended plenoptic camera for three-dimensional imaging of dusty plasmas. Japanese Journal of Applied Physics. 2017;**56**:080305. DOI: 10.7567/JJAP.56.080305/meta

[27] Sun Z, Baldwin JK, Wei X, Wang Z, Jiansheng H, Maingi R, et al. Initial results and designs of dual-filter and plenoptic imaging for high-temperature plasmas. The Review of Scientific Instruments. 2018;**89**(10):E112. DOI: 10.1063/1.5036633

[28] Hunter RS. Photoelectric color difference meter. Journal of the Optical Society of America. 1958;**48**:985-995. DOI: 10.1364/JOSA.48.000985

[29] Ng R, Levoy M, Bredif M, Duval G, Horowitz M, Hanrahan P. Light field photography with a hand-held plenoptic camera. Stanford CTSR, Tech. Rep. No. 2005-02; 2005

[30] Lumsdaine A, Georgiev T. The focused plenoptic camera. In: IEEE Int.

Conf. Computational Photography; 2009. p. 1. DOI: 10.1109/ICCPHOT.2009.5559008

[31] Hayashi Y, Kawano M, Sanpei A, Masuzaki S. Mie-scattering ellipsometry system for analysis of dust formation process in large plasma device. IEEE Transactions on Plasma Science. 2016;**44**:1032. Available from: https://ieeexplore.ieee.org/document/7475947

[32] Hayashi Y, Sanpei A. Mie-scattering ellipsometry. In: Ellipsometry—Principles and Techniques for Materials Characterization. Rijeka, Croatia: Faustino Wahaia, IntechOpen; 2017. DOI: 10.5772/intechopen.70278

[33] Lakowicz JR. Principles of Fluorescence Spectroscopy. 3rd ed. Boston, MA: Springer; 1999. DOI: 10.1007/978-0-387-46312-4

[34] Sanpei A, Kigami T, Kanaya H, Hayashi Y, Sampei M. Levitation of microorganisms in the sheath of an RF plasma. IEEE Transactions on Plasma Science. 2018;**46**:718. Available from: https://ieeexplore.ieee.org/document/8063380

[35] Sanpei A, Kigami T, Hayashi Y, Himura H, Masamune S, Sampei M. First observation of crystallike configuration of microorganisms in an RF plasma. IEEE Transactions on Plasma Science. 2019;**47**:3074. Available from: https://ieeexplore.ieee.org/document/8726398

[36] Richardson WH. Bayesian-based iterative method of image restoration. Journal of the Optical Society of America. 1972;**62**:55. DOI: 10.1364/JOSA.62.000055

[37] Lucy LB. An iterative technique for the rectification of observed distributions. Astronomical Journal. 1974;**79**:745. DOI: 10.1086/111605

[38] Bolan J, Johnson KC, Thurow BS. Preliminary investigation of three-dimensional flame measurements with a plenoptic camera. In: Proceeding in 30th AIAA Aerodynamic Measurement Technology and Ground Testing Conference; 2014. DOI: 10.2514/6.2014-2520

[39] Onchi T, Fujisawa A, Sanpei A. A prototype diagnostics system to detect ultraviolet emission for plasma turbulence. The Review of Scientific Instruments. 2014;**85**:113502. DOI: 10.1063/1.4900660

[40] Onchi T, Ikezoe R, Oki K, Sanpei A, Himura H, Masamune S. Tangential soft-x ray imaging for three-dimensional structural studies in a reversed field pinch. The Review of Scientific Instruments. 2010;**81**:073502. DOI: 10.1063/1.3455216

[41] Nishimura K, Sanpei A, Tanaka H, Ishi G, Kodera R, Ueba R, et al. 2-D electron temperature diagnostic using soft-x ray imaging technique. The Review of Scientific Instruments. 2014;**85**:33502. DOI: 10.1063/1.4867076

[42] Feng Y, Goree J, Liu B. Accurate particle position measurement from images. The Review of Scientific Instruments. 2007;**78**:053704. DOI: 10.1063/1.2735920

[43] Rogers SS, Waigh TA, Zhao X, Lu JR. Precise particle tracking against a complicated background: Polynomial fitting with Gaussian weight. Physical Biology. 2007;**4**:220. DOI: 10.1088/1478-3975/4/3/008

4

Flow and Segregation of Granular Materials during Heap Formation

Sandip H. Gharat

Abstract

Segregation during flow of granular materials is important from an industrial point of view. Granular materials segregate during flow due to their physical properties (such as size, shape, and density). A considerable work has been done on granular segregation in the past (two decades). This chapter is divided into three parts. In the first part, a review of work done on heap formation is presented. Experimental work during heap formation by intermittent feeding is reported in the second part. The system used is a simplified model for the feeding of raw materials to a blast furnace, which is widely used for the manufacture of iron and steel. Experiments carried out using 2-D system and steel balls of size 1 and 2 mm are used as model granular materials. Image analysis is done to detect the position of each particle using an in-house computer code. Accuracy and efficiency of image analysis techniques were found to be good enough as we have used 1 and 2 mm spherical steel balls for all the cases studied. The chapter ends with concluding remarks.

Keywords: granular materials, segregation, heap, granular flow, blast furnace

1. Introduction

Granular materials are a collection of solid particles in the size range of a few millimeters to a few centimeters. They are present everywhere in nature and are the second most manipulated materials after water [1]. Commonly used granular materials are salt, sugar, coffee, and rice in day-to-day life and sand, gravel, coal, powders, fertilizers, and pharmaceutical pills in various industries. They are made up of different constituents, ranging from 1 μm to 1 km (theoretically) and having various shapes. Here, we are mainly interested in the above-mentioned size range as most of the materials used in the industry are in this range. Coke and sinter particles used in the blast furnace are also in the above-mentioned size range. Electrostatic charges and surface forces are negligible for such particles. The properties of granular materials are very different from the properties of constituent particles. They primarily depend on the nature of interstitial fluid (air, water, etc.) and forces acting between the particles. The important forces are van der Waals forces, interaction forces, frictional forces, and capillary forces when some interstitial fluid is present. If the influence of the interstitial fluid (like air) is neglected, then we called it as dry granular materials. Otherwise, the system will become more complex. Granular materials behave like a solid when they are packed in a container or in case of a heap. They can also withstand strong shear force as they can support our body weight while walking on the beach. They can flow through a hopper and can take the shape of a container like liquids. But unlike liquids, the pressure at the base of a bin filled with granular materials does not increase after a certain height of filling. In each of the states (both solid and liquid), they exhibit different properties that are unique in nature. This dual behavior made them an interesting and complicated system to study the basic of granular materials. For the size range of millimeters to few centimeters, we can ignore electrostatic forces. Also, the focus of this study is on simple system (quasi-two-dimensional rectangular bin: heap formation) where particles are uniform in shape (spherical) and air (whose viscosity and density are very less as compared to those of SS balls) is the interstitial fluid that has negligible effect.

The flow of granular material is encountered in various industries during transportation of materials, mixing of rotary kiln, storage of materials in hopper and silos, crushing and grinding, mechanical separations, etc. It is also encountered in nature during transportation at river bed, during formation of sand dunes, in lava flows, and in avalanches. The properties of granular materials during flow are different, which gives problems in various industries. Granular materials have a tendency to segregate. Segregation occurs due to small differences in either size or density when they flow or are shaken or vibrated. But this is not true with liquids and gases. Fluids often mix with each other after agitation in industrial processes. Segregation of granular materials during transport and handling is an important problem in the chemical, pharmaceutical, and food industry wherever mixing is required. Segregation is the fundamental characteristics of granular flow. During flow, granular matter segregates.

Segregation mechanisms have been studied experimentally [2–6] as well as by simulations [7–9]. Particle dynamics [10, 11] or discrete element method [12] and Monte Carlo simulations [10, 13] are used as computational techniques. If a bed of grains of different sizes is sheared parallel to the top surface, the larger grains segregate to the surface of the bed and the smaller ones to the bottom. Brazil nut effect is the most celebrated example to understand the mechanism of segregation where large particles remain at the top.

In this chapter, the problem of segregation has been studied during flow in a blast furnace using a simplified system. A binary mixture of spherical particles of different sizes is used in a quasi-two-dimensional system, with a finite mass of the mixture poured periodically. Although this problem has not been studied in detail previously, some works have reported a related system in which the material is poured continuously. The objective of the present work is to study the segregation of a binary mixture of different diameter spheres in a quasi-two-dimensional rectangular bin (two vertical glass plates separated by a gap of 10 mm) during heap formation by intermittent pouring of a finite mass of the mixture at one edge of the bin. The concentration distribution in the system is obtained by image analysis for different system parameters including the mass of the material poured, the initial concentration of the mixture, the height of pouring, and the size ratio of particles in the mixture. The experimental details are presented next followed by the results and conclusions.

1.1 Blast furnace

The blast furnace is charged with coke and sinter particles by means of an inclined chute rotated about its vertical axis (**Figure 1**). For each rotation of the chute, the particles on the sloping burden surface form a circular ring. The particles in the ring then flow along the burden surface and form a new surface profile. These particles are distributed over a range of sizes. During the flow, the particles have a tendency to separate due to differences in size and density as shown in **Figure 2**.

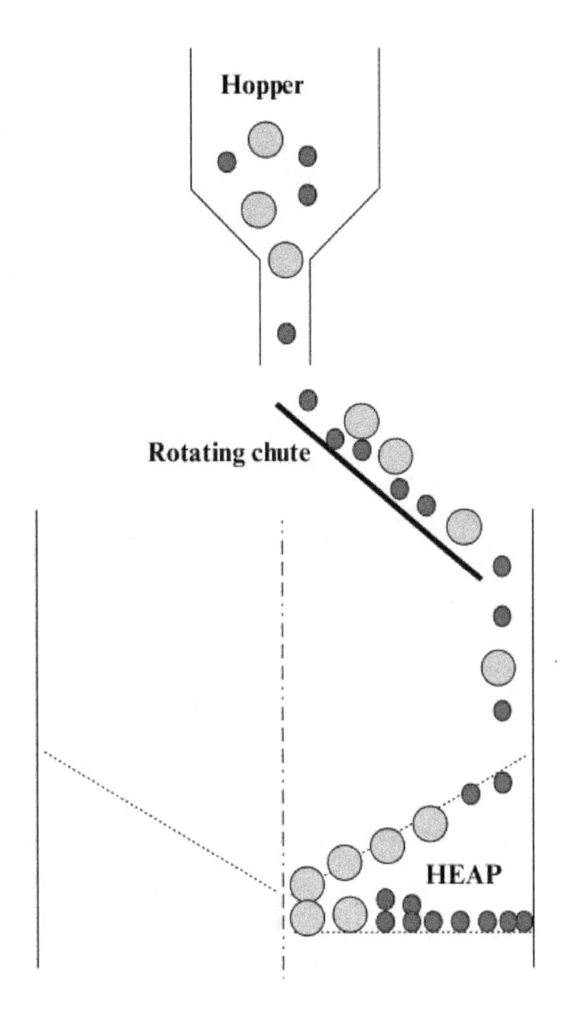

Figure 1.
Schematic view of blast furnace.

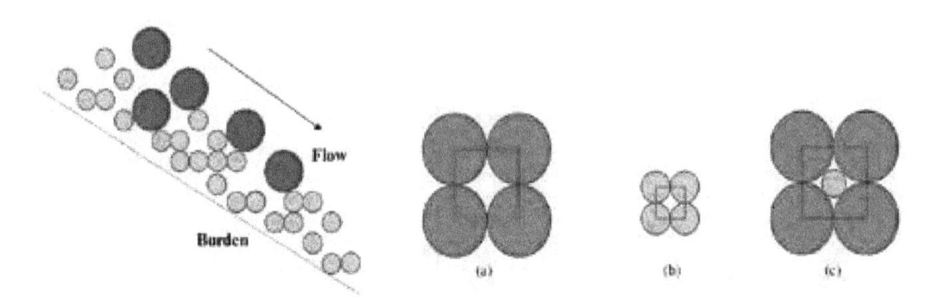

Figure 2.
Schematic view of a partially segregated mixture of particles flowing down along the burden surface (left). Packing of particles: (a) monosize large, (b) monosize small, (c) mixed.

This results in smaller particles being deposited first and larger particles rolling more distance before they stop. According to this mechanism, larger particles should be concentrated near the center of the blast furnace and smaller particles closer to the radius at which the material stream impinges on the burden surface.
Due to limited time of flow, segregation of granular material is generally not complete. This affects the packing of particles and thus the gas flow resistance and

distribution of gas through blast furnace. In the case of complete segregation, the void fraction in the region containing only large particles is roughly the same as void fraction in the small particle region (**Figure 2**). However, the region with larger particles has a large average interstitial pore diameter and thus a lower flow resistance. In a mixed region, the void fraction is lower than that of pure component case and the average pore diameter is lower than the larger particle case. Thus, the extent of segregation affects both void fraction and average pore diameter; both affect flow resistance and consequently gas distribution through the blast furnace.

2. Experimental details

2.1 Experimental setup

Experiments are carried out in quasi-two-dimensional rectangular bins (two vertical glass plates separated by a gap of 10 mm). Stainless steel (SS 316) balls of different size of diameters (1 and 2 mm) are used as model granular materials. The images taken are analyzed using computer code to detect the particles. Image analysis technique is used to detect the position and size of the particles. The profiles of the number fraction of big (2 mm) particles along the flow direction across the depth in the layers are plotted separately for the top and bottom regions of the heap. Flowing layer decreases toward the end. We have assumed flowing depth constant for simplicity. A schematic view of heap formation is shown in **Figure 3**.

The experimental setup consists of two glass plates (52 cm × 29 cm), L-shaped aluminum divider, and an auxiliary hopper. The side walls (glass plates) are transparent to facilitate imaging. The gap between two vertical glass plates is 1 cm, and the L-shaped aluminum spacer is placed. The L-shaped divider is a chute used in the blast furnace with the angle of the chute close to 40°. Aluminum divider forms an inclined chute so that materials will easily flow and then settle at the bottom to form a heap. The auxiliary hopper is placed on the top of the bin for supplying granular materials as a feed. The exit of the chute is designed so as to provide minimum

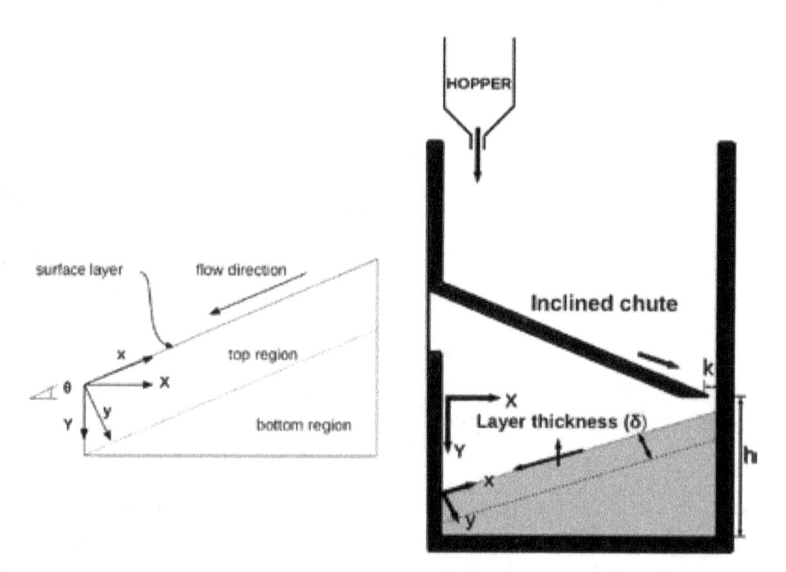

Figure 3.
Schematic view of heap formation, comprising top and bottom regions (left) along with setup used (right). The coordinate system shows x = 0, y = 0 at the low end of the top surface of the pile. The coordinate system employed in the analysis is shown (left).

disturbance to the flow. The tip of the chute is smoothened by reducing the sharpness of the exit (the point at which materials fall from the gap k and then fall onto the heap), which in turn reduces the disturbance to the flow. A black paper is placed at the back of the glass plate so as to ensure good-quality images after experiments. The position of the aluminum divider can be adjusted to control the flow rate and the height of fall to the top of the heap. Halogen lamp (1000 W) is used as light source for better image analysis. The setup is leveled properly to ensure that side walls are vertical and the base is horizontal. The distance between divider exit and heap (forms at the bottom after experiments) is sufficient so that materials will flow easily and get enough time to settle at the bottom.

2.2 Experimental method

A known weight of the binary mixture of a specified composition is poured intermittently into the auxiliary hopper and then allowed to pass through the gap (k) between the divider and the plate and finally settle onto the heap. This forms a layer on the surface of the heap. The process is repeated to form a number of layers until the top of the heap is close to the end of the chute. The heap thus formed is imaged to determine the concentration distribution of the different size particles in the heap. To obtain a sufficiently high resolution, the heap is imaged in eight parts. A transparency of known scale is placed while capturing images for each part. Images taken with and without transparency give more accuracy and also help in combining all parts of the heap. A source of light is placed parallel to the camera and in front of the heap so that each particle has a single reflection of the light, the size of which depends on the particle diameter. Each part is captured using a digital camera (Nikon D3100) at 1/100 shutter speed and intensity of 800/1600. Captured images have a resolution of 4608 $\times$ 3072 pixels; one pixel is equal to 0.020616 mm. Data (in terms of pixel) obtained for the detected images are converted into mm. The images are analyzed to determine the position of each particle (small and big) and number fraction of big particles across the layer. Each experiment is repeated eight times to get average data. The details of image analysis are explained in the next section.

2.3 Image analysis technique

Image analysis technique is used to detect the position and size of the particles from captured images using an in-house computer code (*c programming*). The detection of particles (small and big) is done based on the number of pixels obtained in each cluster (white spots in **Figure 4**). As we have used binary mixture (1 and 2 mm) particles while performing experiments, analysis became easy. Therefore, the detection of particles is accurate and efficiency is high. These data are used to plot profiles of the number fraction of big particles vs. component x (flow direction) with depth y (mm) in the layer. This will be useful to develop the segregation model based on it.

2.3.1 Detection of particles

The images (4608 $\times$ 3072 pixels) are captured using Nikon D3100 camera. In all images, particles appear as tiny bright spots on a black background. The typical captured image is shown in **Figure 4**.

Thresholding is used to convert gray scale images into black and white images. The threshold value is chosen such that the less illuminated particles from the inner layers are eliminated and particles from the front layer appear as clusters of white pixels on a black background. During the thresholding process, individual pixels in

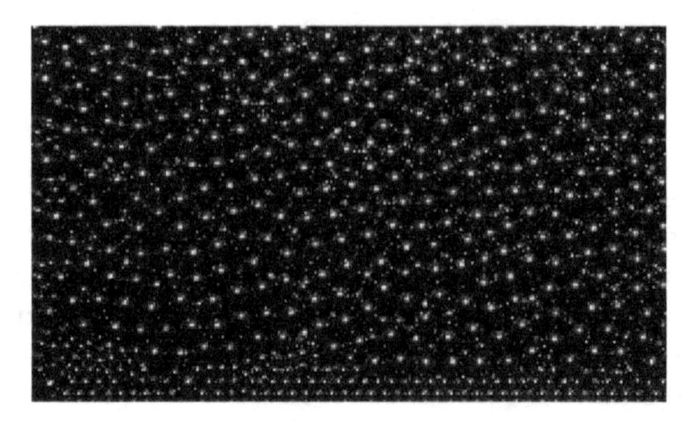

Figure 4.
Typical captured image: the particles are seen as bright spots.

Figure 5.
Detected particle as a white cluster after thresholding.

an image are marked as "object" pixels if their value is greater than some threshold value (assuming an object to be brighter than the background) and as "background" pixels otherwise. This convention is known as threshold above. If the chosen threshold value is 200, it means all the pixel value (range 0–255) that are above 200 will be marked as white (255) and less than 200 will be marked as black (zero value). Also, we can use threshold below, which is opposite of threshold above. In this way, the binary image is created by coloring each pixel white or black, depending on a pixel value. We next scan the thresholded image for a white pixel, starting from top left corner, in both right and downward directions. Once a white pixel is found, then the neighboring pixel should be scanned to detect all the white pixels that form the cluster. Similarly, all the white clusters are identified in that image. The captured image after thresholding shows individual white clusters of pixels (**Figure 5**).

2.3.2 Position and size of particle

The position (X and Y coordinates) of individual cluster is determined based on mean position of all the pixels present in that cluster. The radius of gyration (R_g) is used to calculate the size of the cluster. It is calculated as the root mean square distance of the object's parts from either its center of mass or an axis. Here, we have used this parameter to characterize the size of individual particle (cluster) based on center of mass.

$$\text{Radius of gyration}\left(R_g\right) = \sqrt{\frac{\sum_{i=1}^{N}\left[(X_i - X_0)^2 + (Y_i - Y_0)^2\right]}{N}} \quad (1)$$

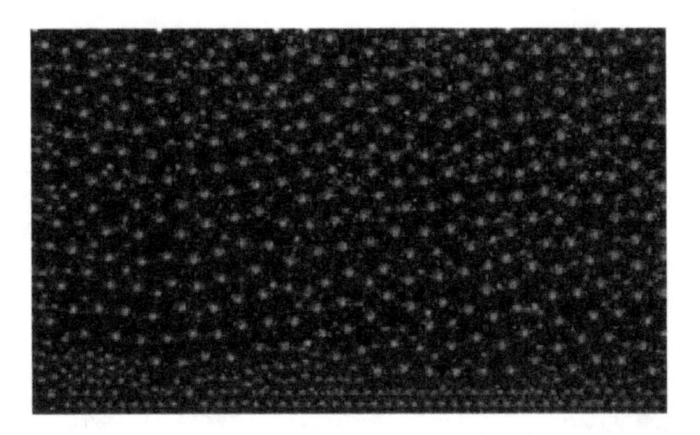

Figure 6.
Detected particle: red (2 mm) and blue (1 mm).

where X_i and Y_i are the X and Y coordinates of each pixel in the cluster. X_0 and Y_0 are center of mass of each cluster of X and Y coordinates, respectively, and they are calculated as

$$X_0 = \sum_{i=1}^{N} \frac{X_i}{N} \tag{2}$$

$$Y_0 = \sum_{i=1}^{N} \frac{Y_i}{N} \tag{3}$$

The particles detected in **Figure 5** as a white cluster are marked as red and blue based on the radius of gyration. The values of R_g greater than or equal to 3.5 are marked as red or as blue. Detected particles are shown in **Figure 6** in red (big particle—2 mm) and blue (small particle—1 mm).

For analysis, we have chosen a coordinate system with its origin (x,y) at the top left corner of the free surface layer, such that the +ve x-axis is opposite the flow direction and + ve y-axis is in the downward direction. The default and transformed coordinate axis is shown in **Figure 3**. The new position of the particles is calculated using the following transformation:

$$x = X\cos\theta + Y\sin\theta \tag{4}$$

$$y = Y\cos\theta - X\sin\theta \tag{5}$$

where x, y and X, Y are the position according to the new and old coordinate system respectively and θ is the angle the free surface makes with the horizontal.

3. Results and discussion

3.1 Heap formation and layer thickness

Figure 7 shows a schematic view of the layers formed due to repeated pouring of the mixture. There is no time internal between two pouring. Once the first layer settles and forms a heap, then another layer is poured over the heap form in earlier layer. In each pouring layer, the upper particle (mostly large in size) gets eroded in the next layer cycle. Several repeated pouring (7–8 layers) forms a heap, which is analyzed to get the data. One pouring corresponds to layer thickness (δ). The initial layers formed do not reach the edge of the bin since a fixed amount of material is poured and the angle of repose is fixed. As more layers are poured, the height of the heap increases and the layer spans the bin. When there is no movement of particles

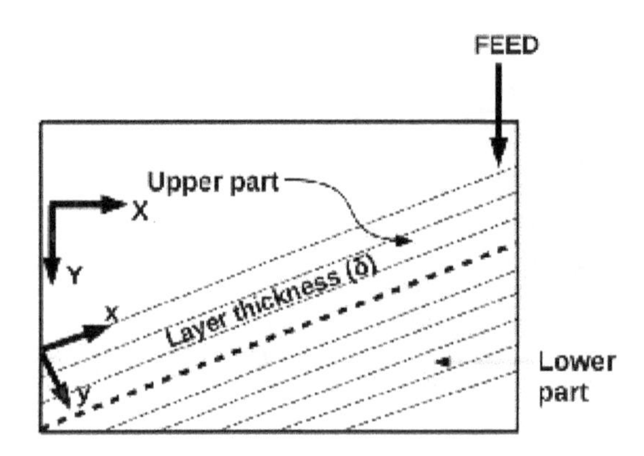

Figure 7.
Schematic diagram of layer formation showing top region, bottom region, and layer thickness.

in the heap form, then stable state heap is achieved. We denote the region in which the layer spans part of the bin as the bottom region and the region with the full layers the top region. The regions are demarcated by a dashed line in **Figure 7**. The layer thickness (δ) is measured experimentally as well as estimated theoretically using measured bulk densities of the mixtures.

The parameters varied in the experiments are the volume (mass) of mixture and height of the divider from the bottom for fixed gap (k = 5 mm) between the plate and the divider. Therefore, results presented below are based on these parameters. The profiles of the number fraction of big (2 mm) particles along the flow direction across the depth in the flowing layer for top and bottom regions are shown below. Each flowing layer is divided into bin of size (layer thickness × 10 mm). Layer thickness depends upon the amount of volume poured in each step. Layer thickness is calculated experimentally as well as theoretically for known volume (mass) of mixture, which is explained below. Practically, thickness of layer reduces with distance but here we assumed as same from initial (pouring point) to the end point. Experimentally, it is measured before and after pouring using Image J software in terms of pixels and then converted to mm. Theoretical calculations are done as follows: amount of material poured (i.e., mass) converted into volume using density of material. Then using setup geometry, the distance of one layer in x-direction (flow direction) is known, which is typically the distance of the surface profile. The gap between two glass plates is known as 10 mm (1 cm). Finally, the following formula is used: volume of material (mm^3) = layer thickness (δ) × gap (k) × distance in flow direction (x). We have found that the obtained values of theoretical layer thickness and experimental are the same. In **Figure 9**, we have obtained layer thickness of 10 mm for 150 g of mixture and average layer thickness taken for each layer is 5 mm (in the middle of layer which is half of 10). Therefore, the depth of layer from surface of heap is 15 mm (we have neglected top layer in all cases studied), which is consider as sampling depth in **Figure 9**, and similar calculation is done for subsequent layers. We also checked number fraction profile at the back side of heap (opposite to front side) and found similar results compared to front plate/side. The error bars denote the standard deviation over eight experiments.

3.1.1 Effect of mass of pouring

Figure 8 shows the profile for 50% mixture of small (1 mm) and big (2 mm) particles with equal density for different volume of pouring for top region (**Figure 8a–c**) and bottom region (**Figure 8d–f**). Equal percent of small and big particles is used for all cases of mass of pouring. All other parameters (composition,

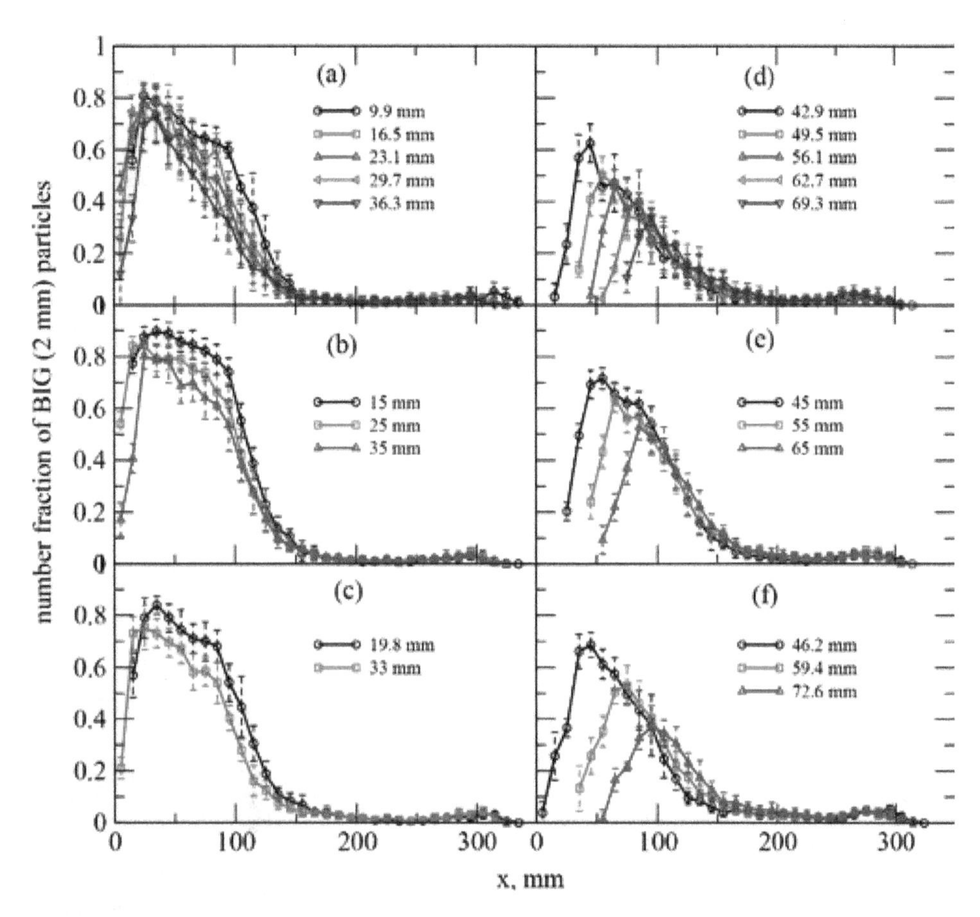

Figure 8.
Profiles of number fraction of big (2 mm) particles along the flow direction, x (mm) across the depth in the flowing layer, y (mm) for equal % of mixture. Volume of pouring (a) 100, (b) 150, and (c) 200 g for top region and similarly for bottom region (d), (e), and (f). Height of the divider from the bottom (h) is: 22 cm. Error bars show the standard deviation over eight experiments.

height, and size ratio) are the same. The number fraction profiles of big particles for bottom region overlap with each other and are almost identical for all the cases of volume of pouring (**Figure 8d–f**). The pattern formation for three different pouring is somewhat similar and is apparent. The profiles of top region **Figure 8a–c** do not show significant variation in the number fraction across the depth in flowing layer, and it indicates that segregation in flowing layer is independent of volume of pouring. Though we increase the mass of particles (i.e., number of particles), percent fraction of small and big particles in the mixture remains the same (i.e., 50%). Also corresponding value of layer thickness varies with the mass of pouring, which is calculated theoretically as well as experimentally. Therefore, in each bin, the variation in number fraction of particles along the flow direction and flow depth is less for all three cases studied (i.e., 100, 150, and 200 g). The data indicate segregation (partial) with big particles at the far end of the heap and small particles settle near the heap. The nature of number fraction profile is almost equal and will overlap with each other if we combined all the plots for top region as well as bottom region.

3.1.2 Effect of height variation

Figure 9 shows profiles of 50% by volume of mixture of small and big particles with equal density for different height of the divider from the bottom. The top

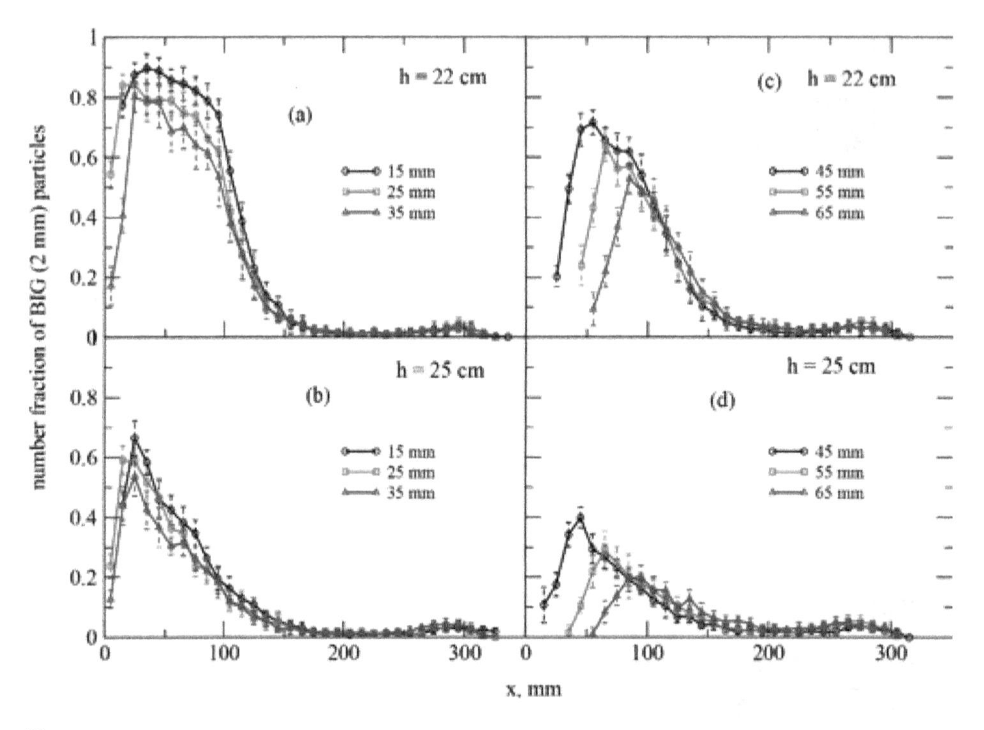

Figure 9.
Profiles of number fraction of big (2 mm) particles along the flow direction, x (mm) across the depth in the flowing layer, y (mm) for equal % of mixture. The height of the divider from the bottom is (a) 22 and (b) 25 cm for top region and similarly for bottom region (c) and (d). The volume of pouring is 150 g. Error bars show the standard deviation over eight experiments.

region (**Figure 9a** and **b**) and bottom region (**Figure 9c** and **d**) show large variations in number fraction profile plotted for different layer thickness. This indicates that mixing is more in flowing layer if height of the divider from the bottom increases. Also increase in height creates more energy for particles during flow due to which small particles may bounce and settle at the far end of the heap. This variation in mixing/segregation pattern also depends on the height of the heap and other working parameters, which have to be studied further. Here, we have considered only two values of height while studying the effect on segregation pattern. Also, the system studied here is 2D; therefore, the effect of wall also plays a vital role in heap formation and mixing/segregation pattern. Wall effect will be neglected if we perform similar study in 3D heap formation setup.

4. Concluding remarks

Segregation of binary granular mixtures, differing in size and equal density, during heap formation by intermittent feeding in a quasi-two-dimensional rectangular bin has been investigated. Experiments were performed by varying the mass (volume) of the mixture and the height of the feeding point above the bottom of the bin. Data show large particles travel more distance than small particles as smaller ones can easily fit into void spaces created during flow of materials. The profiles of the number fraction decreases along the flow path with depth in the deposited layer for both the parameters studied (mass of pouring and height). During heap formation, the small particles concentrate near the upper part of the heap and large particles travel more distance and settle at the bottom. The profiles for different volumes of pouring do not show significant variation and this indicates that segregation in flowing layer is independent of volume of pouring. This is due to mere increase in mass (volume) of pouring, but number fraction in each layer remains the same. Therefore, they overlap with each other as observed. In all cases studied, there is a peak in number fraction profile near the pouring end. Increase in height (h) decreases the number fraction of big particles, which in turn affects the extent of segregation. This indicates that mixing is more in flowing layer if we increase height of the divider above the bottom of the bin for 22 and 25 cm height. Also increase in height creates more energy for particles during flow due to which small particles may bounce and settle at the far end of the heap. The impact for height of fall is clearly apparent near the pouring point of the heap. We have investigated size segregation of particles during heap formation. It has direct implications in engineering applications, and one of the examples of such granular process is blast furnace. These findings may be useful for the optimization of the design of blast furnace and its accessories. Also, the methodology followed in this chapter can be implemented to minimize particle segregation in an actual blast furnace.

Acknowledgements

The experimental work was done at IIT Bombay during a visit by the author and it is gratefully acknowledged.

Author details

Sandip H. Gharat[1,2]

1 Department of Chemical Engineering, Shroff S.R. Rotary Institute of Chemical Technology, Ankleshwar, Gujarat, India

2 Department of Chemical Engineering, Indian Institute of Technology Bombay, Mumbai, India

*Address all correspondence to: sandipgharat78@gmail.com

References

[1] Patrick R, Nicodemi M, Delannay R, Ribiere P, Bideau D. Slow relaxation and compaction of granular systems. Nature Materials. 2005;**4**:121-128

[2] Ding YL, Forster R, Seville JPK, Parker DJ. Segregation of granular flow in the transverse plane of a rolling mode rotating drum. International Journal of Multiphase Flow. 2002;**28**:635-663

[3] Khakhar DV, Orpe AV, Hajra SK. Segregation of granular materials in rotating cylinders. Physica A. 2003;**318**: 129-136

[4] Makse HA, Havlin S, King PR, Stanley HE. Spontaneous stratification in granular mixtures. Nature. 1997;**386**: 379-382

[5] Ottino JM, Khakhar DV. Fundamental research in heaping, mixing, and segregation of granular materials: Challenges and perspectives. Powder Technology. 2001;**121**:117-122

[6] Savage SB, Lun CKK. Particle size segregation in inclined chute flows of dry cohesionless granular solids. Journal of Fluid Mechanics. 1988;**189**:311-335

[7] Shinbrot T, Zeggio M, Muzzio FJ. Computational approaches to granular segregation in tumbling blenders. Powder Technology. 2001;**116**:224-231

[8] Tripathi A, Khakhar DV. Rheology of binary granular mixtures in the dense flow regime. Physics of Fluids. 2011;**23**: 113302

[9] Yaowei Y, Saxen H. Experimental and dem study of segregation of ternary size particles in a blast furnace top bunker model. Chemical Engineering Science. 2010;**65**:5237-5250

[10] Khakhar DV, McCarthy JJ, Ottino JM. Radial segregation of granular mixtures in rotating cylinders. Physics of Fluids. 1997;**9**:3600-3614

[11] Vargas WL, Hajra SK, Shi D, McCarthy JJ. Suppressing the segregation of granular mixtures in rotating tumblers. AIChE. 2008;**54**: 3124-3132

[12] William RK, Curtis JS, Wassgren CR, Hancock BC. Modeling granular segregation in flow from quasi-three-dimensional, wedge-shaped hoppers. Powder Technology. 2008;**179**: 126-143

[13] Rosato A, Prinz F. Monte Carlo simulation of particulate matter segregation. Powder Technology. 1986; **49**:59-69

5

Observation and Analyses of Coulomb Crystals in Fine Particle Plasmas

Yasuaki Hayashi

Abstract

Observations of crystal-like ordering of fine particles in plasmas were first reported in 1994, when we succeeded to observe it by growing carbon fine particles in a methane plasma. Video cameras and Mie-scattering ellipsometry were applied for the analyses of fine particles and their crystal ordering. 3D and 2D crystal structures were observed for smaller and larger particles, respectively. The former structures were fcc, fco, and bct, but bcc structure was not observed. The result is due to the fact that the rearrangement from fcc to fco or bct occurs with both constant particle density in horizontal planes and constant interplane vertical distance. Behaviors of fine particles under microgravity were observed and analyzed using ready-made and injected fine particles. Its experimental result showed that the resultant force composed of electrostatic and ion drag forces pushes fine particles outward from the center forming a void.

Keywords: fine particle plasma, fine particle, plasma, Coulomb crystal, ellipsometry, Mie scattering, Mie-scattering ellipsometry, dusty plasma, microgravity, one-component plasma, strongly coupled plasma

1. Introduction

Observations of crystal-like ordering of fine particles in plasmas were first reported in 1994 [1–4]. They were performed by growing carbon particles in a methane plasma [1], silica particles growing in plasma [2], and ready-made polymer particles in argon plasmas [3, 4]. The possibility of observation of crystal-like ordering of fine particles, which was called Coulomb solid or Coulomb crystal, in a low-pressure plasma was predicted through calculation by Ikezi [5]. It was similar to those in the solution of colloids generally negatively single-charged, while micron-sized fine particles get thousands of electrons or more on their surface in a plasma. They can form solid state easily because of larger value of the Coulomb-coupling parameter, Γ, as follows:

$$\Gamma = \frac{(eQ)^2/4\pi\varepsilon_0 a}{k_B T}, \tag{1}$$

where e, Q, a, and T are the unit charge, the number of electron charge per one particle, the Wigner-Seitz radius, and absolute temperature, respectively. The numerator and denominator of Eq. (1) are the average Coulomb energy and the kinetic energy of a fine particle, respectively. In the case that $a \gtrsim \lambda_D$, where λ_D is the Debye length, the Yukawa-type interaction works and the Coulomb-coupling parameter is replaced with Γ^* defined as

$$\Gamma^* = \Gamma e^{-a/\lambda_D} = \frac{(eQ)^2/4\pi\varepsilon_0 a}{k_B T} e^{-a/\lambda_D,} \quad (2)$$

Charge neutrality is satisfied in a fine particle plasma as

$$-N_d Q + (n_i - n_e) = 0, \quad (3)$$

where N_d, n_i, and n_e are the density of fine particles, ions, and electrons, respectively. The first term of the left side of Eq. (3), $-N_d Q$, is negative and $N_d Q$ is balanced with the positive second term, $(n_i - n_e)$. Since an ion and an electron have much smaller mass and higher mobility than a fine particle, it is suggested that the part $(n_i - n_e)$ acts as background field equivalently. This means that negatively charged fine particles are surrounded by the positive background. Such a system is called one-component plasma [6]. As with charged particles, neutral atoms or molecules act on fine particles as background temperature environment. The friction to fine particles by neutral atoms or molecules decreases the temperature T in Eq. (1) or Eq. (2).

The physics of one-component plasma or strongly coupled plasma has been studied mainly by computer simulation such as the Monte Carlo method and molec-ular dynamics since 1957, when the liquid–solid phase transition by hard spheres in a confined condition was discovered by Alder and others [7–9]. Following this study, the phase transitions were confirmed using other repulsive potentials such as Coulomb [10], soft sphere [11], and Yukawa [12, 13].

Although theoretical or computer-simulation studies on one-component plasma and strongly coupled plasma have been proceeded, there are not many experimental studies except for the use of ion trap or colloidal solutions. Ions confined in a Pawl or Penning trap can form crystal ordering [14, 15]. Colloids in electrolyte solution form crystal structures [16]. However, special skills and time are required for the formation. Compared to that, Coulomb crystals in fine particle plasma are formed more easily and quickly.

In this chapter, the formation, observation, and analysis of Coulomb crystals in fine particle plasmas are presented. Current status of their application to study of solid state physics phenomena is also described.

2. Formation and observation of Coulomb crystal in fine particle plasmas

2.1 Formation of Coulomb crystal

Coulomb crystals in fine particle plasmas have been formed typically under the conditions as shown in **Table 1**.

Figure 1 shows an example of a plasma system for Coulomb crystal formation [17]. The system is a parallel-plate 13.56-MHz radio-frequency (RF) plasma device, which includes eight pieces of permanent magnets in the RF electrode to generate planar-magnetron plasma. Discharge gas is introduced and evacuated far from the area of center of vacuum chamber in order that fine particles are suspended under a stagnant condition to reduce the effect of gas drag. A side view of suspended fine particles above the RF electrode is shown in **Figure 2**. In this case, fine particles of spherical divinylbenzene polymer 2.74 μm in diameter were injected into argon plasma under the pressure of 65 Pa and RF power of 2 W.

Fine particles	Material	Polymer, silica, carbon
	Shape	Sphere
	Size	1–10 μm
	Size distribution	Monodisperse
Discharge device	Power supply	RF (13.56 MHz), DC
	Electrode type	Parallel-plate
Plasma conditions	Discharge gas	Argon, helium, methane
	Pressure	10–100 Pa
	Discharge power	As low as limit of particle suspension

Table 1.
Typical conditions for Coulomb crystal formation in plasma.

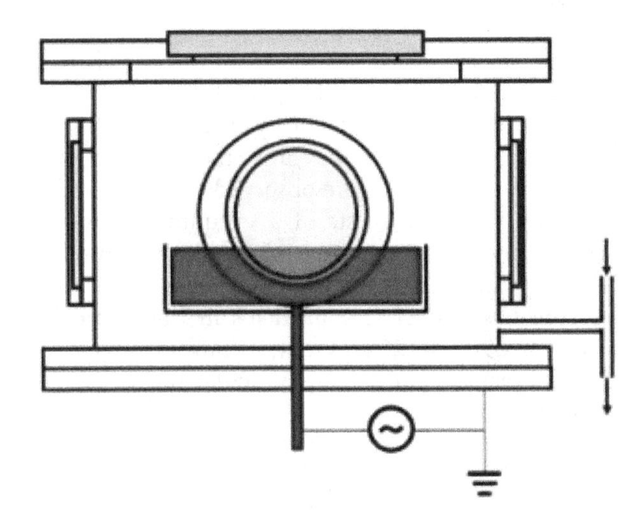

Figure 1.
Schematic view of plasma system for Coulomb crystal formation. Dark red and light blue parts show RF electrode and viewing glass windows, respectively. Other parts are grounded.

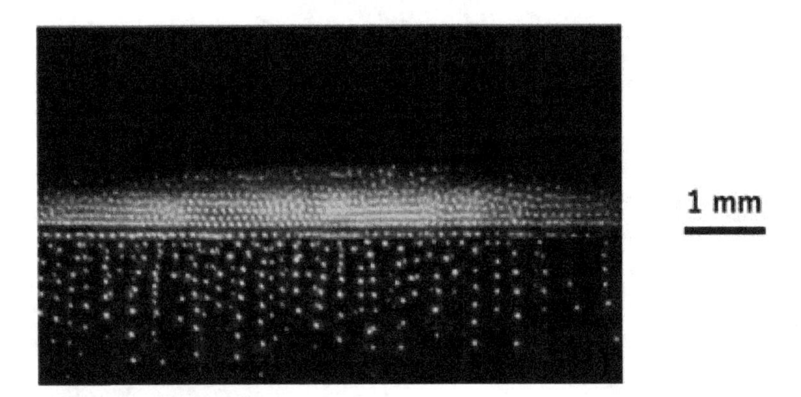

Figure 2.
Side view of arrangement of fine particles above RF electrode.

The arrangement of fine particles in **Figure 2** is divided up and down. The upper part of fine particles forms horizontal planes, while the lower part forms vertical strings. It is judged that the border of the two particle arrangements corresponds to the plasma-sheath boundary. It was explained by theoretical analysis [18] and computer simulation [19] that the vertical particle string of fine particles in the lower part is created by ion-wake field, which is effective in the region of ion speed more than Mach sound velocity, that is, in the sheath region. When the particle strings are closely packed, they form a two-dimensional (2D) hexagonal structure. Meanwhile, the horizontal layers of fine particles in the upper part are formed under the one-dimensional (1D) compressive force in the vertical direction [20]. Fine particles are arranged closely packing in each plane. To minimize the free energy in three-dimension (3D), such particles tend to be staggered with those in the next layer forming 3D close-packed structures: hexagonal close-packing (hcp) or face-centered cube (fcc). In this way, fine particles can form a 2D hexagonal structure and 3D structures such as fcc and hcp. The possibility of formation of body-centered cubic (bcc) structure will be described in Section 2.2.

2.2 Observation of Coulomb crystal

Coulomb crystals can be observed by laser light scattering. When a visible light laser is used, light scattering from micron-sized fine particle is in the Mie-scattering regime. In such a regime, fine particles can be individually observed by the scattered laser light of output power of even a few mW when they are in the arrangement of solid-like state.

3D structures of Coulomb crystal were observed by the use of plasma system as shown in **Figure 3**. The system consists of a vacuum reaction chamber, a blue argon-ion laser (wavelength: 488 nm, 100 mW output power), and two CCD video cameras, and a rotating analyzer for Mie-scattering ellipsometry [21]. In case of confirmation of the position of fine particles in the vertical direction, a red diode laser (wavelength: 690 nm, 20 mW output power) was also used. The light beam of the diode laser was expanded perpendicular to the RF electrode with the use of a cylindrical lens.

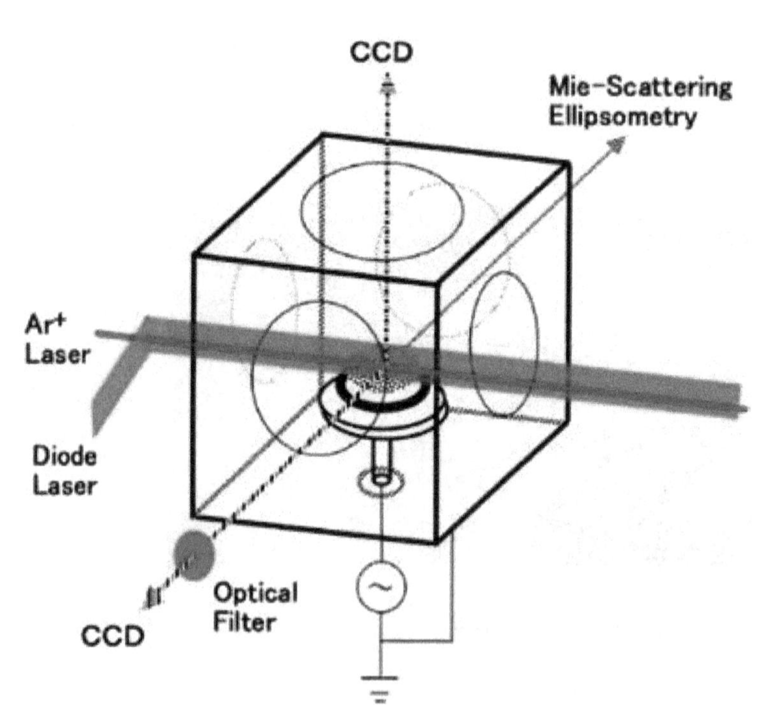

Figure 3.
Schematic of plasma system for observation of 3D structure of Coulomb crystal [21].

The vacuum chamber is a cube with a side length of 10 cm. Viewing windows are provided in the five faces of the chamber. Two are for the incidence and outgoing of laser light. Two others are for the observation of Coulomb crystals from the side and the top. The other one is for the monitoring of the particle diameter by Mie-scattering ellipsometry. RF of 13.56 MHz was applied to an electrode of 4 cm diameter with the chamber wall grounded. A ring of 3 cm inner diameter was put on the electrode in order to effectively trap fine particles above it by forming a potential bucket. Gas inlet and exhausting ports were provided close to each other so that particles were not transported by gas flow.

Fine particles were prepared by film growth on the seeds of ultrafine carbon particles, which were injected at the first stage of the growth, in a 20% methane/argon plasma under the conditions of 5 W RF power and 40 Pa (0.3 Torr) pressure. Spherical and monodisperse carbon particles can be grown through coating of hydrogenated amorphous carbon on the seeds by the dissociation of methane gas in the plasma [22]. Using Mie-scattering ellipsometry, the diameter of growing particles was monitored, and the growth was stopped by the decrease of RF power to 1 W for the particles of diameter of 1.4 μm, with which three-dimensional Coulomb crystals were formed in the experiments that will be shown in Section 3.2.

Fine particles spread over the entire region in the potential bucket are suspended about 5 mm above the electrode in the luminous plasma region of the negative glow. The top and side images, which were taken at the same position and at the same time by irradiation with blue laser light only, of a 3D Coulomb crystal formed by 1.4-μm carbon particles are shown in **Figure 4** [23]. The relative horizontal positions of fine particles in the top view were adjusted so as to agree with those in the side view at a crystal-grain boundary. Bright spots in the top view image indicate particles in the lowest layer and comparatively dimmed or small spots indicate those in the second lowest layer, because particles were illuminated in the off axis of the laser beam of the Gaussian distribution under them. From the correspondence of particle arrangement in the top view with that in the side view, particles are found to be aligned in the perpendicular direction of

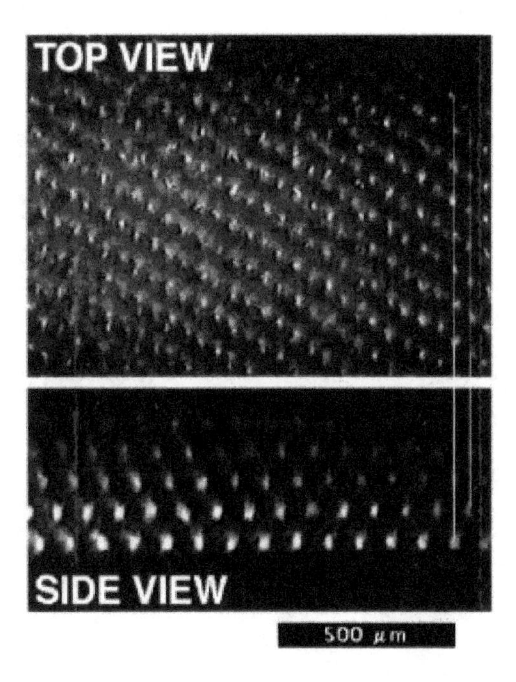

Figure 4.
Top and side video images of a 3D Coulomb crystal taken at the same time and the same position [23].

the side view plane. This means that each bright spot in the side view shows particles piled up in the direction (see the lines on the right in the figure).

It can be seen that the crystal structure in **Figure 4** is similar to the body-centered cubic (bcc) structure with (110) planes parallel to the electrode. The three-dimensional structure of a unit cell of the crystal is depicted in **Figure 5**. The lattice constants a, b, and c are determined from the average in the image to be 106, 157, and 165 mm, respectively. The relation among these sizes is $\sqrt{2}\,a < b \leq c$; therefore, the crystal structure is not strictly body-centered cubic (bcc) but body-centered tetragonal (bct) or generally face-centered orthorhombic (fco). The top view and the side view in **Figure 4** show (001) planes and planes perpendicular to [110] axes of the structure, respectively.

3D observation of the change of structure of Coulomb crystal was carried out. **Figure 6** shows the top and side views of change of fine particles arrangement every 1/3 s. As the blue argon-ion laser beam runs through the lower region of fine particles and the scattered blue light is more intensive than the scattered red diode-laser light, the particle arrangement in a few lowest layers was distinguished in top views. Side views were taken by the use of another CCD video camera with an optical band-pass filter for the red light (690 nm) put in front of it. Therefore particles in a few vertical layers were observed by the latter CCD camera.

The crystal structure of stage "A" in **Figure 6** was analyzed to be fco or bct, while that of stage "D" was fcc [21]. Change from "A" to "D" in the side view indicates that from the crystal observation axis of bct[100] to fcc[110] for, and change from "A" to "D" in the top view shows that from bct[110] to fcc[111]. The change of inclination of particle rows in the vertical direction is clearly seen in the side view. Corresponding to the side view, the particle arrangement in horizontal layers shown in the top view changes in the way that particles in the third lowest layer, which are indicated by very dimmed spots, gradually separate from the particle positions in the lowest layer. The change of crystal structure is illustrated in **Figure 7**.

The results of the observation of the time and space changes of 3D Coulomb crystal structures suggest that the structural transitions occurred by the slip of crystal planes parallel to the electrode. The slip planes were fcc(111) and bct(110). In the martensitic transformation of metallic iron, the slip of crystal planes generally occurs between fcc(111) and bcc(110) or bct(110) [24]. The structural transitions of 3D Coulomb crystals in dusty plasmas well matches those of real metallic crystals.

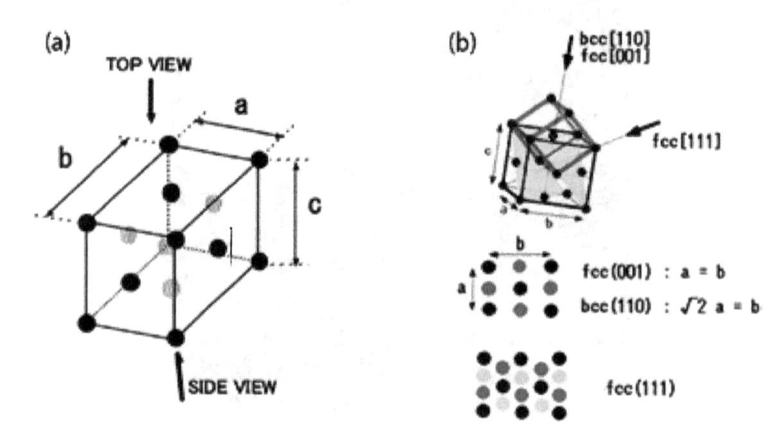

Figure 5.
3D crystal structures of bcc, bct, or fco (a) [23]; bcc and fcc (b) [21].

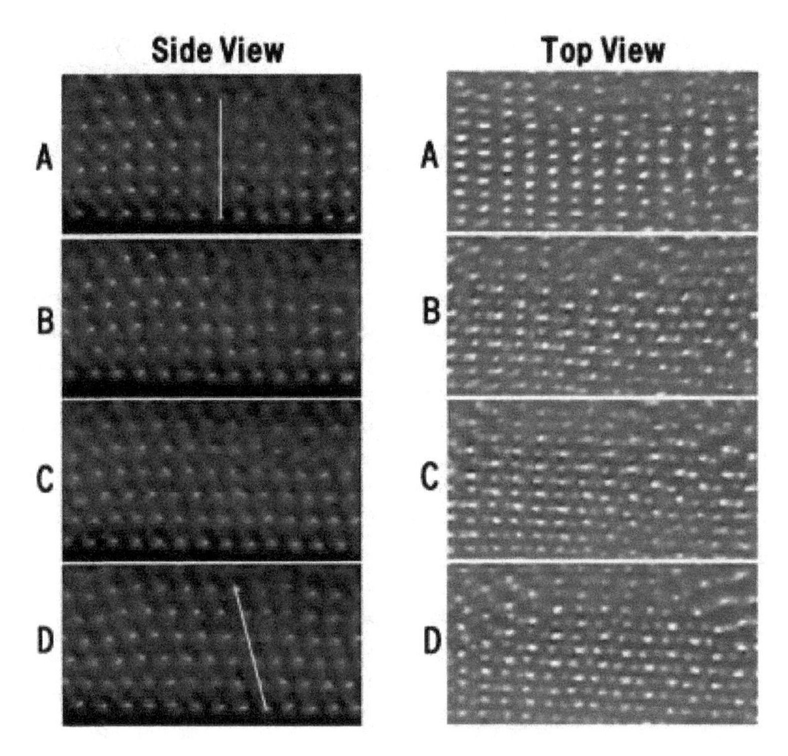

Figure 6.
Structural transition of Coulomb crystal every 1/3 s from A to D. Side and top views were taken at the same time. Fine particles in the red colored regions in the top view correspond to those in the side view [21].

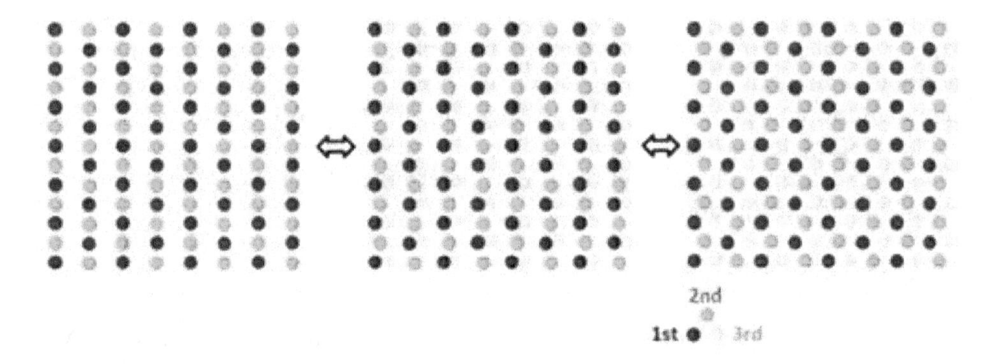

Figure 7.
Illustration of structural change of Coulomb crystal between bct (fco) and fcc from top view. Fine particles in the first, second, and third lowest layers are indicated by dark (blue), pale (green), and light (yellow) circles, respectively.

When a crystal structure simply changes by slip of crystal planes from fcc to another cubic structure, the relation between two horizontal lattice constants should hold as $a : b = 1 : \sqrt{3}$ (=1.73). If the crystal rearranges its structure for 3D minimum Coulomb energy, the structure changes to bcc with (110) planes parallel to an electrode for $a : b = 1 : \sqrt{2}$ (=1.41). The b/a ratio of the crystal in **Figure 4** is about 1.5. This discrepancy can be explained by the fact that the particle arrangement is reconstructed for 2D minimum Coulomb energy with constant particle density in horizontal planes and with interplane vertical distance constant [21, 23]. It is suggested that the crystal structure was formed from the pile of the plain layers perpendicular to the direction of an external force with constant interplane distance, which was also shown by molecular dynamics simulation [20], being arranged by the secondary effect of Coulomb force of particles in nearby layers to minimize average Coulomb energy [23].

3. Analyses of behavior of fine particles in plasmas

3.1 Analysis by polarized light scattering

The intensity of light scattered by a single spherical particle is calculated through the Mie-scattering theory [25, 26] when the diameter, D, and refractive index, m, of a particle are given. For multiple monodisperse particles, the scattered light intensity is proportional to the number of particles, N_d. Thus the scattered light intensity depends on both D and N_d. In order to determine D and N_d separately, the measurement of polarization of scattered light is required simultaneously.

The complex scattering amplitude functions, $\boldsymbol{S}_s$ of a polarization component perpendicular to the scattering plane and $\boldsymbol{S}_p$ of a parallel one, are defined by the following equations:

$$\boldsymbol{S}_s = \sum_{n=1}^{\infty} \frac{2n+1}{n(n+1)} \{\boldsymbol{a}_n \pi_n(cos\theta) + \boldsymbol{b}_n \tau_n(cos\theta)\}, \quad (4)$$

$$\boldsymbol{S}_p = \sum_{n=1}^{\infty} \frac{2n+1}{n(n+1)} \{\boldsymbol{b}_n \pi_n(cos\theta) + \boldsymbol{a}_n \tau_n(cos\theta)\}, \quad (5)$$

where $\boldsymbol{a}_n$ and $\boldsymbol{b}_n$ are expressed with the Bessel functions including the diameter D and complex refractive index $\boldsymbol{m}$ of a fine particle; π_n and τ_n are expressed with the Legendre polynomials as functions of scattering angle θ [27, 28]. The absolute value of ratio, $|\boldsymbol{S}_p/\boldsymbol{S}_s|$, can be determined for monodisperse fine particles with known refractive index with the use of three polarizing elements: one placed in position before scattering and the other two placed after scattering with polarization azimuth angles parallel and perpendicular to the scattering plane [29]. The value of $|\boldsymbol{S}_p/\boldsymbol{S}_s|$ does not depend on particle number N_d but only on diameter D. Thus D is determined first and then N_d is determined from the intensity of scattered light after calibration.

3.2 Analysis by Mie-scattering ellipsometry

In an analogous way to the reflection ellipsometry, angle parameters for Mie scattering are defined from the ratio between two complex scattering amplitude in the direction parallel and perpendicular to the scattering plane. When particles are spherical and monodisperse, the ellipsometric parameters Ψ and Δ are defined by the ratio of the scattering amplitude functions of $\boldsymbol{S}_s$ and $\boldsymbol{S}_p$ [30], which are complex numbers, as

$$\tan\Psi \, e^{i\Delta} = \frac{\boldsymbol{S}_p}{\boldsymbol{S}_s}. \quad (6)$$

It is obvious that $\tan\Psi = |\boldsymbol{S}_p/\boldsymbol{S}_s|$, which can be determined also by the method described in Section 3.1. Another value of parameter, Δ, is added as information about fine particles in Mie-scattering ellipsometry. When particles are polydisperse in the size range of Mie scattering, the scattered light generally results in some depolarization even if the incident light is fully polarized [31]. In this case, the Stokes vector and the Mueller matrix are required for the calculation of polarization state. Details about the analysis of fine particles by Mie-scattering ellipsometry are described in Refs. [30, 32, 33].

Figure 8 shows a (Ψ, Δ) trajectory as a result of Mie-scattering ellipsometry measurement for spherical carbon particles growing by coating in a plasma including methane (**Figure 8(A)**). The time evolution of diameter and density of fine particles are obtained from correspondence of graphs of (a) and (b) in **Figure 8(B)** as shown in **Figure 9** [1].

The transitions of arrangement of fine particles during the growth in a methane/helium plasma under the conditions of pressure of 106 Pa (0.8 Torr) and RF power of 5 W are shown in **Figure 10** [34]. By Mie-scattering ellipsometry, particle diameter was evaluated as 1.3, 1.7, 2.2 μm at 30, 45, and 80 min, respectively.

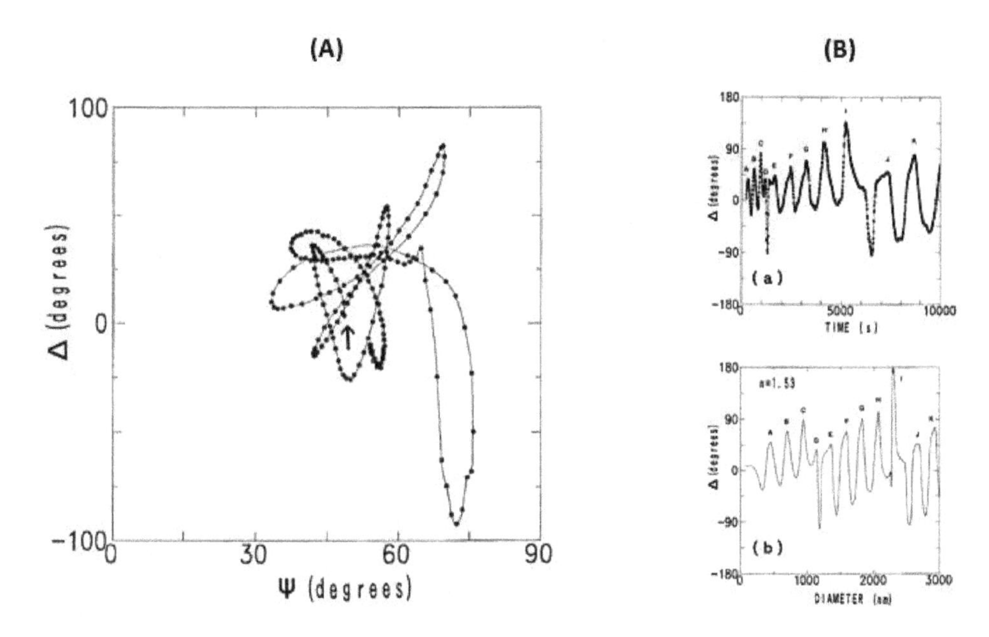

Figure 8.
Experimental result of evolution of ellipsometric parameters, (Ψ,Δ), during carbon particle growth. (A); (a) time evolution of Δ by experiment, (b) size evolution of Δ by simulated calculation (B) [1].

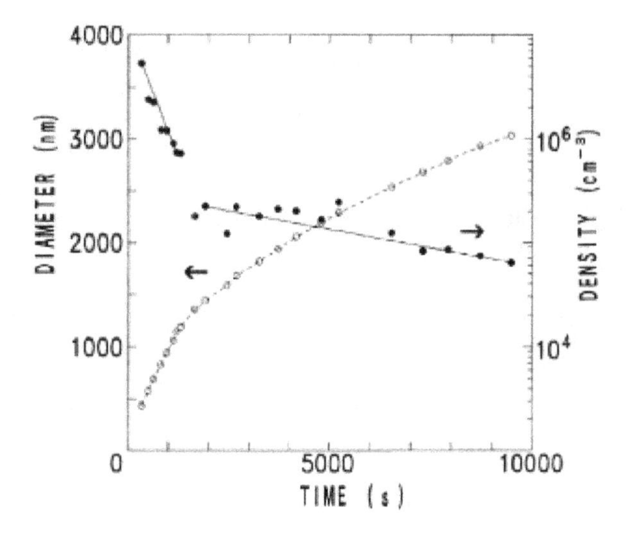

Figure 9.
Time evolution of particle diameter (open circles and broken curves) and density (closed circles and solid lines) [1].

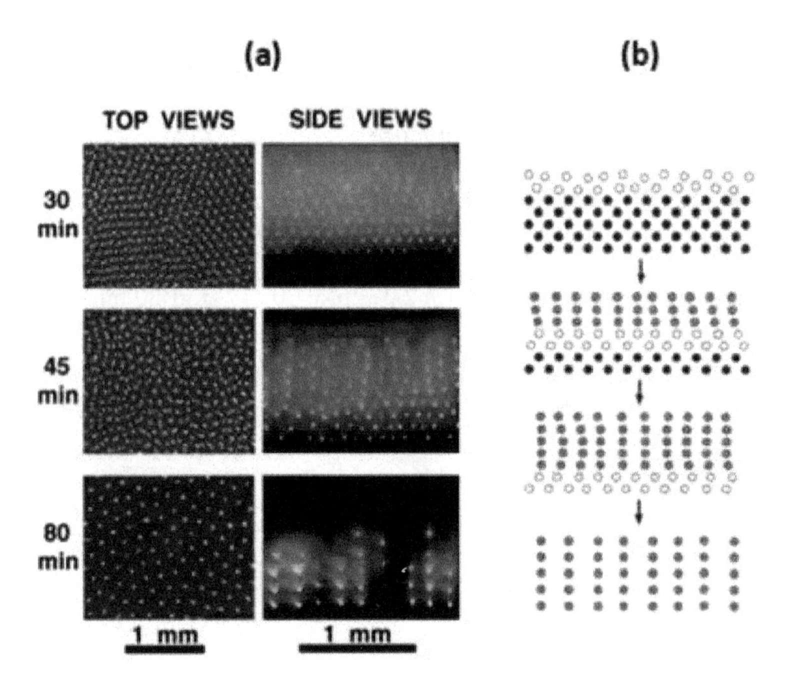

Figure 10.
Changes of top and side views of particle arrangement (a). The particles of top views are in the two lowest layers and those of side views in several lowest layers. Illustrations of change of side-view particle arrangement are shown (b) [34].

The particle arrangement at 30 min shows 3D regularity in the lowest four or five horizontal layers, especially crystal-like ordering in the lowest two layers. It is seen at 45 min that particles tend to be weakly chained in the vertical direction in the upper region. In its video movie, the chains were slowly swinging from right to left. The top view shows that particles in the lowest layer are randomly arranged. At 80 min, the particles are aligned like stiff rods in the vertical direction. They seem to form a close-packed 2D crystal, that is, hexagonal crystal. The structures of arrangement of smaller and larger particles correspond to those of upper and lower parts in **Figure 2**, respectively. However, the crystal edge of the 2D crystal is seen at the lowest side in the side view in **Figure 10**, while it is seen at the top side of 2D structure, which is the boundary between 2D and 3D structures, in **Figure 2**.

The spatial distribution of size of arranged particles under the same conditions as shown in **Figure 2** was analyzed by Mie-scattering ellipsometry using an image sensor instead of a photo-detector [35, 36]. Determined ellipsometric parameters are plotted on the Ψ-Δ coordinate plane, as shown in **Figure 11**, along with the trajectory obtained by calculation for spherical particles of refractive index of 1.56, the scattering angle of 88°, and diameter of 2650–2830 nm for each 10 nm. It should be noted that the trajectory shows a significant change with the diameter. The calculated trajectory was obtained for the best fitting to the determined values. By the comparison of the determined values Ψ and Δ with calculation, the size was evaluated to be 2.70, 2.74, 2.75, and 2.77 μm for particles in upper to lower layers (L4 to L1). Thus larger particles sank in lower position. However, although interlayer distance was observed to be about 100 μm, the distance between force balance positions for isolated 2.74 and 2.75 μm fine particles calculated in a way as

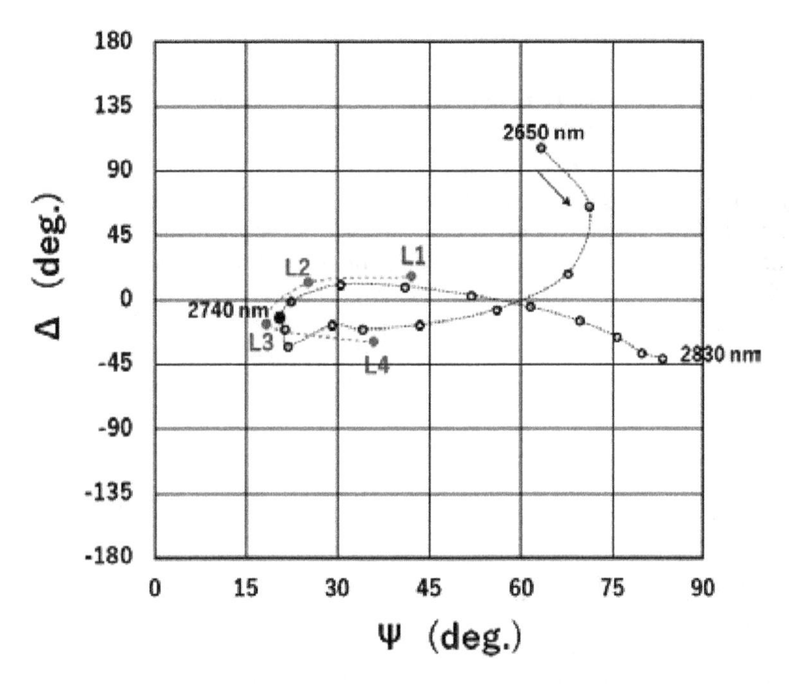

Figure 11.
Ellipsometric parameters of measurement for fine particles in layers upper to lower: L1, L2, L3, L4 (closed circles and dotted line) and calculation for spherical particles of refractive index of 1.56 and diameter of 2650–2830 nm for each 10 nm (open circles and solid line).

shown in Section 4.2 is one order of magnitude smaller than the observed interlayer distance. The interlayer distance of 100 μm must be due to mutual Coulomb repulsion. The reason why larger particles exist in lower layers is explained as follows.

The total chemical potential $\mu = \mu_{\text{int}} + \mu_{\text{ext}}$, where μ_{int} is the internal chemical potential and μ_{ext} is the external potential mainly due to gravity and electrostatic fields, has a constant value in the system of thermal equilibrium. Since μ_{int} is proportional to the logarithm of concentration ratio,

$$c = c_0 \exp(-\mu_{\text{ext}}/k_B T), \tag{7}$$

where c and c_0 are particle concentration and a constant, respectively. The gravitational force is proportional to the third power of the diameter D of fine particle, while electrostatic force is proportional to the diameter. Then, μ_{ext} above the plasma-sheath boundary is related to the equation (see Section 4),

$$\begin{aligned} \mu_{\text{ext}} &= m\,g\,z - \int (eQE)dz \\ &\cong a\,z\,D^3 - b\left(\int Edz\right) D \\ &\cong a\,z\,D^3, \end{aligned} \tag{8}$$

where m, g, z, and E are particle mass, the gravitational constant, vertical coordinate value, and the electrostatic field strength, respectively; a and b are constants independent of z and D. The external potential is larger for larger particles. Thus, larger particles moved to lower position according to the thermal equilibrium in several minutes after injection.

4. Forces acting on fine particle

4.1 Suspension of fine particle in plasma by force balance

Fine particles are suspended mainly at the balanced position of electrostatic force (f_E), ion drag force (f_I), and gravity (f_G). When they are distributed between two parallel-plate electrodes with central axes in vertical direction, the upward resultant force of $f_E - f_I - f_G$ acts on them near the lower electrode while that of $f_I - f_E - f_G$ acts on them near the upper electrode. These three forces are related with particle diameter, D, by the following equations,

$$f_E = e(rD)E, \tag{9}$$

$$f_I = n_i V_s m_i V_i \pi(b_c^2 + 4b_R^2 \ln\Lambda), \tag{10}$$

$$f_G = \pi\rho D^3/6, \tag{11}$$

where r is the particle charge number per unit size of particle diameter D: rD is the particle charge number, E is the electric field, n_i is the ion density, V_i is the ion drift velocity, $V_s = \sqrt{(V_i^2 + V_{iT}^2)}$ is the ion velocity (V_{iT} is the ion thermal velocity), m_i is the mass of ion, b_c is the impact parameter of ion orbit grazing particle, b_R is the impact parameter of right-angle-scattering ion orbit, $\ln\Lambda$ is the Coulomb logarithm, and ρ is the mass density [37].

The three forces were calculated under the following simple assumptions: $r = 10^9$ /m; $E = 500$ V/m; $n_i = 10^{14}$ /m^3; $V_i = 2000$ m/s; $V_{iT} = 1500$ m/s; $m_i = 10^{-26}$ kg; $b_c \approx D/2$; $b_R \approx 4D$, $\ln\Lambda \approx 10$; $\rho = 2*10^3$ kg/m^3. The variations of the forces with particle diameter are indicated by a log-log graph in **Figure 12.** f_E, f_I and f_G are proportional to D, D^2 and D^3, respectively.

Ion drag force could be larger than that calculated by Eq. (10) [37] because of more widespread ion-particle interaction beyond the Debye-length, which will be described in Section 4.2.

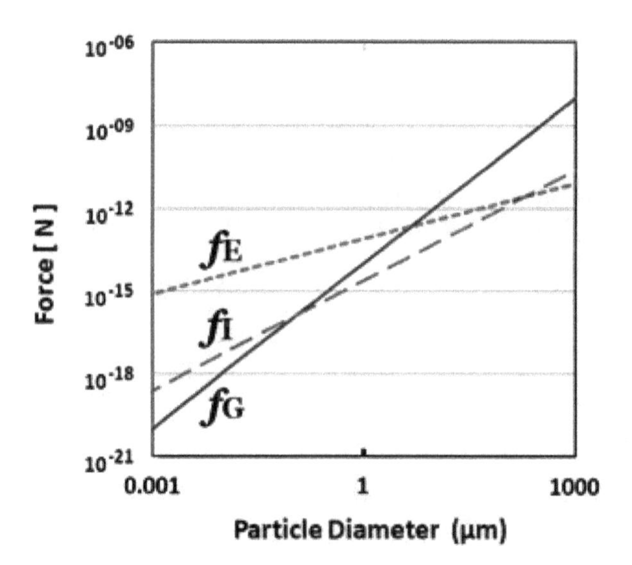

Figure 12.
Variation of electrostatic force (fE), ion-drag force (f_I), and gravity (f_G) with particle diameter.

4.2 Fine particle behavior under microgravity

As was described in Section 4.1, the electrostatic, ion drag, and gravitational forces act on fine particles. The gravitational force pulls fine particles downward. When fine particles are suspended in a plasma generated between two parallel plates placed horizontally, they sink around the lower plasma-sheath boundary. If gravitational force is extremely decreased or lost, the position of force balance changes. Thus, microgravity experiments should be useful for analyzing forces acting on fine particles in a plasma.

A parallel-plate RF plasma system was used for the experiments (**Figure 13**) [38]. A piezoelectric vibrator for the injection of fine particles into a plasma was contained in an RF electrode, which was put at the bottom of vacuum chamber of the system. The top of the electrode is covered with a grid so that fine particles can be pushed up. A grounded counter electrode is placed at the upper side of the RF electrode at the distance of 14 mm with a vertically symmetric structure. The outer diameter of the two electrodes is 40 mm and the gap space of the electrodes is surrounded by a transparent plastic cylinder with an inner diameter of 35 mm. The vacuum chamber is shaped octagonal and has six viewing windows: one for the entrance of the laser light, another for the exit, and the other four for observation of the arrangement of fine particles using scattered laser light. Using two charge-coupled device (CCD) video cameras placed in front of two of the four windows, the XYZ-3D position of each fine particle can be determined.

Spherical divinylbenzene fine particles 2.27 ± 0.10 μm in diameter were put into the piezoelectric vibrator. The density of the fine particle is 1.19 g/cm^3, and the weight is 7.29 × 10^{-12} gw. Helium gas was introduced at the pressure of 133 Pa (1 Torr). After plasma was generated with the RF power of 2 W, fine particles were pushed up into the plasma. When a sufficient quantity of fine particles was intro-duced into the plasma, the vibrator was switched off and then the experimental system was placed in a microgravity environment.

Figure 14 shows images taken using the Y-axis video camera before and during the microgravity condition of less than 10^{-4} G (gravity constant) that was generated in a capsule of drop tower for 4.5 s at the Micro-Gravity Laboratory (MGLAB) at Toki in Japan. The number of injected fine particles was approximately 30,000. A diode laser, which radiates light 690 nm in wavelength, was used for the observation of fine particles by scattered laser light. The image obtained immediately before the

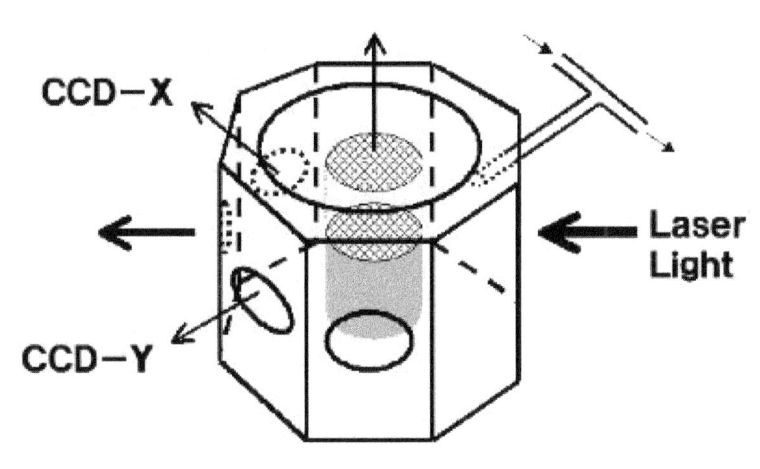

Figure 13.
Schematic of parallel-plate RF plasma system for microgravity experiment [38].

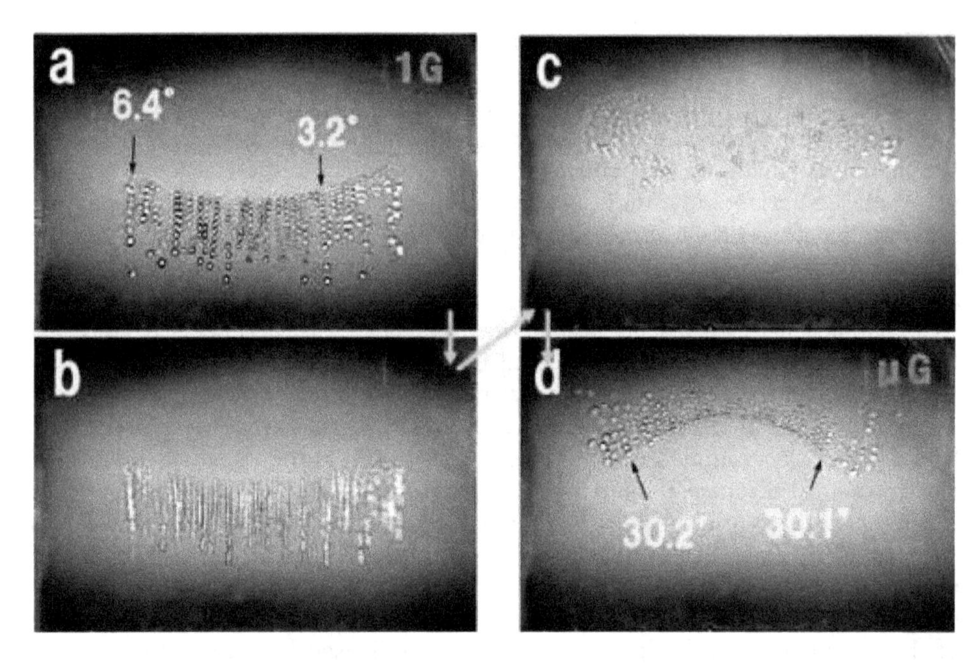

Figure 14.
Evolutions of ordering of fine particles during drop experiment: (a) just before drop under 1 g; (b) 2/30 s after beginning of drop; (c) 6/30 s after; (d) 3 s after. Directions of fine-particles alignment and tilt angles are indicated in (a) and (d). The height and width of the figures are 14 mm and 21 mm, respectively [38].

drop under gravity shows fine particles aligning in a nearly vertical direction. The image obtained 2/30 s after the beginning of the drop shows the fine particles going straight up in the vertical direction, while the 6/30 s image shows the particles radially moving upward. In 1.5-s movie, the movement slowed down and the alignment of the fine particles was observed again, however, it was not vertical but radial. Fine particles aligned in radial directions are seen in the image shown at 3 s in **Figure 14**.

The alignment of fine particles in the sheath should be caused by the effect of the formation of the wake field [18, 19]. The reason why the fine particles moved up after the beginning of the drop is due to the change in the balance of forces. Fine particles are suspended around the balance position of the resultant force (F for upward direction), which is composed of ion drag (f_I), electrostatic (f_E), and gravitational (f_G) forces under gravity, while they are around the position of balance induced by the former two forces (f_I and f_E) under zero gravity. Thus fine particles remain around the lower plasma-sheath boundary under gravity with the condition $F = f_E - f_I - f_G = 0$, and they are pushed upward there when gravity is diminished with the condition $F = f_E - f_I > 0$. The ion drag force directed toward the electrodes is larger than the electrostatic force toward the center of the plasma under the conditions of this experiment, when the ion-particle cross section is calculated with the impact parameter beyond the Debye length [39–41]. Because the mean free path of helium atoms or ions at the pressure of 133 Pa is 130 μm, which is much longer than the ion Debye length, the collisionless limit [41] is applicable to this experiment. As a result, the upward resultant force F changes and fine particles are pushed upward to the position near the upper grounded electrode passing over the center as shown in **Figure 15**. The force was calculated for a particle of the same size and weight as those used in the experiment and under similar plasma conditions [40]. In the calculation, the gravitational force f_G was 0.7×10^{-13} N, and the ion drag and electrostatic forces at the balance positions under gravity (1 G) were about 9.6×10^{-13} N and 10.3×10^{-13} N, respectively, for the charge of the isolated fine particle of 7340 e, which was evaluated under the

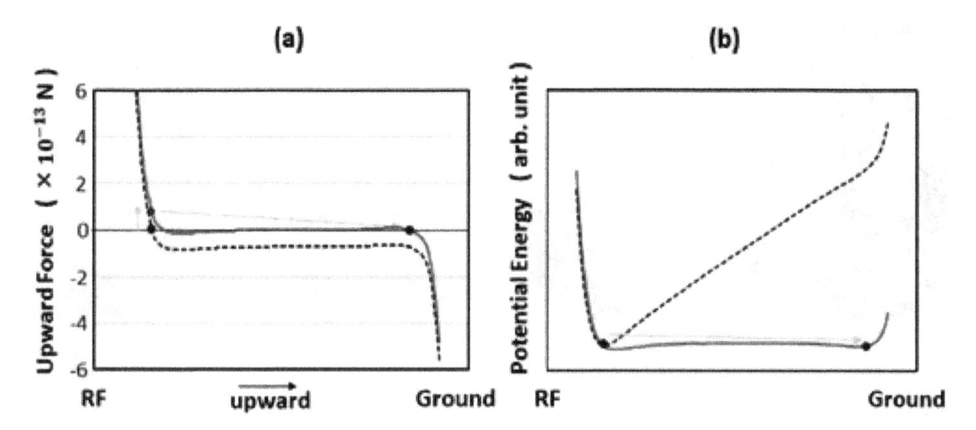

Figure 15.
Illustration of upward resultant force composed of ion drag, electrostatic and gravitational forces and behavior of fine particle in drop experiment under 1 G (black broken curve) to 0 G (red solid curve) (a); potential energy calculated from integral of curves of the resultant force in (a) (b). Closed circles show force balance position under 1 G and upper position under 0 G; open circle shows upward force under 0 G at the position of force balance under 1 G.

same plasma conditions as this experiment. The difference between the electrostatic and ion drag forces, $f_E - f_I = f_G = 0.7 \times 10^{-13}$ N, pushed the fine particles upward at the moment of gravitational change. Fine particles moved to a position around the upper plasma-sheath boundary beyond the unstable zero-resultant position if the upward force was sufficiently large for fine particles to pass through the center of the plasma.

Using the images shown in **Figure 14** along with images taken at the same time with the X-axis video camera, three-dimensional particle position coordinates were obtained. Three-dimensional tilt angles were determined for six inner particles to be 6.4°and 3.2°for the left and right rows in **Figure 14(a)**, and 30.2° and 30.1° for those in **Figure 14(d)**, respectively. Interparticle distances between the inner first and second particles were 320 and 230 μm for the left and right rows in **Figure 14(a)**, and 280 and 180 μm for those in **Figure 14(d)**, respectively, and the Mach numbers of ion speed were calculated to be in the range from 1.2 to 1.6.

Since the structure of the RF plasma system is symmetric both vertically and horizontally, the shape of plasma generated in the plastic cylinder is supposed to be symmetric, that is, spheroidal, when the sheath thickness is comparable to the order of plasma size. Therefore, under zero gravity, fine particles are subjected to a resultant force in the spheroidal plasma, that is, they are affected by the positive ion flow and electric field whose directions are radial. If the wake field theory is accepted here, the radial alignment of fine particles can be reasonably understood as being due to the formation of rows in the direction of ion flow. On the other hand, the vertical alignment of fine particles under gravity cannot be explained with the assumption of a spheroidal plasma. In order to determine the shape of the plasma, we analyzed the density distribution of the plasma containing fine particles and drew 16 contours through the division of optical emission brightness on CCD images. As is shown in **Figure 16**, the plasma was lowered and flattened under gravity compared to that under microgravity. The results suggest that fine particles suspended around the lower plasma-sheath boundary deformed the plasma.

By the effect of plasma wake [18, 19, 42], a negatively charged fine particle forms a positive potential in the downstream position of ion flow, and another fine particle is attracted to the position of maximum potential. At the same time, the gravitational force pulls the downstream fine particle down from its position [43]. The third downstream fine particle that is downstream to the second one is also

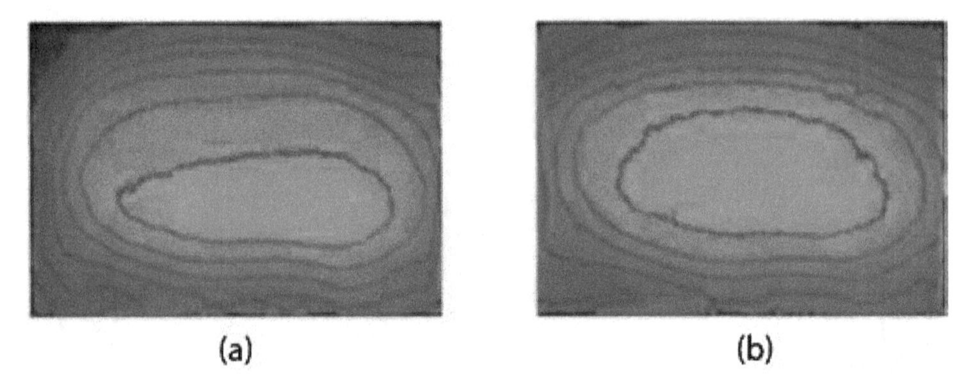

Figure 16.
Contours of intensity distribution of optical emission from plasma containing fine particles under 1 G (a) and microgravity (b). The height and width of the figures are 14 mm and 19 mm, respectively. [38].

pulled down by gravity to a position lower than that of maximum potential. Such force actions are repeated on other particles further downstream. Due to the sinking of the fine particles under gravity, ion flow is affected by the negative potential formed by fine particles in turn and tends to be vertically directed lowering the free energy. As a result, under gravity, the mutual influence between the positive ion flow and the negative fine particles results in the alignment being more vertical and in the deformation of the plasma. The reason why fine particles moved straight up in the vertical direction, as shown in the 2/30 s image of **Figure 14(b)**, is considered to be that the resultant force due to ion flow and the electric field was directed vertically upward just immediately after the diminishment of gravity.

Using the same plasma system that was used for drop experiment, a microgravity experiment by parabolic flight was carried out under the condition of less than 0.04 G in the vertical direction and less than 0.01 G in the horizontal direction for approximately 20 s at the Diamond Air Service Inc. in Japan. The plasma system was installed in a standardized rack in a jet plane. **Figure 17** shows the images of fine particle behavior taken by the Y-axis video camera during the changes in gravitational conditions from 2 G (a) to microgravity (c) and to 1.5 G (e). Under micro-gravity, in **Figure 17(c)**, fine particles are also observed around the lower balance position forming a void in the center as was reported [44]. The number of injected fine particles was approximately 10 times greater than that for the drop experiment. Most of the fine particles were pushed to the upper balance region at the moment of gravitational change, however, some remained behind due to their mutual Coulomb repulsive forces. In addition, fine particles injected continuously during the microgravity condition were also suspended around the lower balance position. Fine particles at the inner boundary were arranged parallel to the electrodes under gravity more than 1 G, while they formed an arc under microgravity even around the lower plasma-sheath boundary. In transition from 2 G to microgravity (b), the fine particles moved vertically upward. On the other hand, in transition from micro-gravity to 1.5 G (d), they moved downward through the outer side of the plasma and a void was formed in the center. The nonsymmetrical behavior of fine particles in transition can be explained by the difference in the state of plasma containing fine particles. As explained above concerning the results of the drop experiment, the resultant of ion drag and electrostatic forces pushed the fine particles vertically upward at the moment shown in **Figure 17(b)**. However, the particles were obliquely pulled down during the increase of gravity as shown in **Figure 17(d)** by the oblique force that is composed of the gravitational force and the outward radial force, which is derived from the ion drag force stronger than the electrostatic force.

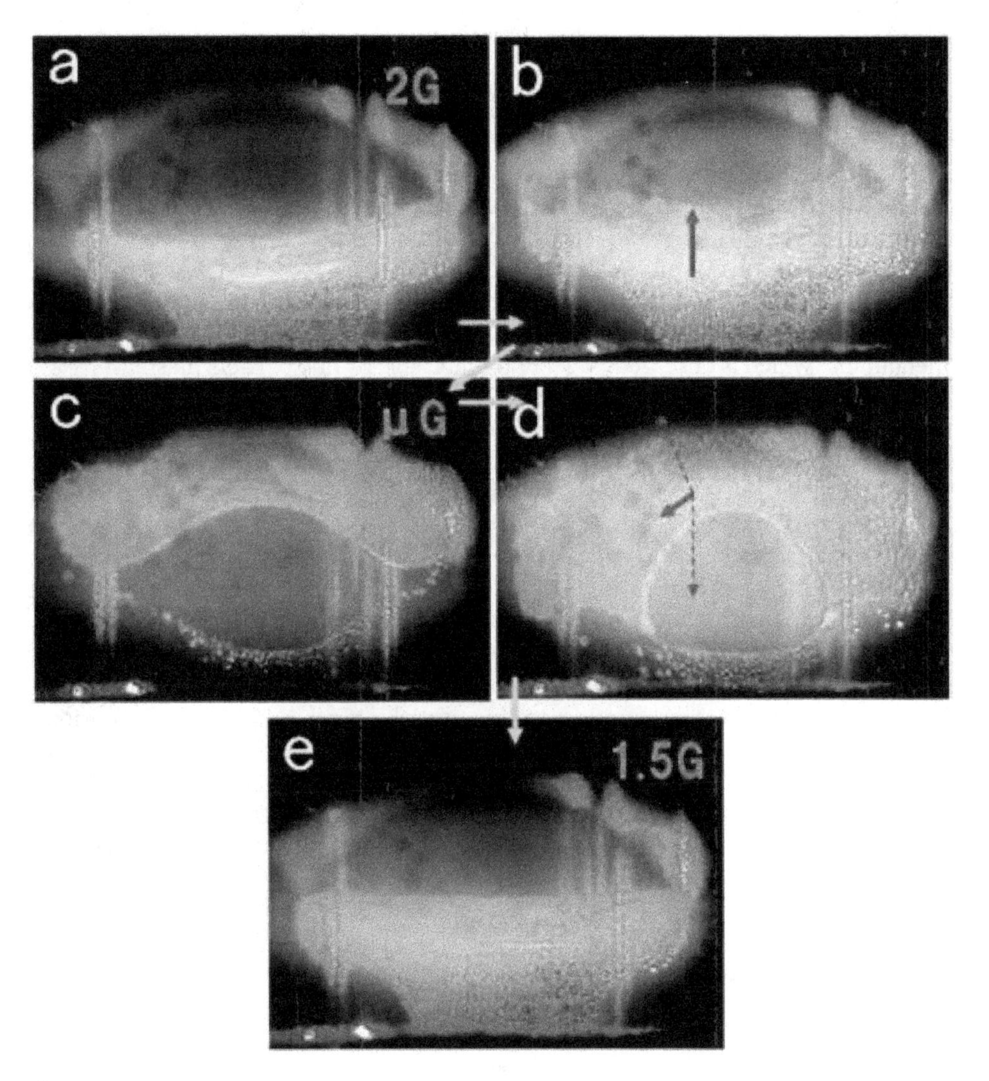

Figure 17.
Video images of fine particle behavior in parabolic flight under gravitational conditions: (a) 2 G, (b) transition from 2 G to microgravity, (c) microgravity, (d) transition from microgravity to 1.5 G and (e) 1.5 G. In the case of (a), fine particles were accelerated by 0.1–0.2 G in the right direction of the image [38].

Such "void" was observed in the center of plasma under microgravity condition performed on the International Space Station (ISS) [45]. The cause of "void" formation is generally considered to be due to strong ion drag force at the center of plasma [39–41].

5. Conclusions

The observation of Coulomb crystals in fine particle plasmas was presented in this chapter. Their 3D crystal structures were fcc, fco, and bct, but bcc structure have not been observed. The latter result is due to the fact that the rearrangement from fcc to fco or bct occurs with constant particle density in horizontal planes and with interplane vertical distance constant. It is explained that the crystal structure was formed from the pile of the plain layers perpendicular to the direction of an external force with constant interplane distance being arranged by the secondary effect of the Coulomb force of particles in nearby layers to minimize the average Coulomb energy.

The structure change of Coulomb crystal during the growth of carbon particles was observed and analyzed with the use of Mie-scattering ellipsometry. The arrangement of fine particles 1.3 μm in diameter showed 3D regularity, while that of diameter 2.2 μm formed 2D close-packed crystal structure.

The spatial distribution of size of particles forming horizontal layers was analyzed by Mie-scattering ellipsometry using an image sensor. By the comparison of the determined values Ψ and Δ with calculation, particle diameter was evaluated to be 2.70, 2.74, 2.75, and 2.77 μm for particles in upper to lower layers. Its diameter was determined with accuracy as low as 0.01 μm (10 nm). Such measurement method may be useful for the analysis of phase separation phenomena, which is observed in general materials, in fine particle plasmas.

3D Coulomb crystals in fine particle plasmas can be good models of real atomic crystal and its formation and melting processes. Thus physical phenomena in solid, liquid, and gas states, which are, for example, the liquid-to-solid phase transition [46, 47], critical phenomena [48], soliton [49], chaos [50], can be simulated and analyzed on the basis of observation of behavior of individual fine particles. Such an experimental method enables the kinetic analyses of component behavior in the study of statistical properties of a system, along with the complementary uses of the computer simulations of molecular dynamics (MD) or Monte-Carlo method (MC).

In order to analyze a 3D Coulomb crystal as a model of real atomic crystal, it is preferable to make its conditions of environment close to the real crystal. Spatially isotropic force field acting on fine particles is required for the conditions. Microgravity environment can eliminate unidirectional gravity; however, the resultant force of electrostatic and ion drag forces pushes fine particles outward from the center forming the void of them [44, 45]. The fabrication of a plasma system that does not generate such a void is yet an unresolved issue and that with homogeneous conditions is under development [51, 52].

Acknowledgements

The author thanks Prof. Kunihide Tachibana, Prof. Kazuo Takahashi, Prof. Akio Sanpei, Prof. Yukio Watanabe, Prof. Noriyoshi Sato, and members of the project of Plasma Control Science Research in Kyoto Institute of Technology for useful discussions. He also thanks laboratory students of Kyoto Institute of Technology for experimental assistance. This work was partly supported by the Ministry of Education, Science, Sports and Culture of Japan.

Author details

Yasuaki Hayashi
Kyoto Institute of Technology, Sakyo-ku, Kyoto, Japan

Address all correspondence to: hayashiy@kit.ac.jp

References

[1] Hayashi Y, Tachibana K. Observation of Coulomb-crystal formation from growing particles grown in a methane plasma. Japanese Journal of Applied Physics. 1994;**33**:L804-L806. DOI: 10.1143/JJAP.33.L804

[2] Chu JH, Lin I. Direct observation of Coulomb crystals and liquids in strongly coupled rf dusty plasmas. Physical Review Letters. 1994;**72**: 4009-4012. DOI: 10.1103/PhysRevLett.72.4009

[3] Thomas H, Morfill GE, Demmel V, Goree J, Feuerbach B, Möhlmann D. Plasma crystal: Coulomb crystallization in a dusty plasma. Physical Review Letters. 1994;**73**:652-655. DOI: 10.1103/PhysRevLett.73.652

[4] Melzer A, Trottenberg T, Piel A. Experimental determination of the charge on dust particles forming Coulomb lattices. Physics Letters A. 1994;**191**:301-308. DOI: 10.1016/0375-9601(94)90144-9

[5] Ikezi H. Coulomb solid of small particles in plasmas. The Physics of Fluids. 1986;**29**:1764-1766. DOI: 10.1063/1.865653

[6] Ichimaru S. Strongly coupled plasma. High-density classical plasmas and degenerate electron liquids. Reviews of Modern Physics. 1982;**54**:1017-1059. DOI: 10.1103/RevModPhys.54.1017

[7] Alder BJ, Wainwright TE. Phase transition for a hard sphere system. Journal of Chemical Physics. 1957;**27**:1208-1209. DOI: 10.1063/1.1743957

[8] Wood WW, Jacobson JD. Preliminary results from a recalculation of the Monte Carlo equation of state of hard spheres. The Journal of Chemical Physics. 1957;**27**:1207-1208. DOI: 10.1063/1.1743956

[9] Alder BJ, Hoover WG, Young DA. Studies in molecular dynamics. V. High-density equation of state and entropy for hard disks and spheres. The Journal of Chemical Physics. 1968;**49**:3688-3699. DOI: 10.1063/1.1670653

[10] Brush SG, Sahlin HL, Teller E. Monte Carlo study of a one-component plasma. I. Journal of Chemical Physics. 1966;**45**:2102-2119. DOI: 10.1063/1.1727895

[11] Hoover WG, Gray WG, Johnson KW. Thermodynamic properties of the fluid and solid phases for inverse power potentials. Journal of Chemical Physics. 1971;**55**:1128-1136. DOI: 10.1063/1.1676196

[12] Robbins MO, Kremer K, Grest GS. Phase diagram and dynamics of Yukawa systems. Journal of Chemical Physics. 1988;**88**:3286-3312. DOI: 10.1063/1.453924

[13] Farouki RT, Hamaguchi S. Thermodynamics of strongly-coupled Yukawa systems near the one-component-plasma limit. II. Molecular dynamics simulations. Journal of Chemical Physics. 1994;**101**:9885-9893. DOI: 10.1063/1.467955

[14] Waki I, Kassner S, Birkl G, Walther H. Observation of ordered structures of laser-cooled ions in a quadrupole storage ring. Physical Review Letters. 1992;**68**:2007-2010. DOI: 10.1103/PhysRevLett.68.2007

[15] Mitchell TB, Bollinger JJ, Dubin DHE, Huang X-P, Itano WM, Baughman RH. Direct observations of structural phase transitions in planar crystallized ion plasmas. Science. 1998;**282**:1290-1293. DOI: 10.1126/science.282.5392.1290

[16] Dosho S et al. Recent studies of polymer latex dispersions. Langmuir.

1993;**9**:394-411. DOI: 10.1021/la00026a008

[17] Hayashi Y, Mizobata Y, Takahashi K. Behaviors of fine particles in a planar magnetron plasma. Journal of Plasma and Fusion Research Series. 2009;**8**:298-301

[18] Nambu M, Vladimirov SV, Shukla PK. Attractive forces between charged particles in plasmas. Physics Letters A. 1995;**203**:40-42. DOI: 10.1016/0375-9601(95)00380-L

[19] Melandso F, Goree J. Polarized supersonic plasma flow simulation for charged bodies such as dust particles and spacecraft. Physical Review E. 1995;**52**:5312-5326. DOI: 10.1103/PhysRevE.52.5312

[20] Totsuji H, Kishimoto T, Totsuji C. Structure of confined Yukawa system (dusty plasma). Physical Review Letters. 1997;**78**:3113-3116. DOI: 10.1103/PhysRevLett.78.3113

[21] Hayashi Y. Structural transitions of 3-dimensional Coulomb crystals in a dusty plasma. Physica Scripta. 2001;**T89**:112-116. DOI: 10.1238/Physica.Topical.089a00112

[22] Hayashi Y, Tachibana K. Analysis of spherical carbon particle growth in methane plasma by Mie-scattering ellipsometry. Japanese Journal of Applied Physics. 1994;**33**:4208-4211. DOI: 10.1143/JJAP.33.4208

[23] Hayashi Y. Structure of a three-dimensional Coulomb crystal in a fine-particle plasma. Physical Review Letters. 1999;**83**:4764-4767. DOI: 10.1103/PhysRevLett.83.4764

[24] Nishiyama Z. X-ray investigation the mechanism of transformation from face-centered-cubic lattice to body-centered cubic. Science Report Tohoku University. 1934;**23**:637-664

[25] Mie G. Beiträge zur Optik trüber Medien, speziell kolloidaler Metallösungen. Annalen der Physik. 1908;**25**:377-445. DOI: 10.1002/andp.19083300302

[26] Born M, Wolf E. Principles of Optics. 5th ed. Oxford: Pergamon; 1975. Section 13.5

[27] van de Hulst HC. Light Scattering by Small Particles. New York: Dover; 1981. 470 p

[28] Tachibana K, Hayashi Y, Okuno T, Tatsuta T. Spectroscopic and probe measurements of structures in a parallel-plates RF discharge with particles. Plasma Sources Science and Technology. 1994;**3**:314-319. DOI: 10.1088/0963-0252/3/3/012

[29] Watanabe Y, Shiratani M, Yamashita M. Observation of growing kinetics of particles in a helium-diluted silane rf plasma. Applied Physics Letters. 1992;**61**:1510-1512. DOI: 10.1063/1.107532

[30] Hayashi Y, Tachibana K. Mie-scattering ellipsometry for analysis of particle behaviors in processing plasmas. Japanese Journal of Applied Physics. 1994;**33**:L476-L478. DOI: 10.1143/JJAP.33.L476

[31] Born M, Wolf E. Principles of Optics. 5th ed. Oxford: Pergamon; 1975. Section 10.8

[32] Hayashi Y, Sanpei K. Mie-scattering Ellipsometry. In: Wahaia F, editor. Ellipsometry. Rijeka, Chapter: IntechOpen; 2017. p. 1. DOI: 10.5772/65558

[33] Hayashi Y. Measurement of fine-particles by Mie-scattering ellipsometry. Journal of The Vacuum Science of Japan. 2001;**44**:617-623. DOI: 10.3131/jvsj.44.617

[34] Hayashi Y, Takahashi K. Structural changes of Coulomb crystal in a carbon fine-particle plasma. Japanese Journal of Applied Physics. 1997;**36**:4976-4979. DOI: 10.1143/JJAP.36.4976

[35] Hayashi Y, Kawano M, Sanpei A, Masuzaki S. Mie-scattering ellipsometry system for analysis of dust formation process in large plasma device. IEEE Transactions on Plasma Science. 2016;**44**:1032-1035. DOI: 10.1109/TPS.2016.2542349

[36] Hayashi Y, Sanpei A, Mieno T, Masuzaki S. Analysis of spatial distribution of fine particles in plasma by imaging Mie-scattering ellipsometry. In: 8th International Conference on the Physics of Dusty Plasmas; May 20-25, 2017; Prague, Czech Republic. 2017

[37] Barnes MS, Keller JH, Forster JC, O'Neil JA, Coultas DK. Transport of dust particles in glow-discharge plasmas. Physical Review Letters. 1992;**66**:313-316. DOI: 10.1103/PhysRevLett.68.313

[38] Hayashi Y. Radial ordering of fine particles in plasma under microgravity condition. Japanese Journal of Applied Physics. 2005;**44**:1436-1440. DOI: 10.1143/JJAP.44.1436

[39] Kilgore MD, Daugherty JE, Porteous RK, Graves DB. Ion drag on an isolated particulate in a low-pressure discharge. Journal of Applied Physics. 1993;**73**:7195-7202. DOI: 10.1063/1.352392

[40] Hayashi Y, Takahashi K, Sato H. Analyses of particle behaviors and formation of Coulomb crystals in plasmas. Journal of The Vacuum Science of Japan. 1998;**41**:782-789. DOI: 10.3131/jvsj.41.782

[41] Khrapak SA, Ivlev AV, Morfill GE, Thomas HM. Ion drag force in complex plasmas. Physical Review E. 2002;**66**: 1-4. DOI: 10.1103/PhysRevE.66.046414

[42] Takahashi K, Oishi T, Shimomai K, Hayashi Y, Nishino S. Analyses of attractive forces between particles in Coulomb crystal of dusty plasmas by optical manipulations. Physical Review E. 1998;**58**:7805-7811. DOI: 10.1103/PhysRevE.58.7805

[43] Takahashi K, Oishi T, Shimomai K, Hayashi Y, Nishino S. Simple hexagonal Coulomb crystal near a deformed plasma sheath boundary in a dusty plasma. Japanese Journal of Applied Physics. 1998;**37**:6609-6614. DOI: 10.1143/JJAP.37.6609

[44] Morfill GE, Thomas HM, Konopka U, Rothermel H, Zuzic M, Ivlev A, et al. Condensed plasmas under microgravity. Physical Review Letters. 1999;**83**:1598-1601. DOI: 10.1103/PhysRevLett.83.1598

[45] Nefedov AP et al. PKE-Nefedov: Plasma crystal experiments on the international Space Station. New Journal of Physics. 2003;**5**:33.1-33.10

[46] Thomas HM, Morfill GE. Melting dynamics of a plasma crystal. Nature. 1996;**379**:806-809. DOI: 10.1038/379806a0

[47] Krapak SA et al. Fluid-solid phase transition in three-dimensional complex plasmas under microgravity conditions. Physical Review E. 2012;**85**:1-11. DOI: 10.1103/PhysRevE.85.066407

[48] Totsuji H, Takahashi K, Adachi S, Hayashi Y, Takayanagi M. Strongly coupled plasmas under microgravity. International Journal of Microgravity Science and Application. 2011;**28**:S27-S30

[49] Sheridan TE, Nosenko V, Goree J. Experimental study of nonlinear solitary waves in two-dimensional dusty plasma. Physics of Plasmas. 2008;**15**:1-6. DOI: 10.1063/1.2955476

[50] Lin I, Juan W, Chiang C, Chu JH. Microscopic particle motions in strongly coupled dusty plasmas. Science. 1996;**272**:1626-1628. DOI: 10.1126/science.272.5268.1626

[51] Arp O, Block D, Piel A, Melzer A. Dust Coulomb balls: Three-dimensional plasma crystals. Physical Review Letters. 2004;**93**:1-4. DOI: 10.1103/PhysRevLett.93.165004

[52] Thomas HM, Knapek C, Huber P, Mohr DP, Zähringer E, Molotkov V, et al. Ekoplasma—The Future of Complex Plasma Research in Space. 2017. Available from: https://core.ac.uk/display/86640180

6

Massive Neutrinos and Galaxy Clustering in $f(R)$ Gravity Cosmologies

Jorge Enrique García-Farieta
and Rigoberto Ángel Casas Miranda

Abstract

Cosmic neutrinos have been playing a key role in cosmology since the discovery of their mass. They can affect cosmological observables and have several implications being the only hot dark matter candidates that we currently know to exist. The combination of massive neutrinos and an adequate theory of gravity provide a perfect scenario to address questions on the dark sector that have remained unanswered for years. In particular, in the era of precision cosmology, galaxy clustering and redshift-space distortions afford one of the most powerful tools to characterise the spatial distribution of cosmic tracers and to extract robust constraints on neutrino masses. In this chapter, we study how massive neutrinos affect the galaxy clustering and investigate whether the cosmological effects of massive neutrinos might be degenerate with $f(R)$ gravity cosmologies, which would severely affect the constraints.

Keywords: massive neutrinos, gravity theories, structure formation, galaxy clustering

1. Introduction

From first principles, it is well known that a theory of gravity is needed to describe the spatial properties and dynamics of the large-scale structures (LSS) of the universe. The observational data collected for several decades provide strong support to the concordance model Lambda cold dark matter (ΛCDM), which yields a consistent description of the main properties of the LSS [1–4]. However, since cosmological observations have entered in an unprecedented precision era, one of the current aims is to test some of the most fundamental assumptions of the concordance model of the universe. In this sense, the ΛCDM model assumes (i) the general theory of relativity (GR) as the theory describing gravitational interactions at large scales, (ii) the standard model of particles and (iii) the cosmological principle. Moreover, in this framework, the universe is currently dominated by dark energy (DE) in the form of a cosmological constant, responsible for the late-time cosmic acceleration [5–7] and by a cold dark matter (CDM) component that drives the formation and evolution of cosmic structures.

Recently, several shortcomings have been found in the ΛCDM scenario, like a possible tension in the parameter constraint of H_0 and σ_8 when different probes are used [4, 8–10]. This has motivated the interest on theoretical models beyond GR. Among them, the models based on $f(R)$ gravity are the favourite ones because of their generality and rich phenomenology [11, 12]. Moreover, modified gravity (MG) models represent one of the most viable alternatives to explain cosmic acceleration [13] that require satisfying simultaneously solar system constraints and to be consistent with the measured accelerated cosmic expansion and large-scale constraints [14–17]. An extra motivation to study MG models is given by the fact that massive neutrinos, the only (hot) dark matter candidates we actually know to exist, can affect these observables and have several cosmological implications [18]. However, the degeneracy between some MG models and the total neutrino mass [19–21] give rise to a limitation of many standard cosmological statistics [22, 23]. In this context, the clustering analysis and redshift-space galaxy clustering have been proven to be a powerful cosmological probe to discriminate among MG scenarios with massive neutrinos, as will be discussed in the following sections.

Regarding the dynamics of background universe in the standard framework, it is well described by the Friedmann-Lemaître-Robertson-Walker (FLRW) metric, whose line element in natural units $c = 1$ is given by

$$ds^2 = -dt^2 + a^2(t)\left[\frac{dr^2}{1-kr^2} + r^2 d\theta^2 + r^2 \sin^2\theta d\phi^2\right]. \quad (1)$$

where $a(t)$ is the so-called scale factor, (r, θ, ϕ) are dimensionless spherical-polar coordinates and k defines the geometry of the universe under consideration to be flat ($k = 0$), open ($k < 0$) or closed ($k > 0$). The equations of motion that describe the time evolution of $a(t)$ and the dynamic growth of the universe are called *Friedmann equations* and are given by

$$\left(\frac{\dot{a}}{a}\right)^2 = \frac{8\pi G}{3}\rho - \frac{k}{a^2} + \frac{\Lambda}{3}, \quad (2)$$

$$2\frac{\ddot{a}}{a} + \left(\frac{\dot{a}}{a}\right)^2 = -8\pi G p - \frac{k}{a^2} + \Lambda, \quad (3)$$

that can be re-expressed as

$$\dot{\rho} = -3\frac{\dot{a}}{a}(\rho + p), \quad (4)$$

$$\frac{\ddot{a}}{a} = -\frac{4\pi G}{3}(\rho + 3p) + \frac{\Lambda}{3} \quad (5)$$

after eliminating the curvature term. These equations lead to the definition of the Hubble parameter as $H \equiv \frac{\dot{a}}{a}$, which drives the expansion rate of the universe and usually is represented in terms of the dimensionless factor h, defined by the expression $H_0 = 100\ h\ km\ s^{-1}\ Mp\ c^{-1}$ at the present epoch. In order to have a full description of the background universe, it is necessary an equation of state (EoS) of the cosmic fluid, considering that it has three principal components: baryonic matter, dark matter and radiation (see **Figure 1a**). A first approximation consists in assuming a linear relationship between ρ and p; thus, the equation of state can be written as $p = w\rho$, where w is a parameter that in principle can be time-dependent $w = w(t)$, but the simplest approach is to consider it as a constant Under this assumption, Eq. (4) is easy to solve, resulting in $\rho(a) = a^{-3(1+w)}$. The case with $w = -1$ corresponds to the so-called cosmological constant; it is obtained

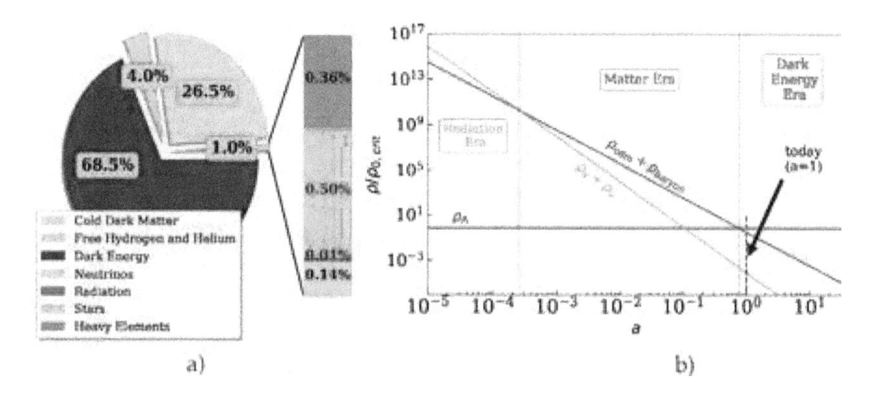

Figure 1.
(a) The current composition of the universe derived from Planck data [4], (b) evolution of the energy density parameter ρ normalized to $\rho_{cr,0}$ as a function of the scale factor a.

assuming a constant density energy so that the corresponding EoS is $p = -\rho$. The case of a flat universe ($k = 0$) is interesting because of its agreement with many observational results [4]; it also implies a special value of the matter density in the universe that allows to introduce naturally a critical density ρ_{cr} in terms of the Hubble parameter. The critical density is also useful to define the dimensionless density parameter Ω_i for the various species i, related to radiation Ω_r, matter Ω_m, cosmological constant Ω_Λ and curvature Ω_k. It is easy to verify that the sum of these parameters is equal to unity, as it can be expected from the Friedmann equations; in fact, $\Omega_m + \Omega_r + \Omega_\Lambda + \Omega_k = 1$ is known as the *cosmic sum rule*. Therefore, a Friedmann universe can be described by the cosmological parameters $(H_0, \Omega_m, \Omega_r, \Omega_k, \Omega_\Lambda)$ such that the expansion rate as a function of the scale factor is given by

$$H^2(a) = H_0^2\left[\Omega_r a^{-4} + \Omega_m a^{-3} + \Omega_k a^{-2} + \Omega_\Lambda\right]. \tag{6}$$

This equation is usually written as a dimensionless function defined by $E(a) = H(a)/H_0$. The last constraints on cosmological parameters obtained by the Planck satellite show that $w = -1.03 \pm 0.03$, consistent with a cosmological constant, and $\Omega_m = 0.3153 \pm 0.0073, \Omega_\Lambda = 0.692 \pm 0.012, \Omega_k = 0.000 \pm 0.005$, where Ω_m contains the density of baryons (Ω_b) and cold dark matter (Ω_{CDM}). Additionally, in the last few years, it became usual to include the energy density of neutrinos Ω_ν; they contribute to the radiation density at early times but behave as matter after the non-relativistic transition at late times [24], so that for a flat universe, the total energy density is given by $\rho = \rho_\gamma + \rho_{\mathrm{CDM}} + \rho_{\mathrm{b}} + \rho_\nu + \rho_\Lambda$. **Figure 1a** shows the percentages, derived from [4] data, in which each species contributes to the total content of the universe. **Figure 1b** shows the evolution of the density parameter ρ (in units of $\rho_{cr,0}$) as a function of the scale factor a. For further details concerning the background universe, see Refs. [25–27].

2. Modified gravity and massive neutrinos

2.1 Modified gravity models

One of the most interesting modifications of GR is that which modifies the Einstein-Hilbert action by introducing a scalar function, $f(R)$, as follows:

$$S = \int d^4x \sqrt{-g}\left(\frac{R + f(R)}{16\pi G} + \mathcal{L}_m\right), \tag{7}$$

where R is the Ricci scalar, G is the Newton's gravitational constant, g is the determinant of the metric tensor $g_{\mu\nu}$ and $\mathcal{L}_m$ is the Lagrangian density of all matter fields. For a classification of $f(R)$ theories of gravity and the assumptions needed to arrive to the various versions of $f(R)$ gravity and GR, see, e.g., [28]. Thus, for a general $f(R)$ model, one can consider a spatially flat FLRW universe with metric $ds^2 = -dt^2 + a^2(t)d\mathbf{x}^2$, so that varying the Einstein-Hilbert action with respect to $g_{\mu\nu}$, one can get a general form of the modified Einstein field equations. Consequently, the corresponding modified Friedmann equations are given by

$$3H^2(1 + f_R) = 8\pi G(\rho_m + \rho_{rad}) + \frac{1}{2}(f_R R - f) - 3H\dot{f}_R \tag{8}$$

$$-2(1 + f_R)\dot{H} = 8\pi G\left(\rho_m + \frac{4}{3}\rho_{rad}\right) + \ddot{f}_R - H\dot{f}_R, \tag{9}$$

where $R = 6(2H^2 + \dot{H})$ and the over-dot denotes a derivative with respect to the cosmic time t. In general, the background evolution of a viable $f(R)$ is not simple as it has been shown by [29–31]. However it is possible to get an approximation in a way that is analogous to the DE models, by neglecting the higher derivative and the non-linear terms. By defining the growth rate as $f \equiv d \ln \delta / d \ln a$, the equation that describes the growth of matter perturbations in terms of the density contrast δ in a $f(R)$ model is approximated by [29, 32]

$$-3\Omega_m(a)(1 - \Omega_m(a))\frac{df}{d\Omega_m(a)} + f^2 + \left(2 - \frac{3}{2}\Omega_m(a)\right)f = \frac{3}{2}\frac{G_{\mathrm{eff}}}{G}\Omega_m(a). \tag{10}$$

where G_{eff} is the effective gravitational constant that can be written as [33]. A plausible $f(R)$ function able to satisfy the solar system constraints, to mimic the ΛCDM model at high-redshift regime where it is well tested by the CMB and, at the same time, to accelerate the expansion of the universe at low redshift but without a cosmological constant [34], suggests that

$$\lim_{R\to\infty} f(R) = \text{const}, \quad \lim_{R\to 0} f(R) = 0, \tag{11}$$

which can be satisfied by a broken power law function such that

$$f(R) = -m^2 \frac{c_1\left(\frac{R}{m^2}\right)^n}{c_2\left(\frac{R}{m^2}\right)^n + 1}, \tag{12}$$

where the mass scale m is defined as $m^2 \equiv H_0^2\Omega_M$ and c_1, c_2 and n are non-negative free parameters of the model [34]. For this $f(R)$ model, the background expansion history is consistent with the ΛCDM case by choosing $c_1/c_2 = 6\Omega_\Lambda/\Omega_M$, where Ω_Λ and Ω_M are the dimensionless density parameters for vacuum and matter, respectively.

Nowadays it is generally accepted that some MG theories such as the [34] $f(R)$ are strongly degenerated in a wide range of their observables with the effects of massive neutrinos; see, e.g., [19–21, 35]. This represents a serious challenge constraining cosmological models from current and future galaxy surveys requiring robust and reliable methods to disentangle both phenomena. Furthermore, for some specific combinations of the $f(R)$ function and of the total neutrino mass $m_\nu \equiv \Sigma m_{\nu,i}$, standard statistics would not distinguish them from the standard ΛCDM expectations (see [21, 22, 36]). In addition, since the degeneracy is mostly driven by the non-linear behaviour of both the MG and the massive neutrinos effects on the LSS, the linear tools are not suitable to properly disentangle the combined parameter space [36].

2.2 Massive neutrinos and the large-scale structure

Motivated by the apparent violation of energy, momentum and spin in β-decay processes, Pauli proposed the existence of neutrinos in 1930 to keep the conservation laws safe. Eventually, 26 years after they have been theoretically postulated, the neutrinos were detected for the first time by Cowan [37]. Neutrinos are classified in three 'flavours' in the standard model of particles; they were considered to be massless for some time until the discovery of the neutrino oscillation phenomena, i.e. related to the change of flavour [38]. Since then, it is known that at least two of the three neutrino families are massive, in contrast to the particle standard model assumption; however, measuring the absolute masses of the neutrinos is not easy, which makes this a very active field of research today, both for cosmology and particle physics. In a cosmological context, the neutrinos leave detectable imprints on observations that can then be used to constrain their properties; in particular, the presence of massive neutrinos impacts the background evolution of the universe and the growth of structures [39]. In the early universe, massive neutrinos are relativistic and indistinguishable from the massless ones, behaving like photons, meaning that their energy density drops like $\propto a^{-4}$. In this stage, neutrinos are in thermal equilibrium, and their momentum follows the standard Fermi-Dirac distribution

$$n(p)\mathrm{d}p = \frac{4\pi g_\nu}{(2\pi\hbar c)^3}\frac{p^2\mathrm{d}p}{e^{\frac{p}{k_B T_\nu}}+1}, \tag{13}$$

where $n(p)$ is the number of cosmic neutrinos with momentum between p and $p+dp, g_\nu$ is the number of neutrino spin states, $T_\nu(z)$ is the neutrino temperature at redshift z and k_B is the Boltzmann constant. In principle, in the momentum distribution function, the chemical potential should be also included; however, it has been shown to be negligible for cosmological neutrinos [40]. The temperature of the cosmic neutrino background and the one from CMB are related by $T_\nu(z=0) = \left(\frac{4}{11}\right)^{1/3} T_\gamma(z=0)$ [41], such that the temperature of the neutrino background at certain redshift z is given by $T_\nu(z) \cong 1.95(1+z)$ K. Then, when the average momentum of neutrinos drops below a certain mass, they become non-relativistic, and their energy density drops like $\propto a^{-3}$, behaving like baryons and cold dark matter. **Figure 2** shows the evolution of the massive neutrino density, normalised to the today's critical density, as a function of the scale factor from its early stage to the late universe.

After neutrinos with mass m_ν decouple from the rest of the plasma at redshift z_{nr}, as shown by Eq. (14)

$$1 + z_{nr}(m_\nu) \simeq 1890\left(\frac{m_\nu}{1\mathrm{eV}}\right), \tag{14}$$

the number density per flavour is fixed by the temperature, so that the universe is currently filled by a relic neutrino background, uniformly distributed, with a density of 113 part/cm^3 per species and an average temperature of 1.95 K. As

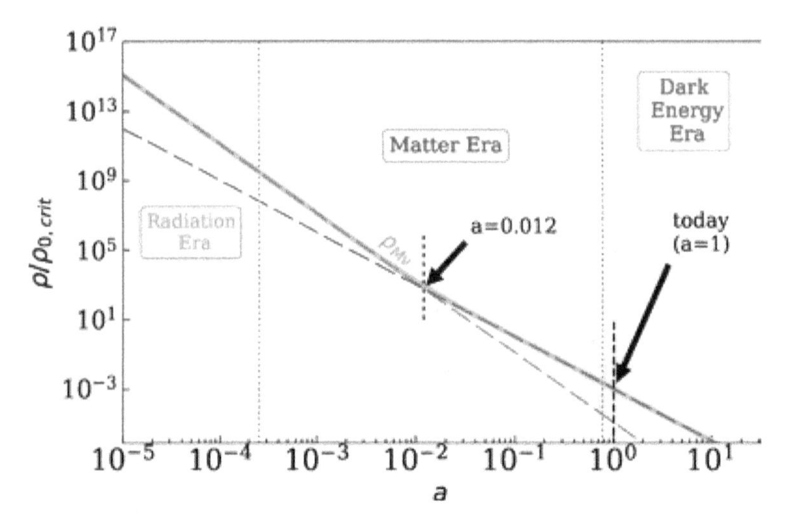

Figure 2.
Evolution of the massive neutrino density, normalised with respect today's critical density, as a function of the scale factor a, *for a neutrino mass* $m_?$ = 0.1 *eV. The pink solid line shows the correct solution for* $\rho_\nu(a)$, *while the green and blue dashed lines show the limit at early and late time, respectively. The early time limit is proportional to* a^4, *underlining that at early times neutrinos are relativistic and behave like radiation. The late time limit, instead, is proportional to* a^3, *following the same evolution of matter. The short vertical dashed line (black) marks the non-relativistic transition for neutrinos of this given mass. The figure was computed by using the CAMB code.*

neutrinos are non-relativistic particles at late times, they contribute to the total matter density of the universe Ω_M, so that $\Omega_M = \Omega_{CDM} + \Omega_b + \Omega_\nu$, where Ω_{CDM}, Ω_b and Ω_ν are the dimensionless density parameters for CDM, baryons and neutrinos, respectively. The density background is affected by massive neutrinos such that a perturbation in the density field is well described by [39, 42] as follows:

$$\delta_m = (1 - f_\nu)\delta_{CDM} + f_\nu \delta_\nu, \quad \text{where } f_\nu \equiv \frac{\Omega_\nu}{\Omega_{CDM} + \Omega_\nu}, \tag{15}$$

δ_ν being the neutrino perturbations and Ω_ν the density contribution related to massive neutrinos that can be expressed in terms of the total neutrino mass, $m_\nu \equiv \sum_i m_{\nu_i}$, as follows:

$$\Omega_\nu = \frac{\Sigma_i m_{\nu_i}}{93.14\, h^2 eV}. \tag{16}$$

Currently, several observations provide limits on the total neutrino mass under the assumption of standard GR [43]. Depending of their mass, neutrinos can affect different quantities such as the matter-radiation equality and at the same time imprint features in cosmological observables like the clustering, the matter power spectrum, the halo mass function and the redshift-space distortions [18, 44–47].

3. Halo mass function and clustering analysis

Considering the impact of MG and massive neutrinos in the clustering, a powerful cosmological test to discriminate among these scenarios is provided by the redshift-space distortions (RSD), that is, the shift in the position of the tracers due to their peculiar motions. For this purpose, cosmological simulations have become a powerful tool for testing theoretical predictions and to lead observational projects.

In this context, the formation and evolution of cosmic structures can be understood as a dynamical system of many particles, which trace the underlying mass distribution in a certain cosmological model. The N-body simulations, methods and algorithms have progressed continuously, achieving a high resolution to resolve finer structures with millions of particles, reducing the gap between theory and observations. For a detailed description on fundamentals of cosmological simulations, see, e.g., [48–51].

Since the formation and evolution of cosmic structures is based on the growth of small fluctuations in the density field, it is expected that the amplitude of these initial perturbations have the correct value at late times to match the observed clustering today. An analytical development based on perturbation theory makes possible to follow the growth of structures to a certain extent using the linear approximation, being valid as long as $\delta\rho \ll \rho$. Nevertheless, these calculations are limited and cannot be extrapolated to explain completely the observational data; they break down on a scale where the density contrast $\delta \sim 1$. Moreover, beyond the linear regime, the observed structures have a density contrast in a wide range from cosmic voids with $\delta \sim -1$ to $\delta \sim 10^6$ and larger. It makes necessary a more elaborated description of the perturbations in the non-linear regime, which can be achieved using higher-order perturbation theory or numerical simulations [48].

In this section we show a complementary analysis to the one performed in [52] in order to investigate the clustering in the context of modified gravity with massive neutrinos. We used a subset of the DUSTGRAIN-*pathfinder* runs [36], which implement the Hu-Sawicki $f(R)$ model including massive neutrinos and whose cosmological parameters are consistent with Planck 2015 constraints [53].

3.1 Halo mass function

As CDM haloes form from collapsing regions that detach from the background density field, their abundance can be related to the volume fraction of a Gaussian density smoothed on a radius R above a critical collapse threshold δ_c [54]. The comoving number density of the haloes is strictly related to underlining cosmological model, such that within a mass interval $[M, M + dM]$, the halo mass function is given by

$$\frac{dn(M,z)}{dM} = f(\sigma(M,z))\frac{\bar{\rho}}{M}\frac{d\ln\sigma(M,z)^{-1}}{dM}, \tag{17}$$

where $f(\sigma)$ is the multiplicity function, σ the RMS variance of the linear density field smoothed on scale $R(M)$ and $\bar{\rho}$ the mean matter density. The product $f(\sigma(M,z))\bar{\rho}$ quantifies the amount of mass contained in fluctuations of typical mass $M = \frac{4}{3}\pi R^3\bar{\rho}$. The simplest argument to compute analytically the multiplicity function $f(\sigma)$ comes from the spherical collapse theory, following the [54] formalism, such that a perturbation is supposed to collapse when it reaches the threshold $\delta_c \simeq 1.68$, by assuming that the probability distribution for a perturbation on a scale M is a Gaussian function with variance σ_M^2, resulting in

$$f(\sigma(M)) = \sqrt{\frac{2}{\pi}}\frac{\delta_c}{\sigma(M)}\exp\left(-\frac{\delta_c^2}{2\sigma^2(M)}\right). \tag{18}$$

Another approach to determine $f(\sigma)$ is given by accurate fitting functions, like the proposed by [55], which extends phenomenologically the results of [54]. For [55] the function $f(\sigma)$ is expected to be universal to the changes in redshift and cosmology and is parameterized as follows:

$$f(\sigma) = A\left[\left(\frac{\sigma}{b}\right)^{-a} + \sigma^{-c}\right] \exp\left(-d/\sigma^2\right), \tag{19}$$

where A is an amplitude of the mass function and a, b, c and d are free parameters that depend on halo definition. The variance σ^2 is usually given by

$$\sigma^2(R(M), z) = \int \frac{P(k,z)}{2\pi^2} W^2(kR(M))k^2 \mathrm{d}k, \tag{20}$$

where $P(k)$ is the linear matter power spectrum as a function of the wave number k and W is the Fourier transform of the real-space top-hat window function of radius R. A fundamental feature of the mass function is that it decreases monotonically with increasing masses; furthermore, its dependency on cosmology is encoded in the variance σ^2, as shown by the integrand of Eq. (20). From the point of view of N-body simulations, an approach to compute the mass function is given straightforward from Eq. (17), by counting the number of haloes N_h above a certain mass threshold M_{min} in a comoving volume V such as

$$N_h = A \int_{z_{min}}^{z_{max}} \int_{M_{min}}^{\infty} \frac{\mathrm{d}V}{\mathrm{d}z} n(M, z) \mathrm{d}M \mathrm{d}z, \tag{21}$$

where A is the area, z_{min} and z_{max} are the redshift boundaries and dV/dz is the comoving volume element.

Figure 3 shows the mass function of CDM haloes measured for all models of the DUSTGRAIN-*pathfinder* runs at six different redshifts $z = 0, 0.5, 1, 1.4, 1.6, 2$. Each panel contains the mass function, per each model as labelled, to track its evolution in redshift. As reference, the black dashed line represents the theoretical expectation by [55] for a flat ΛCDM model. As expected, massive haloes are less abundant with respect to smaller ones in a fixed comoving volume. The mass function decreases with redshift, since at earlier times the density field is smoother than at late times. The plot is logarithmic, meaning that the number density of large mass haloes falls off by several orders of magnitude over the range of redshifts shown. The $f(R)$ models both with and without neutrinos reproduce in very well agreement this pattern, but only at really high masses, significant differences appear.

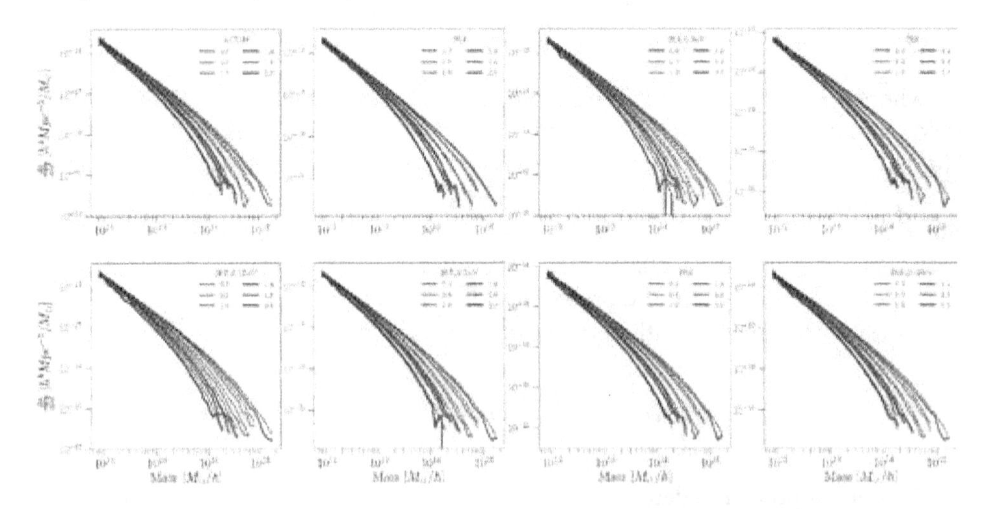

Figure 3.
The mass function of CDM haloes per each model of the DUSTGRAIN-pathfinder project at six different redshifts $z = 0, 0.5, 1, 1.4, 1.6, 2$ *as labelled. Each panel corresponds to one model as labelled, and the dashed line represents the theoretical expectation by [55], assuming a flat ΛCDM model.*

Figure 4 compares the halo mass functions of the different DUSTGRAIN-*pathfinder* simulations, computed at $z = 0$ (left column), $z = 1$ (central column) and $z = 1.6$ (right column). The lower panels show the percentage difference with respect to the ΛCDM model. It is possible to see that the effect of $f(R)$ and massive neutrinos on the dynamical evolution of the matter density field results in different halo formation epochs and different number density of collapsed systems. In particular, the $fR4$ model (blue) is the most deviated model from the standard scenario, whereas the $fR6$, $fR6\_0.06eV$ and $fR6\_0.1eV$ models mimic the ΛCDM behaviour over a wide range of masses.

3.2 Clustering analysis

To quantify the halo clustering, we used the two-point correlation function (2PCF) that can be defined as the joint probability of finding a pair of objects at certain spatial separation given a tracer distribution. To measure the full 2PCF denoted as $\xi(r,\mu)$, we used the Landy-Szalay estimator given by [56]:

$$\hat{\xi}(r,\mu) = \frac{DD(r,\mu) - 2DR(r,\mu) + RR(r,\mu)}{RR(r,\mu)}, \tag{22}$$

where μ is the cosine of the angle between the line of sight and the comoving halo pair separation, r, and DD, RR and DR represent the normalised number of data-data, random-random and data-random pairs, respectively. This estimator is almost unbiased with minimum variance; this is the reason why it is preferred over the other estimators regarding the clustering measurements. Since the possible deviations from GR are more evident on small scales, we consider an intermediate non-linear range from $1\,h^{-1}$Mpc to $50\,h^{-1}$Mpc and random samples ten times larger than the halo ones. Then, in order to examine how significant is the RSD correction, it is convenient to expand the 2D 2PCF in the orthonormal basis of the Legendre polynomials $L_l(\mu)$ [57] such that

$$\xi(s,\mu) \equiv \xi_0(s)L_0(\mu) + \xi_2(s)L_2(\mu) + \xi_4(s)L_4(\mu), \tag{23}$$

where each coefficient corresponds to the lth multipole moment:

$$\xi_l(r) = \frac{2l+1}{2}\int_{-1}^{+1} d\mu\,\xi(r,\mu)L_l(\mu). \tag{24}$$

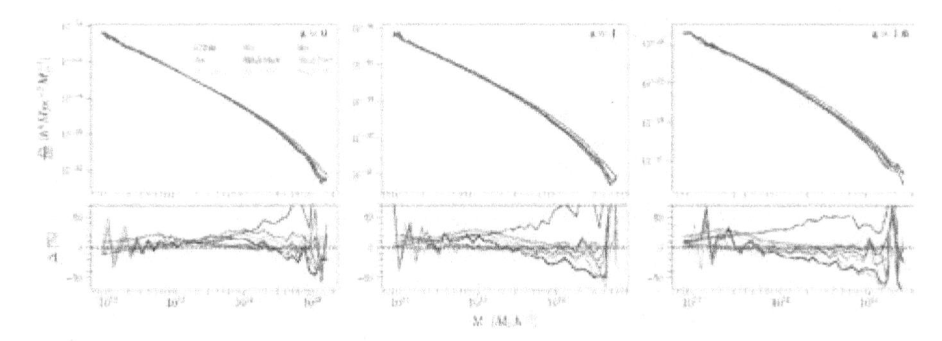

Figure 4.
The mass function of CDM haloes for all the models of the DUSTGRAIN-pathfinder at three different redshifts: $z = 0$ (left column), $z = 1$ (central column) and $z = 1.6$ (right column). The lower panels show the percentage difference with respect to the ΛCDM model. As in **Figure 3** *the dashed line represents the theoretical prediction by [55].*

Figure 5 shows the RSD effects on the iso-correlation curves of the 2D two-point correlation function (2PCF) in the plane $(r_{\perp}, r_{\parallel})$, where $r_{\perp}$ and $r_{\parallel}$ coordinates are, respectively, the perpendicular and parallel components along the line of sight of the observer. The 2PCF has been computed for the ΛCDM catalogue of the DUSTGRAIN-*pathfinder* simulations, in real space (*left panel*), and for the corresponding sample in redshift space (*right panel*). The contours are drawn at the iso-correlation levels $\xi(s_{\perp}, s_{\parallel}) = 0.3, 0.5, 1.0, 1.4, 2.2, 3.6, 7.2, 21.6$. In real space the correlation function is undistorted, describing circular curves in this plane. In redshift space, the effect caused by the RSD is clearly visible on small scales, where the 2PCF is stretched in the direction of $r_{\parallel}$ (Fingers-of-God effect), and in the infall effect on large scales, the contours are squashed along the perpendicular direction (Kaiser effect).

The symmetry of the 2PCF in real space means that the full clustering signal is encoded in the monopole moment $\xi_0(r)$, while the rest of the multipole moments are statistically equal to zero. **Figure 6** shows the monopole moment, $\xi_0(r)$, of the CDM haloes for all models considered in the DUSTGRAIN-*pathfinder* project, at three different redshifts $z = 0, 1, 2$. Subpanels show the percentage difference between MG models [$f(R)$ with and without massive neutrinos] and the ΛCDM model. The monopole moments of the 2PCF of the $fR4$ and $fR4\_0.3eV$ models are the ones that deviate most from ΛCDM at low redshift. This behaviour is also present in the models $fR5$, $fR5\_0.15eV$ and $fR5\_0.1eV$, but it is less significant. In general, it is observed that massive neutrinos increase the clustering signal for all models, especially at high redshift, the $fR6$, $fR6\_0.06eV$ and $fR6\_0.1eV$ models being degenerated with respect to ΛCDM. The quadrupole and hexadecapole moments are consistent with zero at 1σ error bar.

Another important feature to take into account in the clustering analysis is related to the bias that is introduced when the ΛCDM model is wrongly assumed to predict the DM clustering of a $f(R)$ universe with massive neutrinos [58]. This

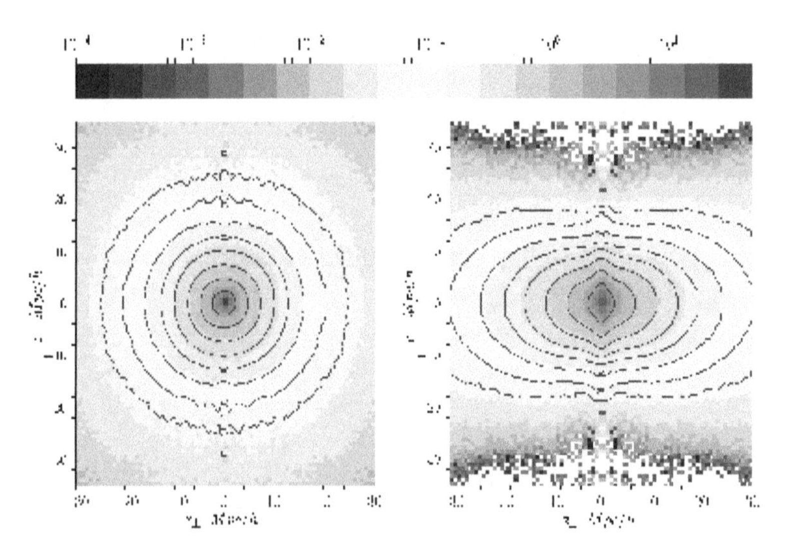

Figure 5.
Contours of the 2D two-point correlation function (2PCF) in the plane $(r_{\perp}, r_{\parallel})$, the $r_{\perp}$ and $r_{\parallel}$ coordinates are, respectively, the perpendicular and parallel components along the line of sight of the observer. The 2PCF has been computed from an N-body simulation, in real space (left panel), and for the corresponding sample in redshift space (right panel). The contours are draw at the iso-correlation levels $\xi(s_{\perp}, s_{\parallel})$= 0.3, 0.5, 1.0, 1.4, 2.2, 3.6, 7.2, 21.6 as indicated by the colour bar. Both the effect of redshift space distortions on small scales (Finger-of-God effect) and the infall effect at large scales (Kaiser effect) are clearly visible in the left panel. On small scales the 2PCF is stretched in the direction of $r_{\parallel}$, and on large scales the contours are squashed along the perpendicular direction.

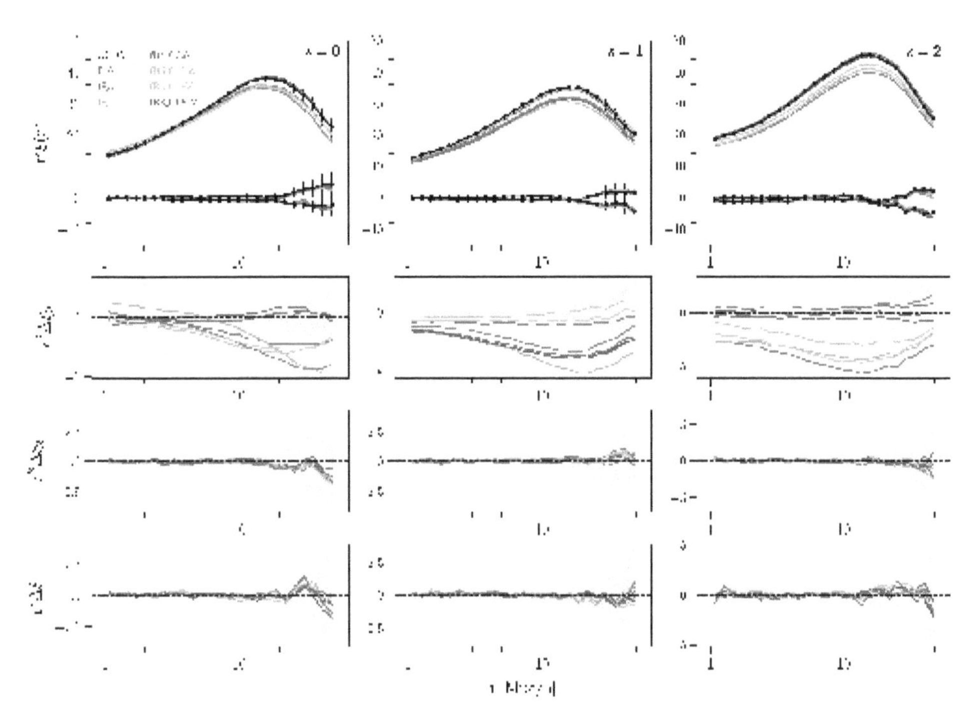

Figure 6.
The real-space 2PCF $r^2\xi_0$ of CDM haloes for all the models of the DUSTGRAIN-pathfinder project at three different redshifts: $z = 0$ (left column), $z = 1$ (central column) and $z = 2$ (right column). From top to bottom, the panels show the fR4, fR5 and fR6 models, respectively, compared to the results of the ΛCDM model. The error bars, shown only for the ΛCDM model for clarity reasons, are the diagonal values of the bootstrap covariance matrices used for the statistical analysis. Percentage differences between f(R), $f(R) + m_\nu$ and ΛCDM predictions are in the subpanels, while the shaded regions represent the deviation at 1σ confidence level.

effective halo bias, $\langle b \rangle$, allows to characterise the relation between the halo clustering and the underlying mass distribution. By using the theoretical effective bias proposed by [59] and its corresponding mass function, it is possible to disentangle the degeneracy with respect to σ_8. The level of agreement with the measurements obtained from the $f(R)$ and $f(R) + m_\nu$ scenarios give us a better understating of the discrepancy with the ΛCDM ones. To compute the theoretical mass function, we consider the linear CDM + baryon power spectrum as have been stated in the CDM prescription [60] and replacing ρ_m with $\rho_{\rm CDM}$ [42]. These quantities can be obtained with CAMB or another Boltzmann code; the linear power-spectrum for CDM, $P^{\rm CDM}_{lin}(k) = T^2_{\rm CDM}/T^2_m P^m_{lin}(k)$, is expressed in terms of $P^{{\rm CDM}+b}_{lin}(k)$ and $P^m_{lin}(k)$, with $T_{\rm CDM}(k)$ and $T_b(k)$ being the transfer functions. It implies, as shown by [61], that the effect of neutrinos on the cluster abundance is well captured by rescaling the smoothed density field such that

$$\sigma^2 \to \sigma^2_{\rm CDM}(z) = \int \frac{P_{\rm CDM}(k,z)}{2\pi^2} W^2(kR)k^2 {\rm d}k \tag{25}$$

with the CDM power spectrum obtained by rescaling the total matter power spectrum with the corresponding transfer functions, T_{CDM} and T_b, weighted by the density of each species so that

$$P_{\rm CDM}(k,z) = P_{\rm m}(k,z)\left(\frac{\Omega_{\rm CDM}T_{\rm CDM}(k,z) + \Omega_{\rm b}T_{\rm b}(k,z)}{T_{\rm m}(k,z)(\Omega_{\rm CDM} + \Omega_{\rm b})}\right)^2. \tag{26}$$

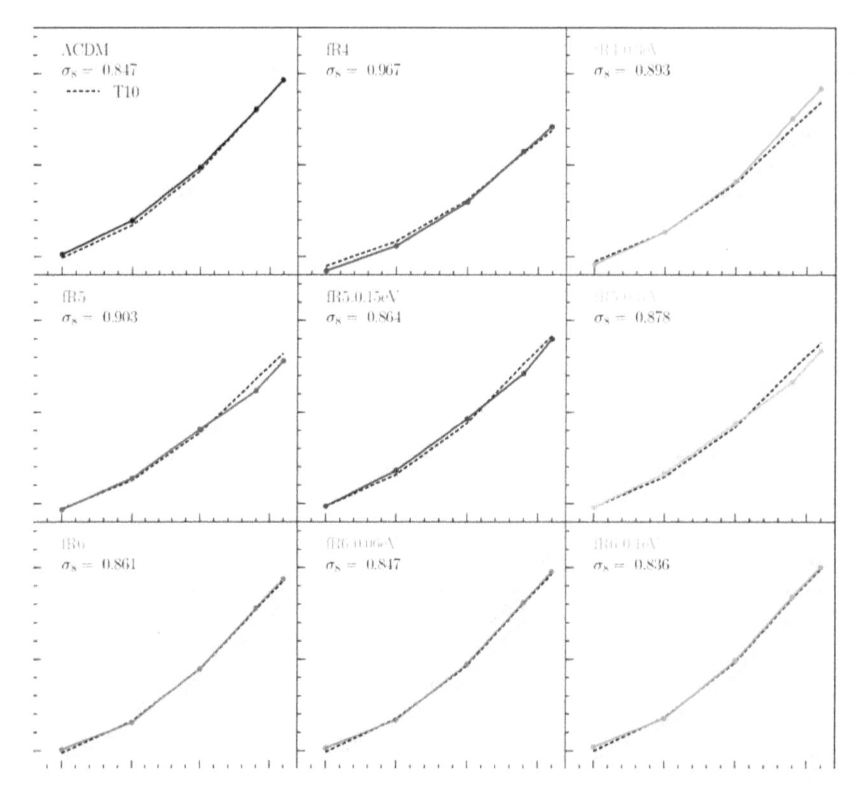

Figure 7.
The coloured solid lines represent the apparent effective halo bias, $\langle b \rangle$, as a function of redshift, averaged in the range $10\ h^{-1}$ Mpc $< r < 50\ h^{-1}$ Mpc. Black lines show the theoretical ΛCDM effective bias predicted by [59] (dashed), normalised to the σ_8 values of each DUSTGRAIN-pathfinder simulation, while the cyan-shaded areas show a 10% error. The upper set of panels shows the results, considering the total power spectrum, while the lower set of panels shows the results when the CDM+baryon power spectrum is used instead.

The impact on the bias when the CDM prescription is not considered can be appreciated in **Figure 7**. In most cases this correction is small with the exception of the $fR4$_0.3 eV model. A detailed discussion on these results can be found in [52].

3.3 Modelling the redshift-space distortions

In a realistic case, spectroscopic surveys observe a combination of density and velocity fields in redshift space. The observed redshift is a combination of cosmological effects plus an additional term caused by the peculiar motions along the line of sight of the observer. This combination makes the redshift-space catalogues appear distorted with respect to the real-space ones, and they can be reproduced from N-body simulations since the positions and velocities are known. Currently, the modelling of the redshift-space distortions provide a powerful tool to test the gravity theory by exploring the spatial statistics encoded in the 2PCF, which is anisotropic due to the dynamic distortions.

We consider the first two even multipoles of the 2PCF, ξ_0 and ξ_2, taking into account that odd multipole moments vanish by symmetry. The analysis consisted of modelling (i) the individual multipole moments and (ii) both multipole moments simultaneously. To derive cosmological constraints from the clustering signal and quantify the effects of $f(R)$ gravity and massive neutrinos on RSD, we performed a Bayesian analysis to set constraints on the linear growth

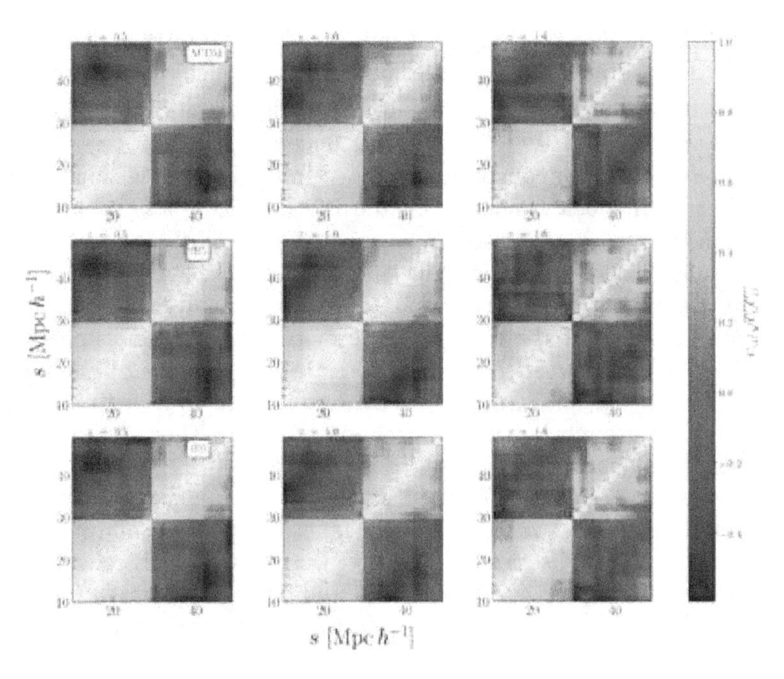

Figure 8.
Covariance matrices from the analysis of the redshift-space monopole and quadrupole moments of CDM haloes with Bootstrap errors. The models correspond to Λ*CDM (upper panels),* fR4 *(central panels) and* fR5 *(bottom panels), at three different redshifts* $z = 0.5, 1.0$ *and* 1.6 *from left to right.*

rate $f(\Omega_m)$ and the linear bias. All numerical tasks were performed with the CosmoBolognaLib[1] [62].

The Kaiser formula is a good description of the RSD only at very large scales, where non-linear effects can be neglected, but it does not describe accurately the non-linear regime. Thus, with the aim of extracting information from the RSD signal at non-linear regime and considering the increasing precision of recent and upcoming surveys, many more approaches have been proposed. There is a vast literature that shows the efforts to model the RSD beyond the linear Kaiser model [63–66], some of them making use of a phenomenological description of the velocity field and others, instead, taking into account higher orders in perturbation theory since, in principle, there is no reason to stop at linear order. Other approaches do a combination of both frameworks. A simple alternative to model the redshift-space 2PCF at small scales consists of extending the Kaiser formula, by adding a phenomenological damping factor that plays the role of a pairwise velocity distribution. It can account for both linear and non-linear dynamics. Therefore, to construct the likelihood, we consider this model sometimes called *dispersion model* [67], which introduces a damping function to describe the distortions in the clustering at small scales (Fingers-of-God). This model is enough accurate to quantify the relative differences between $f(R)$ models with massive neutrinos and ΛCDM. For the Bayesian analysis, the dispersion model is fully described by three parameters, $f\sigma_8$, $b\sigma_8$ and Σ_S, that we constrain by minimising numerically the negative log-likelihood.

Figure 8 shows the normalised covariance matrices $\left(C_{i,j}/\sqrt{C_{i,i}C_{j,j}}\right)$ of the redshift-space monopole and quadrupole moments of CDM haloes with bootstrap errors resampling at three different redshifts $z = 0.5$, 1.0 and 1.6. As it can be appreciated, the covariance matrices represent how the scatter propagates into the

[1] Freely available at the public GitHub repository https://github.com/federicomarulli/CosmoBolognaLib.

likelihood and on the final posterior probabilities of the parameters. Then, to assess the posterior distributions of the three model parameters, we perform a MCMC analysis. The fitting analysis is limited to the scale range $10 \leq r\ [\mathrm{Mpc\,h^{-1}}] \leq 50$, assuming flat priors in the ranges $0 \leq f\sigma_8 \leq 2$, $0 \leq b\sigma_8 \leq 3$ and $0 \leq \Sigma_S \leq 2$. **Figure 9** shows the $f\sigma_8$-$b\sigma_8$ posterior constraints obtained from the MCMC analysis of ξ_0, ξ_2 and $\xi_0 + \xi_2$ for each mock catalogue of the DUSTGRAIN-*pathfinder*. The figure shows the constraints for all models considered in this work at $z = 1.0$ using the monopole (orange), quadrupole (green) and monopole plus quadrupole (blue), while the intersection regions correspond to the joint analysis of the multipoles. Thus, the joint analysis of the redshift-space monopole and quadrupole is able to break the degeneracy between $f\sigma_8$ and $b\sigma_8$ even in the presence of massive neutrinos. In **Figure 10**, we show the theoretical behaviour of the linear distortion parameter and the growth factor as a function of the redshift for each family of models, assuming a flat ΛCDM model.

From the $f\sigma_8$-$b\sigma_8$ posterior contours, at $1-2\sigma$ confidence levels, the results suggest that the clustering information encoded in the two first non-null multipole moments of the 2PCF can discriminate the alternative MG models considered in this work at $z \gtrsim 1$. At low redshifts the $f(R)$ models studied are statistically indistinguishable from ΛCDM, and further studies are required to break this degeneracy.

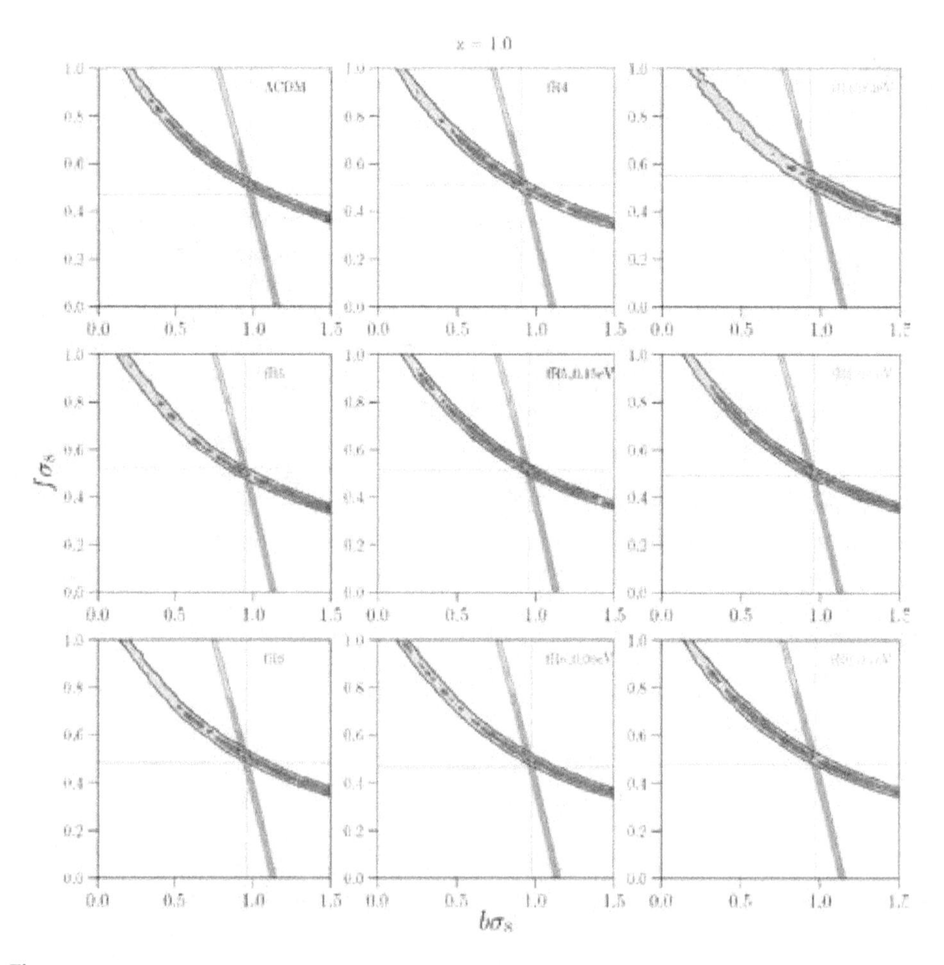

Figure 9.
Contours at $1-2\sigma$ confidence level of the $f\sigma_8 - b\sigma_8$ posterior distributions, obtained from the MCMC analysis in redshift space for 2PCF multipoles of CDM haloes. The contours correspond to $z = 1.0$ with the monopole in orange and the quadrupole in green.

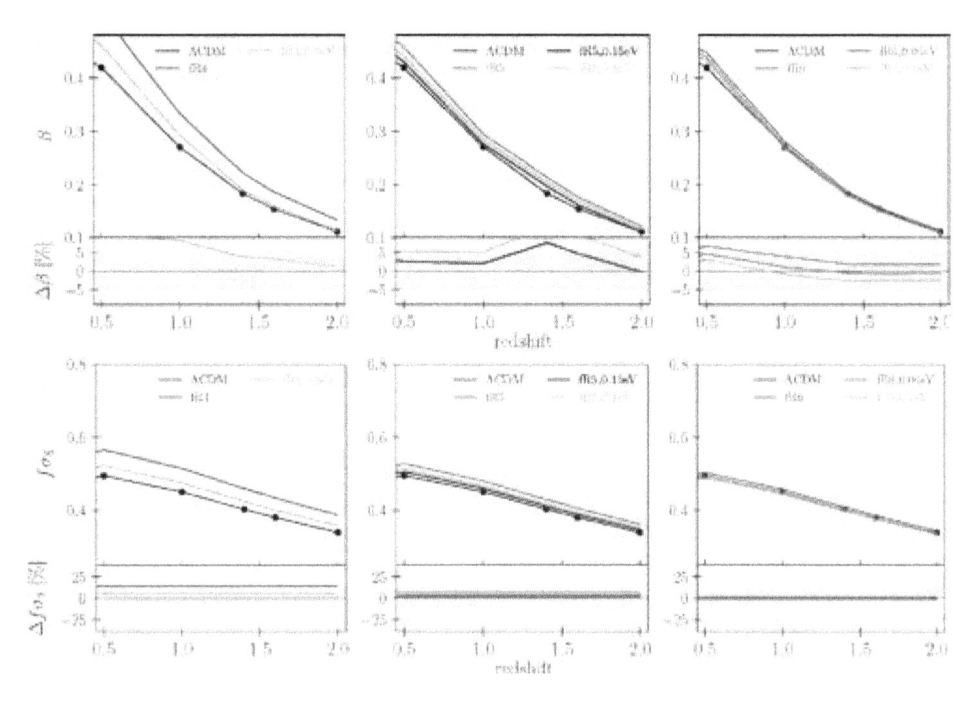

Figure 10.
Theoretical expectation of the linear distortion parameter β (upper panel) and the growth rate $f\,\sigma_8$ (lower panel) as a function of the redshift for each family of mocks from the DUSTGRAIN-pathfinder runs, assuming a flat ΛCDM model.

4. Conclusions

In this chapter we have introduced the theoretical framework of modern cosmology in present massive neutrinos. We emphasize on the structure formation and on the statistical description of the density field as well as the measurements of galaxy clustering and discuss the redshift-space distortions and the differences between clustering in real and redshift space, considering that in the recent years, the spatial distribution of matter on cosmological scales has become one of the most efficient probes to investigate the properties of the universe, such as test gravity theories on large scales, to explore the dark sector and the origin of the accelerated expansion of the universe as well as a probe to constrain alternative cosmological models.

In the context of models based on modified gravity and massive neutrino cosmologies, we investigated the spatial properties of the large-scale structure by exploiting the DUSTGRAIN-*pathfinder* simulations that follow, simultaneously, the effects of $f(R)$ gravity and massive neutrinos. These are two of the most interesting scenarios that have been recently explored to account for possible observational deviations from the standard ΛCDM model. In particular, we studied whether redshift-space distortions in the 2PCF multipole moments can be effective, breaking the cosmic degeneracy between these two effects. We analysed the redshift-space distortions in the clustering of dark matter haloes at different redshifts, focusing on the monopole and quadrupole moments of the two-point correlation function, both in real and redshift space. The deviations with respect to ΛCDM model have been quantified in terms of the linear growth rate parameter. We found that multipole moments of the 2PCF from redshift-space distortions provide a useful probe to discriminate between ΛCDM and modified gravity models, especially at high redshifts ($z\gtrsim1$), even in the presence of massive neutrinos. The linear growth rate constraints that we obtain from all the analysed $f(R) + m_\nu$ mock catalogues are statistically distinguishable from ΛCDM predictions at high redshifts.

Acknowledgements

We thank Lauro Moscardini, Federico Marulli and Alfonso Veropalumbo for the continuous development of the CosmoBolognaLib and their suggestions during this project. We also thank Carlo Giocoli and Marco Baldi for the crucial work performing the DUSTGRAIN-*pathfinder* runs.

Author details

Jorge Enrique García-Farieta* and Rigoberto Ángel Casas Miranda
Universidad Nacional de Colombia, Bogotá, Colombia

*Address all correspondence to: joegarciafa@unal.edu.co

References

[1] Tonry JL et al. Cosmological results from high-zSupernovae. The Astrophysical Journal. 2003;**594**(1):1-24

[2] Hamana T, Sakurai J, Koike M, Miller L. Cosmological constraints from subaru weak lensing cluster counts. Publications of the Astronomical Society of Japan. 2015;**67**(3):34

[3] Abbott TMC et al. Dark energy survey year 1 results: Cosmological constraints from galaxy clustering and weak lensing. Physical Review D. 2018; **98**:043526

[4] Planck Collaboration; Aghanim N et al. Planck 2018 results. VI. Cosmological parameters. arXiv e-prints, arXiv:1807.06209. July 2018

[5] Riess AG et al. Observational evidence from supernovae for an accelerating universe and a cosmological constant. The Astronomical Journal. 1998;**116**(3):1009

[6] Schmidt BP et al. The high-Z supernova search: Measuring cosmic deceleration and global curvature of the universe using type IA supernovae. Astrophysical Journal. 1998;**507**:46-63

[7] Perlmutter S et al. Measurements of ω and λ from 42 high-redshift supernovae. The Astrophysical Journal. 1999;**517**(2):565

[8] Planck Collaboration, Ade PAR, et al. Planck 2015 results—xxiv. Cosmology from Sunyaev-Zeldovich cluster counts. Astronomy & Astrophysics. 2016;**594**:A24

[9] Riess AG et al. A 2.4% determination of the local value of the hubble constant. The Astrophysical Journal. 2016;**826**(1):56

[10] Luis Bernal J, Verde L, Riess AG. The trouble with h_0. Journal of Cosmology and Astroparticle Physics. 2016;**2016**(10):019

[11] Sotiriou TP, Faraoni V. $f(R)$ theories of gravity. Reviews of Modern Physics. 2010;**82**(1):451-497

[12] De Felice A, Tsujikawa S. $f(R)$ theories. Living Reviews in Relativity. 2010;**13**(1):3

[13] Joyce A, Lombriser L, Schmidt F. Dark energy versus modified gravity. Annual Review of Nuclear and Particle Science. 2016;**66**(1):95-122

[14] Uzan J-P. Testing general relativity: from local to cosmological scales. Philosophical Transactions of the Royal Society of London A: Mathematical Physical and Engineering Sciences. 2011;**369**(1957):5042-5057

[15] Will CM. The confrontation between general relativity and experiment. Living Reviews in Relativity. 2014;**17**(1):4

[16] Pezzotta A et al. The VIMOS Public Extragalactic Redshift Survey (VIPERS). The growth of structure at $0.5 < z < 1.2$ from redshift-space distortions in the clustering of the PDR-2 final sample. Astronomy & Astrophysics. 2017;**604**:A33

[17] Collett TE et al. A precise extragalactic test of general relativity. Science. 2018;**360**(6395):1342-1346

[18] Lesgourgues J, Pastor S. Massive neutrinos and cosmology. Physics Reports. 2006;**429**(6):307-379

[19] Motohashi H, Starobinsky AA, Yokoyama J. Cosmology based on $f(r)$ gravity admits 1 eV sterile neutrinos. Physical Review Letters. 2013;**110**: 121302

[20] He J-H. Weighing neutrinos in $f(r)$ gravity. Physical Review D. 2013;**88**: 103523

[21] Baldi M et al. Cosmic degeneracies—I. Joint n-body simulations of modified gravity and massive neutrinos. Monthly Notices of the Royal Astronomical Society. 2014;**440**(1):75- 88

[22] Peel A, Pettorino V, Giocoli C, Starck J-L, Baldi M. Breaking degeneracies in modified gravity with higher (than 2nd) order weak-lensing statistics. Astronomy & Astrophysics. 2018;**619**:A38

[23] Hagstotz S, Costanzi M, Baldi M, Weller J. Joint halo mass function for modified gravity and massive neutrinos. I: Simulations and cosmological forecasts. arXiv e-prints, arXiv: 1806.07400. June 2018

[24] Lesgourgues J, Pastor S. Neutrino mass from cosmology. Advances in High Energy Physics. 2012;**2012**

[25] Peebles PJE. Principles of Physical Cosmology. Princeton, New Jersey: Princeton University Press; 1993

[26] Dodelson S. Modern Cosmology. San Diego, California: Academic Press; 2003

[27] Mukhanov V. Physical Foundations of Cosmology. Cambridge, UK: Cambridge University Press; 2005

[28] Sotiriou TP. $f(R)$ gravity and scalar tensor theory. Classical and Quantum Gravity. 2006;**23**(17):5117-5128

[29] Boisseau B, Esposito-Farèse G, Polarski D, Starobinsky AA. Reconstruction of a scalar-tensor theory of gravity in an accelerating universe. Physical Review Letters. 2000;**85**(11): 2236-2239

[30] Nojiri S, Odintsov SD, Sáez-Gómez D. Cosmological reconstruction of realistic modified F(R) gravities. Physics Letters B. 2009;**681**(1):74-80

[31] Cognola G, Elizalde E, Odintsov SD, Tretyakov P, Zerbini S. Initial and final de Sitter universes from modified $f(R)$ gravity. Physical Review D. 2009;**79**(4): 044001

[32] Narikawa T, Yamamoto K. Characterizing the linear growth rate of cosmological density perturbations in an $f(R)$ model. Physical Review D. 2010; **85**(11):2236-2239

[33] Tsujikawa S. Matter density perturbations and effective gravitational constant in modified gravity models of dark energy. Physical Review D. 2007; **76**(2):023514

[34] Hu W, Sawicki I. Models of $f(r)$ cosmic acceleration that evade solar system tests. Physical Review D. 2007; **76**:064004

[35] Wright BS, Winther HA, Koyama K. Cola with massive neutrinos. Journal of Cosmology and Astroparticle Physics. 2017;**2017**(10):054

[36] Giocoli C, Baldi M, Moscardini L. Weak lensing light-cones in modified gravity simulations with and without massive neutrinos. Monthly Notices of the Royal Astronomical Society. 2018; **481**(2):2813-2828

[37] Cowan CL Jr, Reines F, Harrison FB, Kruse HW, McGuire AD. Detection of the free neutrino: A confirmation. Science. 1956;**124**(3212):103-104

[38] Cleveland BT, Daily T, Davis R Jr, Distel JR, Lande K, Lee CK, et al. Measurement of the solar electron neutrino flux with the homestake chlorine detector. Astrophysics Journal. 1998;**496**:505-526

[39] Lesgourgues J, Mangano G, Miele G, Pastor S. Neutrino Cosmology. Cambridge, UK: Cambridge University Press; 2013

[40] Dolgov AD, Hansen SH, Pastor S, Petcov ST, Raffelt GG, Semikoz DV. Cosmological bounds on neutrino

degeneracy improved by flavor oscillations. Nuclear Physics B. June 2002;**632**:363-382

[41] Weinberg S. Cosmology. Oxford: Oxford University Press; 2008

[42] Castorina E, Sefusatti E, Sheth RK, Villaescusa-Navarro F, Viel M. Cosmology with massive neutrinos. II: On the universality of the halo mass function and bias. Journal of Cosmology and Astroparticle Physics. 2014; **2014**(02):049

[43] Seljak U, Slosar A, McDonald P. Cosmological parameters from combining the Lyman-α forest with CMB, galaxy clustering and SN constraints. Journal of Cosmology and Astroparticle Physics. 2006;**10**:014

[44] Brandbyge J, Hannestad S, Troels H, Wong YYY. Neutrinos in non-linear structure formation—The effect on halo properties. Journal of Cosmology and Astroparticle Physics. 2010;**2010**(09): 014

[45] Saito S, Takada M, Taruya A. Impact of massive neutrinos on the nonlinear matter power spectrum. Physical Review Letters. 2008;**100**: 191301

[46] Villaescusa-Navarro F, Bird S, Peña-Garay C, Viel M. Non-linear evolution of the cosmic neutrino background. Journal of Cosmology and Astroparticle Physics. 2013;**2013**(03):019

[47] Viel M, Haehnelt MG, Springel V. The effect of neutrinos on the matter distribution as probed by the intergalactic medium. Journal of Cosmology and Astroparticle Physics. 2010;**2010**(06):015

[48] Yepes G. The universe in a computer: The importance of numerical simulations in cosmology. In: Martnez VJ, Trimble V, Pons-Bordera MJ, editors. Historical Development of Modern Cosmology, Volume 252 of Astronomical Society of the Pacific Conference Series. 2001. p. 355

[49] Knebe A. How to simulate the universe in a computer. Publications of the Astronomical Society of Australia. 2005;**22**(3):184189

[50] Moscardini L, Dolag K, Colpi M, Gorini V, Moschella U. Cosmology with Numerical Simulations. Dordrecht: Springer Netherlands; 2011. pp. 217-237

[51] Dolag K, Borgani S, Schindler S, Diaferio A, Bykov AM. Simulation techniques for cosmological simulations. Space Science Reviews. 2008;**134**(1): 229-268

[52] Garca-Farieta JE et al. Clustering and redshift-space distortions in modified gravity models with massive neutrinos. Monthly Notices of the Royal Astronomical Society. 2019:07

[53] Planck Collaboration, Ade PAR, et al. Planck 2015 results. xiii. Cosmological parameters. Astronomy & Astrophysics. 2016;**594**:A13

[54] Press WH, Schechter P. Formation of galaxies and clusters of galaxies by self-similar gravitational condensation. The Astrophysical Journal. 1974;**187**:425-438

[55] Tinker J, Kravtsov AV, Klypin A, Abazajian K, Warren M, Yepes G, et al. Toward a halo mass function for precision cosmology: The limits of universality. The Astrophysical Journal. 2008;**688**(2):709-728

[56] Landy SD, Szalay AS. Bias and variance of angular correlation functions. The Astrophysical Journal. 1993;**412**:64-71

[57] Hamilton AJS. Measuring Omega and the real correlation function from the redshift correlation function. The Astrophysical Journal Letters. 1992;**385**: L5-L8

[58] Marulli F, Baldi M, Moscardini L. Clustering and redshift-space distortions in interacting dark energy cosmologies. Monthly Notices of the Royal Astronomical Society. 2012; **420**(3):2377-2386

[59] Tinker JL, Robertson BE, Kravtsov AV, Klypin A, Warren MS, Yepes G, et al. The large-scale bias of dark matter halos: Numerical calibration and model tests. The Astrophysical Journal. 2010;**724**:878-886

[60] Villaescusa-Navarro F et al. Cosmology with massive neutrinos. I: Towards a realistic modeling of the relation between matter, haloes and galaxies. Journal of Cosmology and Astroparticle Physics. 2014;**3**:011

[61] Costanzi M et al. Cosmology with massive neutrinos. III: The halo mass function and an application to galaxy clusters. Journal of Cosmology and Astroparticle Physics. 2013;**2013**(12): 012

[62] Marulli F, Veropalumbo A, Moresco M. CosmoBolognaLib: C++ libraries for cosmological calculations. Astronomy and Computing. 2016;**14**: 35-42

[63] Scoccimarro R, Couchman HMP, Frieman JA. The bispectrum as a signature of gravitational instability in redshift space. The Astrophysics Journal. 1999;**517**(2):531

[64] Scoccimarro R. Redshift-space distortions, pairwise velocities, and nonlinearities. Physical Review D. 2004; **70**(8):083007

[65] Taruya A, Nishimichi T, Saito S. Baryon acoustic oscillations in 2D: Modeling redshift-space power spectrum from perturbation theory. Physical Review D. 2010;**82**(6):063522

[66] Reid BA, White M. Towards an accurate model of the redshift-space clustering of haloes in the quasi-linear regime. Monthly Notices of the Royal Astronomical Society. 2011;**417**(3): 1913-1927

[67] Peacock JA, Dodds SJ. Reconstructing the linear power spectrum of cosmological mass fluctuations. Monthly Notices of the Royal Astronomical Society. 1994;**267**: 1020

7

Dusty Plasmas in Supercritical Carbon Dioxide

Yasuhito Matsubayashi, Noritaka Sakakibara, Tsuyohito Ito and Kazuo Terashima

Abstract

Dusty plasmas, which are systems comprising plasmas and dust particles, have emerged in various fields such as astrophysics and semiconductor processes. The fine particles possibly form ordered structures, namely, plasma crystals, which have been extensively studied as a model to observe statistical phenomena. However, the structures of the plasma crystals in ground-based experiments are two-dimensional (2D) because of the anisotropy induced by gravity. Microgravity experiments successfully provided opportunities to observe the novel phenomena hidden by gravity. The dusty plasmas generated in supercritical fluids (SCFs) are proposed herein as a means for realizing a pseudo-microgravity environment for plasma crystals. SCF has a high and controllable density; therefore, the particles in SCF can experience pseudo-microgravity conditions with the aid of buoyancy. In this chapter, a study on the particle charging and the formation of the plasma crystals in supercritical CO_2, the realization of a pseudo-microgravity environment, and the outlook for the dusty plasmas in SCF are introduced. Our studies on dusty plasmas in SCF not only provide the pseudo-microgravity conditions but also open a novel field of strongly coupled plasmas because of the properties of media.

Keywords: plasma crystal, supercritical fluid, pseudo-microgravity, surface dielectric barrier discharge, dusty plasmas in dense fluids

1. Introduction

The dynamics of statistical phenomena, such as phase transitions and wave propagation, and kinetic phenomena, such as the motion of dislocations in a crystal, are difficult to observe because atoms are too small. For a long time, many models have been developed that imitate crystal structures and can be observed with an optical microscope. For example, in 1947, Bragg, who established X-ray diffraction, and Nye reported the bubble model to understand the dynamics of dislocations [1]. The most extensively studied system is probably a charged particle system. For example, colloidal crystals, ordered structures of microparticles in colloidal dispersion, were developed [2]. Colloidal crystals have been studied not only as a model of crystal structure but also for application to optical materials [3]. Interparticle distances are close to the wavelength of visible light, which results in opalescence.

Dusty plasmas or fine particle plasmas, a system composed of dust particles and plasmas, have also been extensively studied as a model of crystal structure. In 1986,

Ikezi theoretically predicted that microparticles (diameter, 0.3–30 μm) embedded in a commonly used plasma processing possibly form an ordered structure or plasma crystal [4]. Microparticles inside the plasma become negatively charged because of the higher mobility of electrons. The charged particles exert a repulsive Coulombic force on each other. Plasma crystals can be formed when the interparticle electrostatic potential exceeds the kinetic energy of particles. A good measure for the formation of plasma crystals is a Coulomb coupling parameter Γ. Γ is defined as a ratio of the electrostatic potential energy to the kinetic energy of particles. The plasma with $\Gamma > 1$ is defined as a strongly coupled plasma [5], and Monte Carlo simulation suggests that an ordered structure can be formed when $\Gamma > 170$ [6].

In 1994, three independent groups simultaneously reported the experimental observation of plasma crystals [7–9]. However, these crystal structures were strongly affected by gravity. Because the electric field in a plasma sheath can compensate for gravity, the structure of plasma crystals can be maintained in a sheath region. Such compression in a direction of gravitational force gives plasma crystals a two-dimensional (2D) structure. To eliminate the gravitational anisotropy, micro-gravity experiments using the International Space Station and a sounding rocket have been conducted and provided three-dimensional (3D) plasma crystals [10]. The 3D plasma crystals in microgravity experiments show some new phenomena, such as an unexpected void structure and various crystal structures (fcc, bcc, and hcp) [11]. Such microgravity experiments give promising results; however, they are time-consuming and costly. To overcome the time and cost issues, a ground-based "microgravity" experiment is greatly needed. Previously, several concepts have been proposed. Applying thermophoretic force was reported as an effective approach to cancel gravity [12, 13]. The shell structure of dust particles, "Coulomb balls," was found. Another approach is the magnetic field. It was reported that the magnetic field applied on super-paramagnetic particles can compensate for the gravity [14]. In the case of colloidal dispersion, gravity affects the crystal structures in a similar manner. Microgravity experiments were conducted as is for dusty plasmas, which revealed that the crystal structure under microgravity is basically random stacking of hexagonally close-packed planes alone and suggested that fcc, which is often observed in ground-based experiments, is induced by gravity [15]. Buoyancy is employed to compensate for gravity in ground-based experiments [16]. Buoyancy can be tuned by changing the ratio of H_2O and D_2O; the density of media can be matched to that of microparticles.

In the present study, buoyancy in supercritical fluids (SCFs) is proposed as a means for compensating for gravity in dusty plasmas. SCF is a state of matter whose temperature and pressure exceed those of the critical point (T_c, P_c), as described in **Figure 1**. The critical point is the end point of the vapor pressure curve, above which it is impossible to distinguish whether the phase is gas or liquid, and the phase is defined as SCF. SCF has liquid-like solubility, high density, and gas-like low viscosity. Owing to such unique properties, SCF has been applied to fabrication processes of such materials as aerogels and nanoparticles [17]. The critical temperature T_c varies with molecules. CO_2 and Xe are frequently used as SCF media, because they have T_c near room temperature, 304 and 290 K, respectively. The critical pressure P_c of each medium is 7.38 and 5.84 MPa, respectively; therefore, temperature control by a water cooling/heating system, and the application of pressure by a pump can provide SCF states of CO_2 and Xe. The properties of SCF that are significant for the application to dusty plasmas are density and viscosity. **Figure 2a** shows the dependence of the density of CO_2, Xe, and typical liquid (H_2O and ethanol) on pressure. The temperatures of Xe and CO_2 are their own T_c, and those of water and ethanol are set to room temperature (293 K). The densities of H_2O and ethanol show little change against pressure. Meanwhile, those of CO_2 and

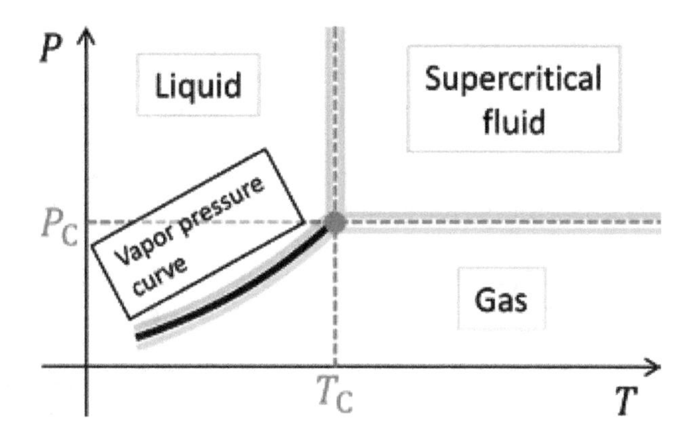

Figure 1.
T-P *phase diagram of a matter focusing around the critical point.*

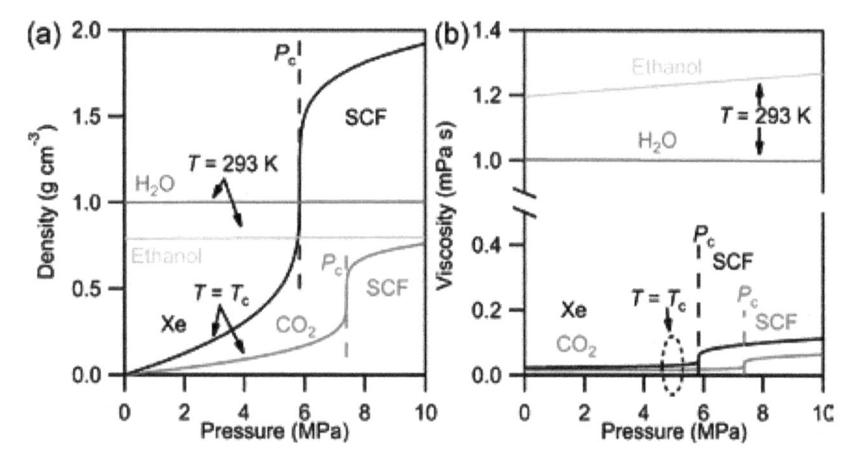

Figure 2.
Dependences of the density (a) and the viscosity (b) of Xe, CO_2, H_2O, and ethanol on pressure: the temperatures of the former two and the latter two are set to their own T_c and room temperature (293 K), respectively (the values were obtained from the NIST database [25]).

Xe successively increase with increasing pressure and reach or exceed those of typical liquids above P_c, where they become SCF. This means that controllable buoyancy can be applied to microparticles in SCF, which possibly results in the realization of pseudo-microgravity conditions for dusty plasmas. **Figure 2b** shows the dependence of the viscosity of each medium on pressure. Despite the large density of SCF, the viscosities are only one-tenth those of typical liquids. Therefore, microparticles in SCF experience little viscous drag, which delays reaching a condition of thermal equilibrium, as is observed in colloidal dispersion [18]. Therefore, SCFs are considered to be attractive media suitable for the generation of dusty plasmas in a pseudo-microgravity condition.

The generation of nonthermal plasmas in SCF is challenging, because the pressure is so high that applying higher voltage is necessary based on Paschen's law. The discharge plasmas in SCFs have been successfully generated by employing electrodes with a gap on the order of micrometers [19]. The possibilities of the plasmas in SCF for application to carbon nanomaterial syntheses and unique phenomena, such as a large decrease of breakdown voltage near the critical point, were shown [20]. For application to the generation of dusty plasmas, surface dielectric barrier discharge (DBD) in the field-emitting regime was employed [21]. The breakdown voltages of CO_2 for the discharges in the "standard regime," in which electrons are dominantly provided by ionizations, increase with increasing pressure, while it was found that field emission plays a major role in generating discharges under high pressure, which results in discharges with breakdown voltages as low as 2 kV. The surface DBD in the field-emitting regime is considered to be suitable for the generation of dusty plasmas in SCF, because the discharge with such low breakdown voltages possibly generates less heat and causes less damage to microparticles and electrodes.

In this chapter, a study on dusty plasmas in supercritical CO_2 ($scCO_2$) is introduced. In Section 2, the first report on the generation of dusty plasmas in SCF, on the formation of plasma crystals in $scCO_2$, and on the estimation of the particle charges is described [22]. Section 3 covers the realization of a pseudo-microgravity environment for dusty plasmas in $scCO_2$ and the 3D arrangement of particles [23]. In Section 4, the outlook for dusty plasmas in SCF, which includes the further applications of pseudo-microgravity conditions and the comparisons with other strongly coupled plasmas, is briefly discussed.

2. Motion of particles in dusty plasmas generated in $scCO_2$

The particle motion in dusty plasmas generated in $scCO_2$ was analyzed. The particles were electrically charged by the surface DBD in the field-emitting regime and showed the formation of an ordered structure above the electrodes. The analysis of the equation of motion revealed that the charge of a particle was on the order of -10^4 to $-10^5 e$ C (e: elementary charge). The kinetic energy of a particle was estimated by recording the motion with a high-speed camera. The estimated Coulomb coupling parameter was 10^2–10^4, from which the formation of strongly coupled plasmas was confirmed.

2.1 Experimental approach

Figure 3 shows a schematic diagram of the experimental setup for the generation of the dusty plasmas in $scCO_2$. As shown in **Figure 3a**, CO_2 pressurized by a high-pressure pump with a cooling circuit was introduced into the high-pressure chamber. The temperature inside the chamber was controlled by a water cooling/heating system. The temperature and pressure of CO_2 were 304.1–305.8 K and 0.10–8.33 MPa, respectively, which includes gaseous, liquid, and SCF states of CO_2. High voltages of up to 10 $kV_{p\text{-}p}$ with a frequency of 0.1–10 kHz were applied to the electrode. The microparticles (divinylbenzene resin; diameter, 30.0 μm; density, 1.19 g cm^{-3}) were placed on the etched region of the electrodes before applying voltages. The density of the particles was larger than that of CO_2 in this experimental condition; therefore, a pseudo-microgravity condition could not be achieved. The interest of this study is the charging and the motion of the particles in $scCO_2$. The motion of microparticles was observed by an optical microscope through a sapphire window under light-emitting diode (LED) illumination, as shown in **Figure 3b**. The high-speed camera was employed for the detection of the fast motion with frame rates up to 1000 fps. The motion in the direction of gravitational force was observed through a mirror. **Figure 3c** shows the detailed structure of the electrodes. The upper, powered electrode consists of a Cu film deposited on polyimide film, etched in a linear fashion, and closed by Ag pastes to confine particles. This rectangular region is referred to as the "etched region." The thicknesses of the Cu films and polyimide films were 30 and 20 μm, respectively. Ag paste was deposited on the reverse side and connected to the grounded chamber.

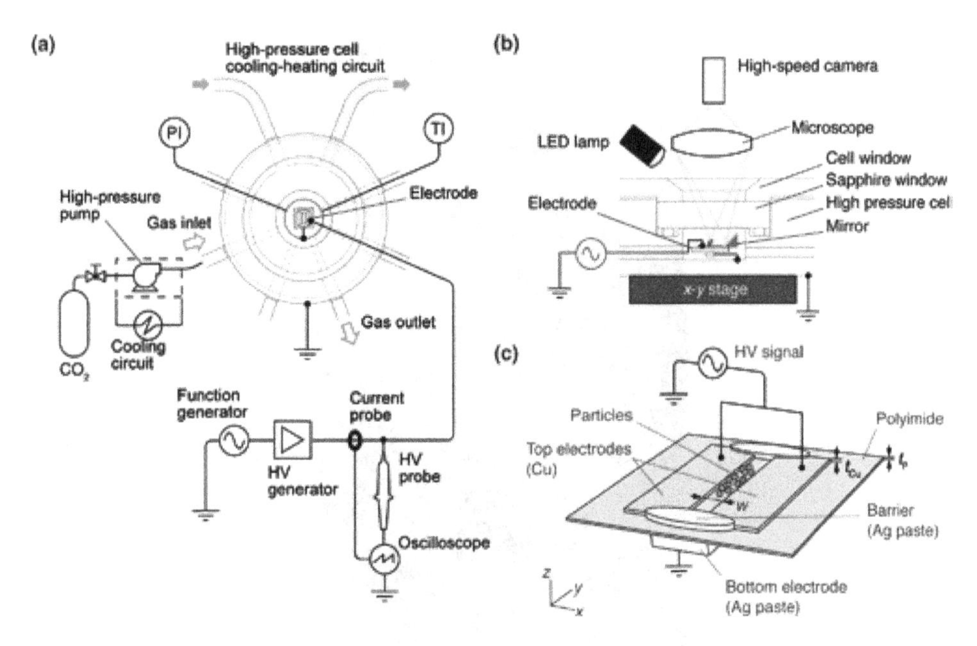

Figure 3.
Schematic of the experimental setup for the generation of the dusty plasmas in $scCO_2$*: (a) top view, (b) side view, and (c) the electrodes (adapted from [22], © IOP Publishing Ltd., all rights reserved).*

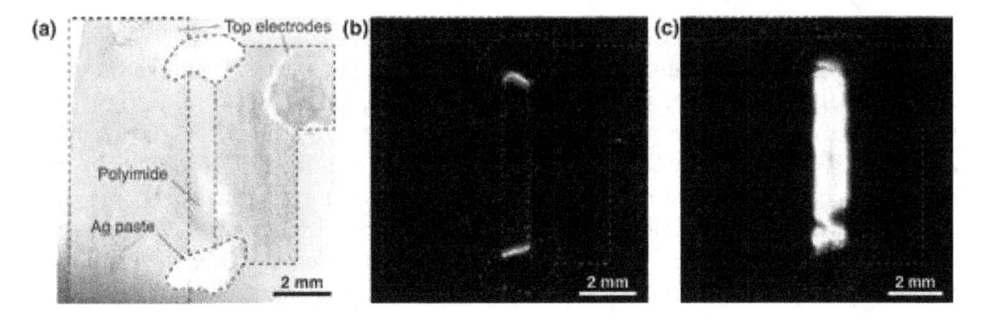

Figure 4.
Images of electrodes (a), surface DBD in field-emitting regime in $scCO_2$ *(b), and in standard-regime in* CO_2 *at 1 atm (c) (adapted from [22], © IOP Publishing Ltd., all rights reserved).*

2.2 Generation of dusty plasmas in $scCO_2$

Figure 4 shows photos of the electrodes and the plasmas generated with them without microparticles placed. **Figure 4a** shows photos of the electrodes whose etching width was 670 μm. **Figure 4b** shows the plasmas generated in $scCO_2$ in the field-emitting regime. The red luminescence is consistent with the previous optical emission spectroscopy measurement, which suggests that this is induced by electron-neutral bremsstrahlung [24]. **Figure 4c** shows the plasmas generated at atmospheric pressure in the standard regime, whose luminescence was blue or white, which might be derived from atomic emission.

In the experiments with particles, the particles in the etched region of the electrodes started to move intensely near the electrode surface when the voltage of ~3.0 $kV_{p\text{-}p}$ with a frequency of 10 kHz was applied, while many particles adhered to the Cu film and the Ag pastes, as shown in **Figure 5**. The moving particles were possibly electrically charged and accelerated by the AC electric field. When the frequency was decreased to 1 kHz, several particles floated above the electrodes, as shown **Figure 6**. **Figure 6a** shows the top view, where it is confirmed that the

Figure 5.
Image of the particles moving near the electrode surface.

particles aligned at the center of the etched region. The particles showed motion along the electrode edge, whose direction is indicated in **Figure 6a**. **Figure 6b** shows the side view. The particles were levitated at a height of 500 μm or more above the electrode surface. These phenomena could be observed in the condition of high-pressure gaseous, liquid, and supercritical CO_2. It was considered that the particles levitated after the applied frequency was lowered, because the particles are likely to follow the AC electric fields with lower frequency, which is discussed in detail later.

2.3 Numerical simulations of the particle motion

The equation of motion of charged particles in $scCO_2$ shown below was numerically solved:

$$m\frac{d^2z}{dt^2} = -mg + \rho Vg - k\frac{dz}{dt} + QE(z)\sin 2\pi ft \quad (1)$$

where m is the particle mass, z is the height of the particle from the electrode surface, t is the time, g is the gravitational constant, ρ is the mass density of CO_2, V is the volume of the particle, $k = 6\pi\eta r$ is the viscous drag coefficient, η is the viscosity of CO_2 and was obtained from the NIST database [25], r is the radius of the particle, Q is the particle charge, $E(z)$ is the electric field along z, and f is the applied frequency. $E(z)$ was obtained from the derivative of the electric potential calculated with the finite-element method using freely available software [26]. **Figure 7** shows the contour map of the electric potential near the electrodes with an applied voltage of 5 kV. Here, z is defined as the opposite direction of the gravitational force so that z = 0 corresponds to the electrode surface (the surface of the polyimide film in the etched region); x and y are defined as the direction parallel to the electrode surface, as indicated in **Figure 6**. **Figure 7** is the cross-sectional view of the electrodes in

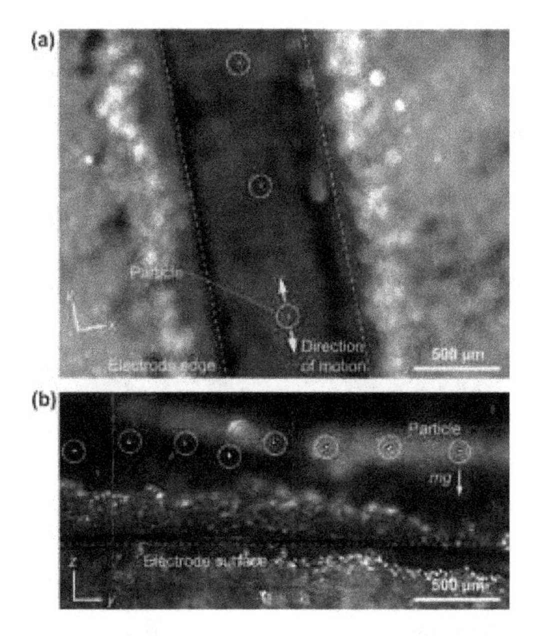

Figure 6.
Images of particles aligned above the electrodes: (a) top view and (b) side view (adapted from [22], © IOP Publishing Ltd., all rights reserved).

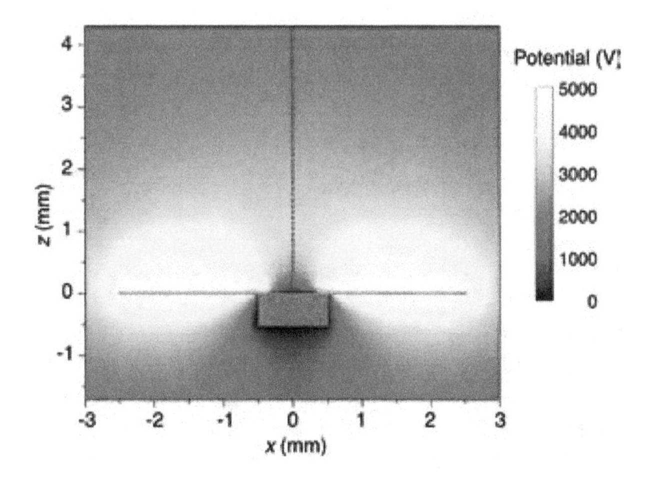

Figure 7.
Contour map of the calculated electric potential near the electrodes with the applied voltage of 5 kV (adapted from [22], © IOP Publishing Ltd., all rights reserved).

the x-z-plane, and it is assumed for the calculation that the y-direction is infinitely extended. The position of interest is $x = 0$, where the particles aligned in the experiments. **Figure 8** shows the electric field along the z-direction $E(z)$ at $x = 0$. However the applied voltage is large, the absolute value of $E(z)$ is the highest at the electrode surface, and $E(z)$ becomes 0 at a certain height from the electrode surface, above which its sign becomes inverted.

The time evolution of the height from the electrode surface of the charged particle is shown in **Figure 9**. The condition is $scCO_2$, where the pressure is 8.07 MPa, temperature is 305.8 K, density is 0.641 g cm^{-3}, product of particle charge and peak voltage QU is $1.5 \times 10^5 e$ C kV, and applied frequency is 1.0 kHz.

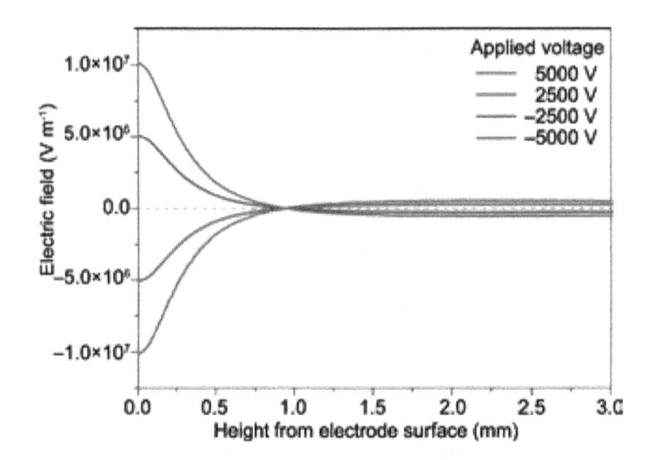

Figure 8.
Electric field along z-direction for each applied voltage at x = 0, which is indicated by the dotted line in **Figure 7** *(adapted from [22], © IOP Publishing Ltd., all rights reserved).*

Three types of initial condition were tried for the calculation. The first one was the particle on the electrode surface ($z = 0$), the second was the particle at the potential valley ($E_z = 0$), and the last was the particle staying far from the electrode surface ($z = 3$ mm). Whichever initial condition was used, the particle settled at a certain z after enough time passed. **Figure 10** shows the magnified graph of the particle position with the initial condition $z = 0$. The particle moves upward very rapidly and reaches the maximum height within 10 ms. After that, it moves downward and settles at z between 650 and 660 μm. The inset shows the particle motion after the settlement. The particle shows oscillation with an amplitude of 5 μm and a period of 1 ms. The frequency of this oscillation corresponds to the applied AC frequency. This oscillation is too small and fast to visualize in the scale of **Figure 9**.

2.4 Estimation of charge and coulomb coupling parameter

The equilibrium positions of the particle were plotted against the product of the particle charge and the applied peak voltage QU for each AC frequency, as shown in **Figure 11**. The condition is the same as the calculation shown in **Figure 9** except for QU. The error bars indicate the oscillation amplitudes. The equilibrium height increases with increasing QU and lowering the AC frequency. A higher QU means stronger electric fields applied on the particle. The particle is likely to follow the AC field with a lower frequency, which is also confirmed by the fact that the oscillation amplitudes are larger for the lower frequency. The experimental result that the particles levitated after the lowering of the AC frequency can be explained by this analysis. The plots in **Figure 11** have important implications for the particle charge. In fact, the particle charges can be estimated from the measured height of the particle from the electrode surface. The range of the particle charge estimated from the experimental results was on the order of $(10^4–10^5)e$ C. The particle charge is reported to be $(1–3) \times 10^3 e$ C with the orbital motion limited theory [27], $(10^3–10^5)e$ C with the analysis of particle motion [28], and on the order of $10^6 e$ C with Faraday cup measurement [29]. Therefore, the estimated particle charge is considered to be of a reasonable order. These analyses lack the information on the sign of the particle charges, although it is a common understanding that particles in a discharge plasma get negatively charged because

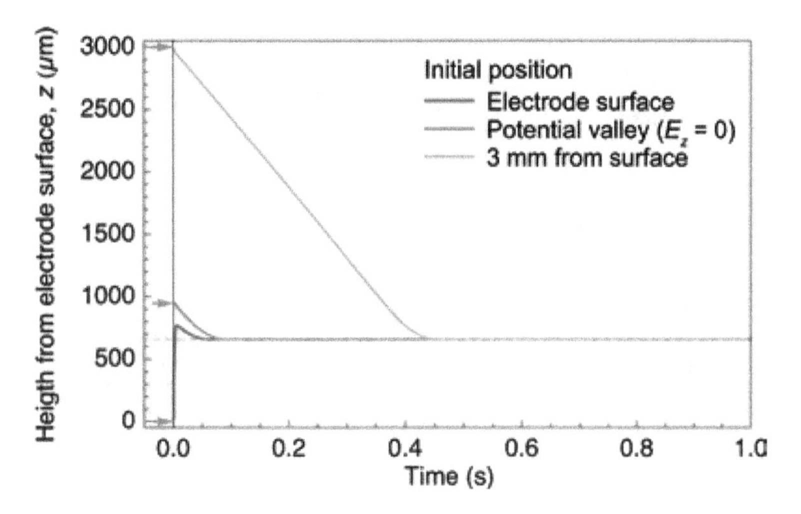

Figure 9.
Calculated time evolution of the height of charged particle for each initial position (adapted from [22], © IOP Publishing Ltd., all rights reserved).

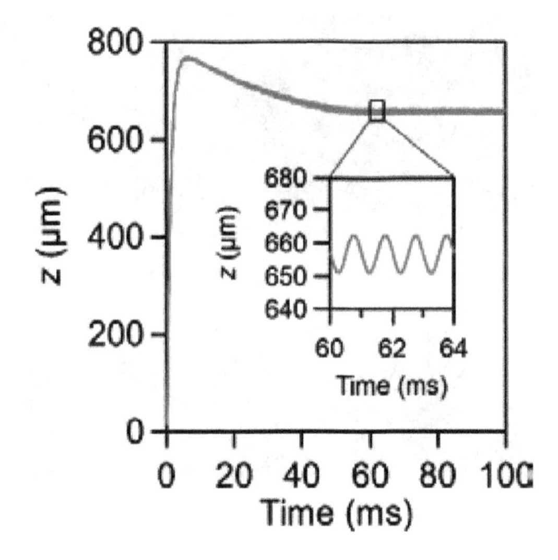

Figure 10.
Magnified graph of **Figure 9** *for the initial position of electrode surface (data adapted from [22]).*

of the higher mobility of electrons. Therefore, DC offset was applied to clarify the sign. **Figure 12** shows the changes of the particle positions with DC offset. When negative bias was applied, the particle moved downward. Meanwhile, the particle moved upward with the positive offset. This behavior is consistent with the numerical simulation results for the negatively charged particle, as shown **Figure 13**. The average position of the particle is 820 μm with the applied offset of +50 V and 656 μm without offset. It is considered that the particles get negatively charged by the discharges in $scCO_2$ because of the flux of the electrons emitted from the electrodes by field emission.

The kinetic energy of a particle was estimated with high-speed imaging. The motion along the *y*-direction is shown in **Figure 6a**. The movie was recorded with a frame rate of 125 fps and a duration of 3.848 s. **Figure 6a** is captured from this movie. The velocities of three particles were in the range of (1.4–2.1) × 10^2 μm s^{-1},

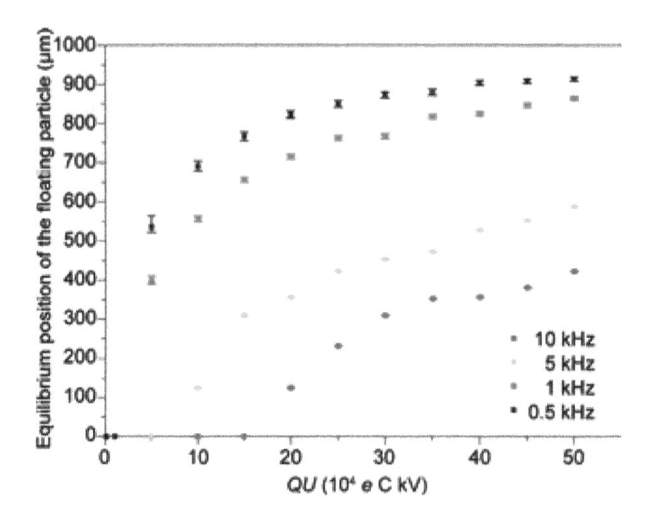

Figure 11.
Equilibrium position of the particle against the product of particle charge and applied peak voltage QU (© IOP Publishing Ltd., all rights reserved).

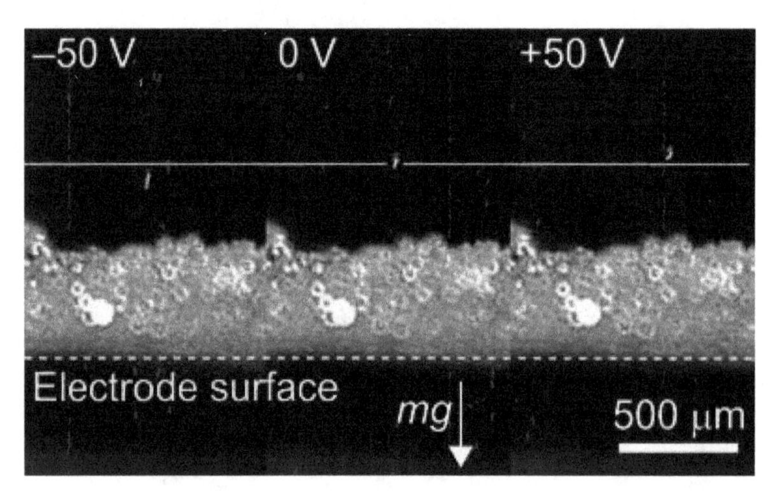

Figure 12.
Change in the height of the particle by applying DC offset.

from which the kinetic energy was (1.8–4.5) × 10^4 K. Assuming that the interparticle distance was 700 mm and the particle charges were $-(10^4\text{–}10^5)e$ C, this system had a Coulomb coupling parameter on the order of 10^2–10^4 and is considered to be a strongly coupled plasma.

To the authors' knowledge, this is the first report on the formation of strongly coupled dusty plasmas in a dense medium. Almost all the reports employ RF plasmas in a vacuum to generate dusty plasmas. There are a few reports on the strongly coupled dusty plasmas generated in thermal plasmas under atmospheric pressure, where CeO_2 particles get positively charged by the thermal emission of electrons [30]. No other studies on the generation of dusty plasmas in a dense medium, such as high-pressure gas, liquid, and SCF, have been reported. Furthermore, this study is the first report on the formation of plasma crystals using DBD. Conventional plasma crystals have been formed in DC or RF glow discharge with metallic electrodes. DBD, which is usually employed for generating low-temperature plasmas under relatively high pressure, such as atmospheric-pressure plasmas, has not been used for plasma crystals up to now.

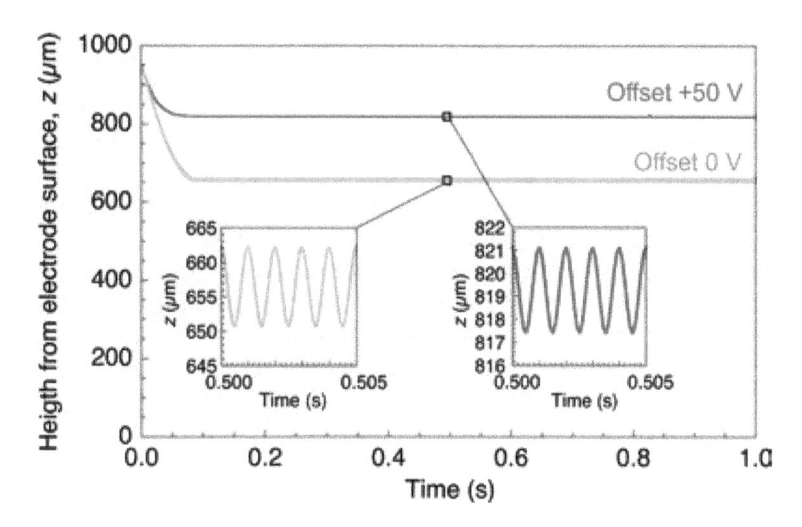

Figure 13.
Time evolution of the height of the particle with and without DC offset (© IOP Publishing Ltd., all rights reserved).

3. Pseudo-microgravity environment for dusty plasmas in $scCO_2$

The charging of particles by discharge in $scCO_2$ was clarified, and the particle charges were successfully estimated, as shown in the previous section. However, the microparticles used in Section 2 were so heavy that the pseudo-microgravity environment was hardly realized. In this section, lighter resin particles were used. Compensation for gravitational force by buoyancy was confirmed by controlling the balance between gravitational force and buoyancy, suggesting the formation of a microgravity environment for dusty plasmas in $scCO_2$. The formation behavior was analyzed by the estimation of the Coulomb coupling parameter.

3.1 Experimental approach

The experimental setup and the simulation method were basically the same as in Section 2. The width of the etched region of the upper electrodes was 2.7 mm. The particles employed were lighter. Their mass density was 0.5 g cm^{-3}. The experimental conditions were $mg > \rho Vg$ for condition (a); $mg \sim \rho Vg$ for conditions (b), (c), and (d); and $mg < \rho Vg$ for condition (e) as shown in **Figure 14**. The density of CO_2 for each condition is shown in **Figure 14**. The temperature was set to 31.7–32.1°C, slightly higher than the critical temperature of CO_2 (31.1°C). Condition (a) has a pressure of 6.92 MPa, where CO_2 is gaseous, and its density is lower than that of the particle. CO_2 is a supercritical condition in conditions (b), (c), (d), and (e).

3.2 Generation of dusty plasmas in $scCO_2$ with varying density of media

As in the previous section, several particles got levitated and trapped above the electrodes after the AC frequency was decreased. The applied AC voltage was 5 $kV_{p\text{-}p}$, and the frequency was initially 10 kHz and decreased to 155 Hz. **Figure 15i** is the side view of the particle arrangement with condition (a). The particles formed a single-layer structure. However, when the CO_2 density was controlled in condition (c), particles were arranged in the gravitational direction, as shown in **Figure 15ii**. The particles were also arranged in the *x-y* (horizontal)-plane, as shown in **Figure 16**, which confirmed the 3D arrangement of particles resulting from the compensation of the gravity. It was observed that some particles exchanged their positions with their neighbors. Furthermore, the structure is not periodic—in other words, it is liquid-like. The particles formed a single- or double-layer structure with condition (e), as shown in **Figure 15iii**. Of all the experimental conditions, 3D structures were formed in the pseudo-microgravity conditions (b), (c), and (d). The window of the density of CO_2 for the formation of 3D plasma crystals was found to be ±0.2 g cm^{-3} compared with that of the particles.

The particles showed oscillation in the gravitational direction with a frequency equal to the applied AC frequency. In condition (c), the particles staying above and below the dashed line indicated in **Figure 15ii** showed oscillations antiphase to each other, the amplitudes of which were 20 and 60 μm, respectively.

3.3 Numerical simulations for pseudo-microgravity condition

To analyze the motion of a particle in the experimental conditions, Eq. (1), explained in the previous section, was applied. The initial position of the particle was set to $z = 0$. The particle charge was assumed to be $-7 \times 10^4 e$ C based on the results explained in the previous section. **Figure 17** shows the calculated time evolution of a particle for each density of CO_2. The particle starts to move upward after $t = 0$ and reaches a steady height in all conditions. The inset shows the magnified graph with the particles staying at a steady position. The steady height increases with increasing CO_2 density and approaches the position where the electric field is zero at any time with ρVg approaching mg. Additionally, the particles oscillate with applied AC frequency. The oscillation amplitude was smaller when $mg = \rho Vg$ than in any other condition.

3.4 Estimation of Coulomb coupling parameter

For the estimation of the Coulomb coupling parameter, the kinetic energies of both the oscillation and thermal motion of the particles were taken into consideration, instead of the thermal energy. The latter was assumed to be the same as the temperature of the chamber because of the small heat generation in the surface DBD in the field-emitting regime [24]. The oscillation energy was calculated as $mA^2(2\pi f)^2/2$ on the supposition that the oscillation is harmonic, where A is the

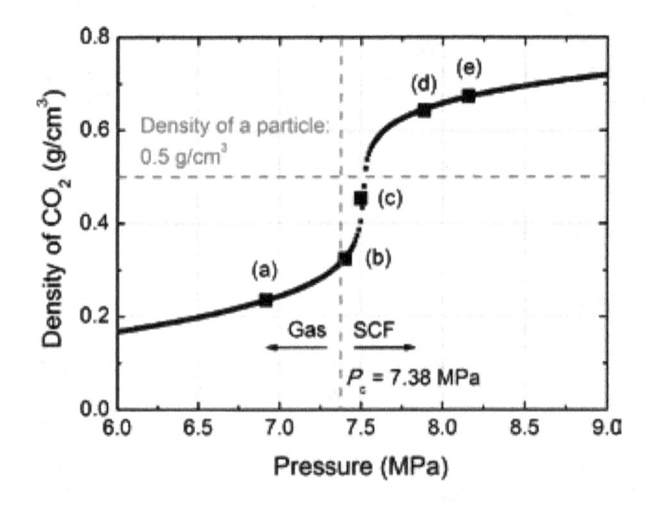

Figure 14.
Dependence of CO_2 density on pressure: the density of the particle is indicated by the red broken line, and the experimental conditions (a)–(e) are plotted (reprinted from [23], with the permission of AIP Publishing).

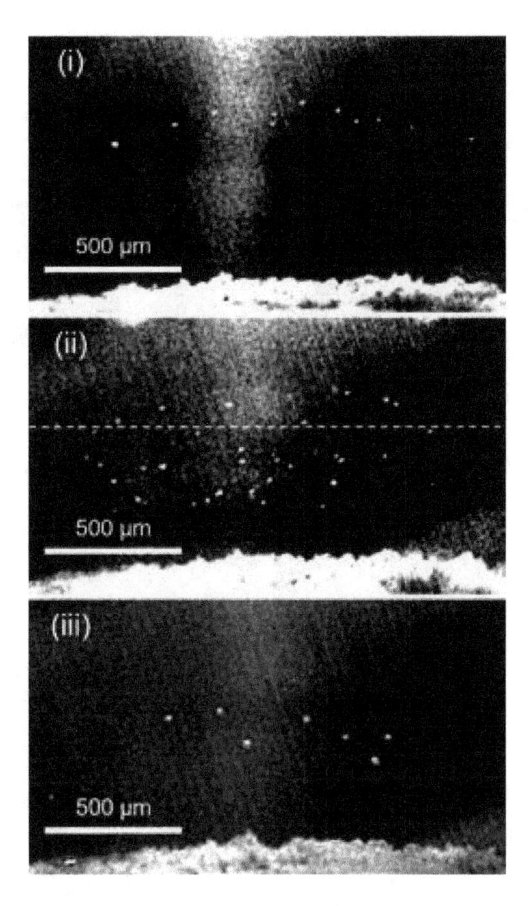

Figure 15.
Side view of the trapped particles above the electrodes with the condition of $mg > \rho Vg$ *(i),* $mg = \rho Vg$ *(ii), and* $mg < \rho Vg$ *(iii) (reprinted from [23], with the permission of AIP Publishing).*

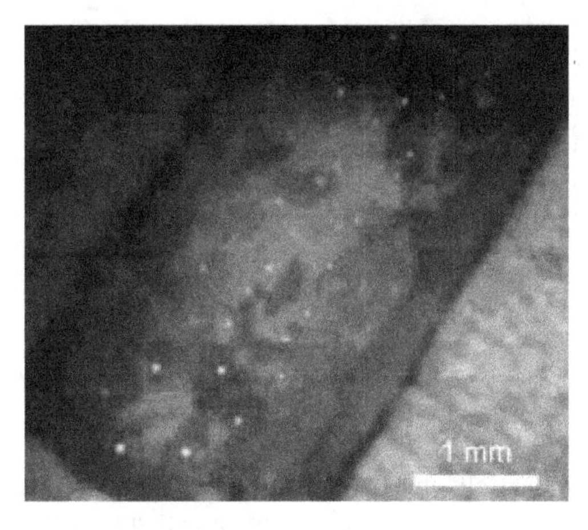

Figure 16.
Top view of the trapped particles with the condition of $mg = \rho Vg$ *(reprinted from [23], with the permission of AIP Publishing).*

oscillation amplitude of a particle. The dependence of the oscillation amplitude on CO_2 density is shown in **Figure 18a**. The amplitude has the minimum value when $mg = \rho Vg$, as suggested in **Figure 17**. The calculated oscillation energy was in the range of 9.4×10^{-24}–3.4×10^{-15} J, while the thermal random motion energy

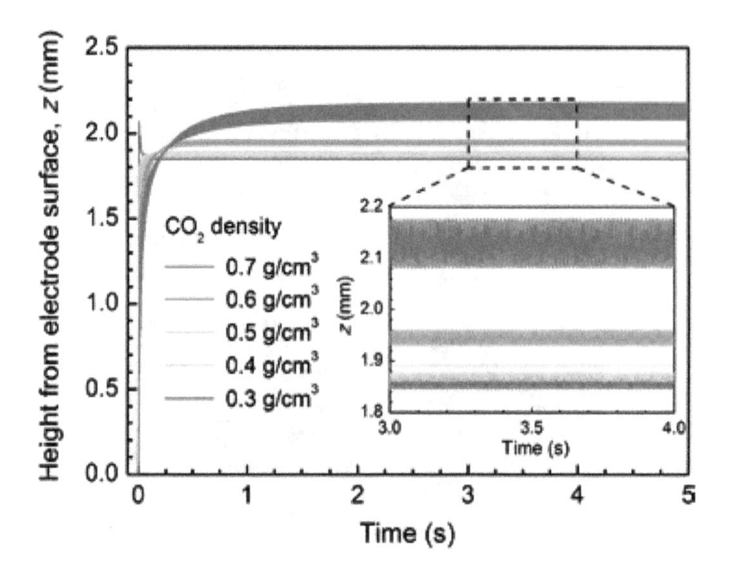

Figure 17.
Calculated time evolution of the height of the particle with each CO_2 density (reprinted from [23], with the permission of AIP Publishing).

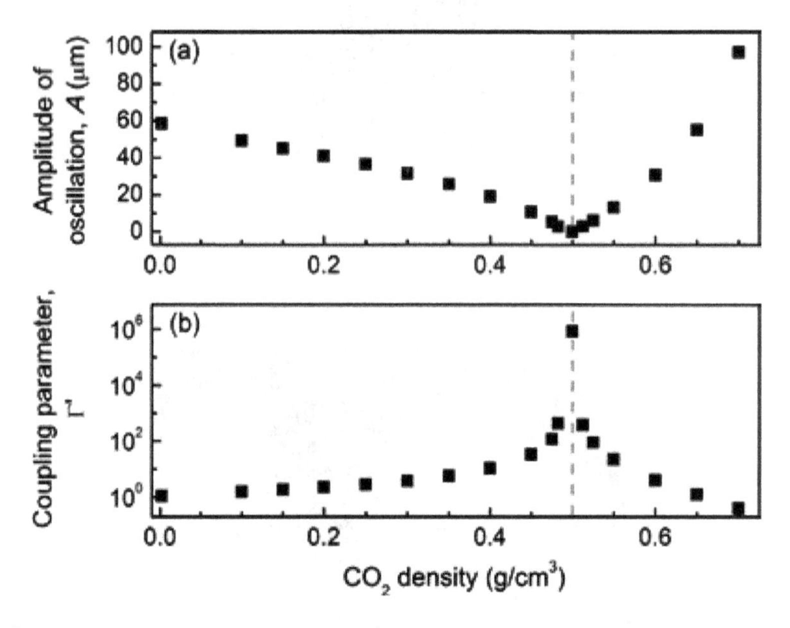

Figure 18.
Dependence of the amplitude of the oscillation of the particle (a) and the Coulomb coupling parameter (b) (reprinted from [23], with the permission of AIP Publishing).

was 4.2×10^{-21} J. The particle charge and the interparticle distance are assumed to be $-7 \times 10^4 e$ C and 250 μm, respectively. **Figure 18b** shows the dependence of the Coulomb coupling parameter on CO_2 density. The coupling parameter shows the maximum value when $mg = \rho Vg$ because of the suppression of the oscillation energy. The coupling parameter is possibly enhanced with the aid of the gravity compensation by buoyancy. The calculated Coulomb coupling parameter for condition (c), which is the closest to the condition with $mg = \rho Vg$ in this study, is less than the previously reported criterion for the transition from liquid to solid ($\Gamma > 170$) [31–33], which is consistent with the observation of liquid-like behavior.

4. Outlook for dusty plasmas in SCF

4.1 Pseudo-microgravity condition

The pseudo-microgravity conditions for dusty plasmas provide the opportunity to study collective phenomena without any anisotropy. **Figure 19** summarizes this study and shows its further application, as discussed below. Microgravity experiments using huge, expensive pieces of apparatus, such as a space station, largely limit the number of experimental trials. Therefore, some exciting discoveries have been possibly missed.

One of the unexplored phenomena is the rotation of a particle on its own axis, which was suggested by Sato at a workshop held at Tohoku University in 2014 [34]. Before going into details, the similarity between dusty plasma physics and solid-state physics should be considered. It is useful for finding novel phenomena in dusty plasmas to learn from solid-state physics. The idea of plasma crystals is similar to that of strongly correlated electron systems in a solid. In commonly used plasmas, ions and electrons have a large kinetic energy (=0.01–1 eV) and small interparticle electrostatic energy owing to small particle charge and large interparticle distance (low number density), which results in a very small Coulomb coupling parameter. Therefore, strongly coupled plasmas ($\Gamma > 1$) can hardly be realized. The introduction of dust particles changes that situation. The particles collect many electrons and have a large electric charge. In addition, they have a large mass to prevent acceleration and small kinetic energy, which results in a large Coulomb coupling parameter and possible realization of strongly coupled plasmas. Meanwhile, electron-electron interaction has great importance in the field of solid-state physics. The conducting electrons in a simple metal, such as Cu and Al, have no or weak interactions between electrons and can move freely. Some transition metal oxides and rare-earth oxides possibly show the localization of electrons and insulating behavior because of large electrostatic repulsion between electrons, even if they have a partially filled electron band in a simple band picture. The so-called Mott insulator shows various phenomena, such as a metal-insulator transition, magnetic transition, and superconducting transition with or without applying temperature change, chemical doping, and so on. Mott insulators have been one of the most extensively studied subjects [35]. There seem to be similarities between dusty plasmas and a strongly correlated electron system, given that, in both cases, novel ordered phases appear with strengthening of the interactions, which was considered to be ignorable. Some of the exciting phenomena observed in a strongly correlated electron system possibly emerge in dusty plasmas in a similar manner.

One of the examples might be the rotation of a particle on its own axis, which is like the spin of an electron. The positions of particles in dusty plasmas have been successfully tracked; however, there are no reports on their rotation to the authors' knowledge. It may be difficult to observe because the particle size is small (several tens of micrometers at largest). Larger particles are likely to sink because they are heavy. However, particle size does not matter in microgravity experiments. Using millimeter-sized particles possibly reveals the details of particle motion. The pseudo-microgravity condition can also make experiments with larger particles possible, because particle size does not matter so long as the density of the media is matched to that of the particle.

The applicability of heavier particles in microgravity experiments provides opportunities to use functional oxide or metal particles. Many studies on dusty plasmas employ resin particles, which possess no notable functional properties except for lightness. Using functional particles might make dusty plasmas functional: susceptible to a magnetic field using magnetic particles [36], photocatalytic using photocatalyst particles, and so on. Such functional dusty plasmas are attractive as a functional fluid whose flow and reactivity can be controlled by a magnetic field or light irradiation. In addition, functional particles, such as magnetic or ferroelectric ones, are expected to show the interparticle interaction via such properties. Such interaction yields intentionally introduced anisotropy, which possibly causes the emergence of the crystal structures that still have not been observed in dusty plasmas or novel phase transitions.

4.2 Dusty plasmas in dense media

Another aspect of dusty plasmas in SCF is a density of media higher than in dusty plasmas in vacuum and lower than or comparable to that in colloidal dispersion. In colloidal dispersion, interparticle interaction is mediated via solvent flow, unlike in the case of dusty plasmas, where particles interact directly with each other via electrostatic force. Dusty plasmas in SCF are considered to be placed at the transient region, as shown in **Figure 20**. Changes of the mode of the interactions might be observable in dusty plasmas in SCF.

In this study, the existence of electrons and ions in the region of trapped particles and their screening effects were ignored. That region is somewhat far from surface

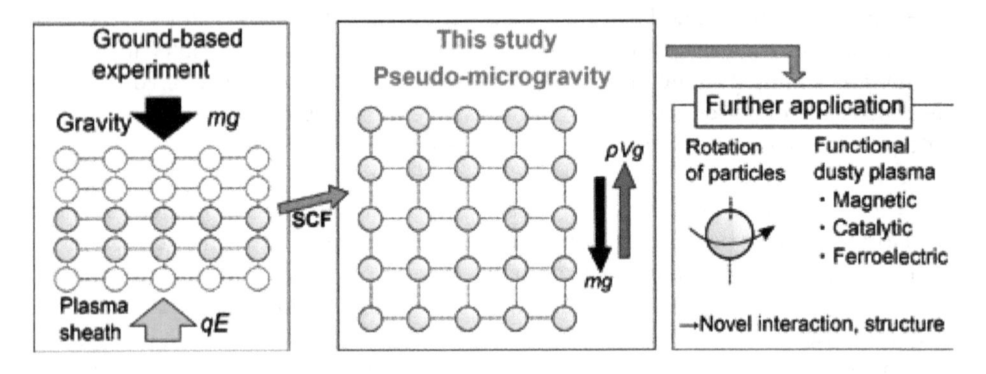

Figure 19.
Overview and further application of this study.

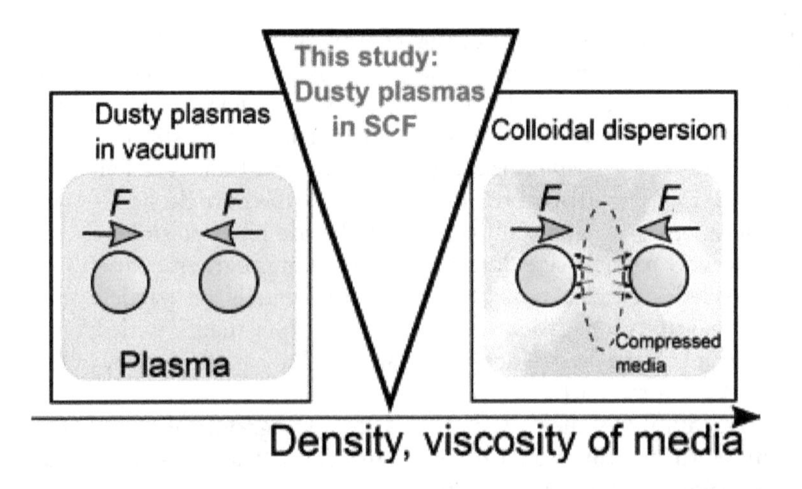

Figure 20.
Interparticle interaction modes in dusty plasmas in vacuum, colloidal dispersion, and the transient region, dusty plasmas in SCF.

DBD; therefore, there seem to be few electrons and ions when SCF is taken as a high-pressure gas, because their lifetime is too short under high pressure owing to frequent collisions. However, SCF also has a liquid-like characteristic. Solvated ions in a liquid have a long lifetime, and it was reported that electrons generated by discharge plasmas can be solvated [37]. The lifetime of ions and electrons in SCF is possibly so long that they can reach where the particles stay. The behaviors of the electrons and ions in plasmas generated in SCF are still not fully understood. Further analysis of the interparticle interaction in dusty plasmas in SCF could serve as a probe to clarify it.

5. Conclusion

Plasma crystals, realized in dusty plasmas, provide opportunities to study statistical phenomena and lattice dynamics in a crystal. However, in ground-based experiments, gravity imposes a large anisotropy on plasma crystals, which results in their 2D structure. This study on the dusty plasmas in SCF aimed at overcoming the problems caused by gravity that hinder research on dusty plasmas. The particles were successfully electrically charged by the surface DBD in $scCO_2$. The estimation of the particle charge and the analysis of the particle motion confirmed the formation of strongly coupled plasmas. With the density of $scCO_2$ matched to that of particles, which means a pseudo-microgravity condition for the particles, 3D-ordered structures were successfully formed. The pseudo-microgravity conditions provide opportunities to find novel phenomena and to develop functional dusty plasmas. In addition, the dusty plasmas in SCF can be considered as the intermediate phase between dusty plasmas in vacuum and colloidal dispersion. SCF is left largely unexplored with regard to the media for the generation of strongly coupled plasmas. Our pioneering works has been opening a novel field of strongly coupled plasmas.

Acknowledgements

This work was supported financially by a Grant-in-Aid for Scientific Research on Innovative Areas (Frontier Science of Interactions between Plasmas and Nano-interfaces, Grant No. 21110002) and the Grant-in-Aid for Challenging Exploratory Research (Grant No. 15K13389) from the Ministry of Education, Culture, Sports, Science, and Technology (MEXT) of Japan. We would like to thank Sekisui Chemical Co., Ltd. for providing us the fine resin particles (HB-2051) used in this study. One of the authors of this chapter, K.T., would like to give special thanks to Professor O. Ishihara (Chubu University) for his valuable suggestions and encouragement.

Author details

Yasuhito Matsubayashi[1*], Noritaka Sakakibara[2], Tsuyohito Ito[2] and Kazuo Terashima[2]

1 Advanced Coating Technology Research Center, National Institute of Advanced Industrial Science and Technology, Tsukuba, Ibaraki, Japan

2 Department of Advanced Materials Science, Graduate School of Frontier Sciences, The University of Tokyo, Chiba, Japan

*Address all correspondence to: y-matsubayashi@aist.go.jp

References

[1] Bragg WL, Nye JF. A dynamical model of a crystal structure. Proceedings of the Royal Society of London. Series A: Mathematical and Physical Sciences. 1947;**190**:474-481

[2] Pieranski P. Two-dimensional interfacial colloidal crystals. Physical Review Letters. 1980;**45**:569

[3] Hayward RC, Saville DA, Aksay IA. Electrophoretic assembly of colloidal crystals with optically tunable micropatterns. Nature. 2000;**404**:56

[4] Ikezi H. Coulomb solid of small particles in plasmas. Physics of Fluids. 1986;**29**:1764-1766

[5] Ichimaru S. Strongly coupled plasmas: High-density classical plasmas and degenerate electron liquids. Reviews of Modern Physics. 1982;**54**: 1017

[6] Slattery WL, Doolen GD, DeWitt HE. Improved equation of state for the classical one-component plasma. Physical Review A. 1980;**21**:2087

[7] Thomas H, Morfill GE, Demmel V, Goree J, Feuerbacher B, Möhlmann D. Plasma crystal: Coulomb crystallization in a dusty plasma. Physical Review Letters. 1994;**73**:652

[8] Hayashi Y, Tachibana K. Observation of Coulomb-crystal formation from carbon particles grown. Japanese Journal of Applied Physics. 1994;**33**:L804

[9] Chu JH, Lin I. Direct observation of Coulomb crystals and liquids in strongly coupled rf dusty plasmas. Physical Review Letters. 1994;**72**:4009

[10] Morfill GE, Thomas HM, Konopka U, Rothermel H, Zuzic M, Ivlev A, et al. Condensed plasmas under microgravity. Physical Review Letters. 1999;**83**:1598

[11] Nefedov AP, Morfill GE, Fortov VE, Thomas HM, Rothermel H, Hagl T, et al. PKE-Nefedov*: Plasma crystal experiments on the International Space Station. New Journal of Physics. 2003;**5**:33

[12] Arp O, Block D, Piel A, Melzer A. Dust Coulomb balls: Three-dimensional plasma crystals. Physical Review Letters. 2004;**93**:165004

[13] Arp O, Block D, Klindworth M, Piel A. Confinement of Coulomb balls. Physics of Plasmas. 2005;**12**:122102

[14] Samsonov D, Zhdanov S, Morfill G, Steinberg V. Levitation and agglomeration of magnetic grains in a complex (dusty) plasma with magnetic field. New Journal of Physics. 2003;**5**:24

[15] Zhu J, Li M, Rogers R, Meyer W, Ottewill RH, Russel WB, et al. Crystallization of hard-sphere colloids in microgravity. Nature. 1997;**387**:883

[16] Ito K, Yoshida H, Ise N. Void structure in colloidal dispersions. Science. 1994;**263**:66-68

[17] Cansell F, Chevalier B, Demourgues A, Etourneau J, Even C, Pessey V, et al. Supercritical fluid processing: A new route for materials synthesis. Journal of Materials Chemistry. 1999;**9**:67-75

[18] Ivlev A, Hartmut LÃ, Morfill G, Royall CP. Complex Plasmas and Colloidal Dispersions: Particle-Resolved Studies of Classical Liquids and Solids. Singapore: World Scientific Publishing Company; 2012

[19] Stauss S, Muneoka H, Terashima K. Review on plasmas in extraordinary media: Plasmas in cryogenic conditions and plasmas in supercritical fluids. Plasma Sources Science and Technology. 2018;**27**:23003

[20] Stauss S, Muneoka H, Urabe K, Terashima K. Review of electric discharge microplasmas generated in highly fluctuating fluids: Characteristics and application to nanomaterials synthesis. Physics of Plasmas. 2015;**22**:57103

[21] Pai DZ, Stauss S, Terashima K. Surface dielectric barrier discharges exhibiting field emission at high pressures. Plasma Sources Science and Technology. 2014;**23**:25019

[22] Matsubayashi Y, Urabe K, Stauss S, Terashima K. Generation of dusty plasmas in supercritical carbon dioxide using surface dielectric barrier discharges. Journal of Physics D: Applied Physics. 2015;**48**:454002

[23] Sakakibara N, Matsubayashi Y, Ito T, Terashima K. Formation of pseudo-microgravity environment for dusty plasmas in supercritical carbon dioxide. Physics of Plasmas. 2018;**25**:10704

[24] Pai DZ, Stauss S, Terashima K. Field-emitting Townsend regime of surface dielectric barrier discharges emerging at high pressure up to supercritical conditions. Plasma Sources Science and Technology. 2015;**24**:25021

[25] Lemmon EW, Huber ML, McLinden MO. NIST standard reference database 23: NIST Reference Fluid Thermodynamic and Transport Properties Version; 9; 2010. p. 55

[26] Meeker D. Finite element method magnetics. FEMM; 4; 2010. p. 32

[27] Fortov VE, Khrapak AG, Khrapak SA, Molotkov VI, Petrov OF. Dusty plasmas. Physics-Uspekhi. 2004;**47**:447

[28] Fortov VE, Nefedov AP, Molotkov VI, Poustylnik MY, Torchinsky VM. Dependence of the dust-particle charge on its size in a glow-discharge plasma. Physical Review Letters. 2001;**87**:205002

[29] Walch B, Horanyi M, Robertson S. Measurement of the charging of individual dust grains in a plasma. IEEE Transactions on Plasma Science. 1994;**22**:97-102

[30] Fortov VE, Nefedov AP, Petrov OF, Samarian AA, Chernyschev AV. Particle ordered structures in a strongly coupled classical thermal plasma. Physical Review E. 1996;**54**:R2236

[31] Farouki RT, Hamaguchi S. Thermal energy of the crystalline one-component plasma from dynamical simulations. Physical Review E. 1993;**47**:4330

[32] Fortov VE, Molotkov VI, Nefedov AP, Petrov OF. Liquid-and crystallike structures in strongly coupled dusty plasmas. Physics of Plasmas. 1999;**6**:1759-1768

[33] Fortov VE, Ivlev AV, Khrapak SA, Khrapak AG, Morfill GE. Complex (dusty) plasmas: Current status, open issues, perspectives. Physics Reports. 2005;**421**:1-103

[34] Sato N. Private communication. Workshop in Tohoku University; 2014

[35] Imada M, Fujimori A, Tokura Y. Metal-insulator transitions. Reviews of Modern Physics. 1998;**70**:1039

[36] Block D, Melzer A. Dusty (complex) plasmas—Routes towards magnetized and polydisperse systems. Journal of Physics B: Atomic, Molecular and Optical Physics. 2019;**52**:63001

[37] Rumbach P, Bartels DM, Sankaran RM, Go DB. The solvation of electrons by an atmospheric-pressure plasma. Nature Communications. 2015;**6**:7248

8

Basic Properties of Fine Particle (Dusty) Plasmas

Hiroo Totsuji

Abstract

Beginning with typical values of physical parameters, basic properties of the system of fine particles (dusts) in plasmas are summarized from the viewpoint of statistical physics. Mutual interactions and one-body *shadow* potential for fine particles are derived, and, by analytic treatments and numerical solutions of drift-diffusion equations, it is shown that one of their most important characteristics is the large enhancement of the charge neutrality in fine particle clouds. This observation leads to simple models for the structures of fine particle clouds under both microgravity and usual gravity. Due to large magnitudes of their charges and relatively low temperatures, fine particles are often in the state of strong coupling, and some interesting phenomena possibly expected in their system are discussed with concrete examples. Also reviewed is the shell model, a useful framework to obtain microscopic structures in strongly coupled Coulomb and Coulomb-like systems with specific geometry.

Keywords: interaction and one-body potential, drift-diffusion equations, charge neutrality enhancement, cloud models under microgravity/gravity, strong coupling, shell model

1. Introduction

Fine particle (dusty) plasmas are weakly ionized plasmas including fine particles (dusts) of micron sizes [1–5]. As the gas species, inert gases such as He, Ne, Ar, etc. are usually (but may be not exclusively) used: one of the reasons is that various phenomena can be analyzed without considering chemical reactions including neutral atoms or ions and values of physical parameters are mostly known or easily estimated. To generate plasmas, the rf or dc discharge is used, and fine particles are intentionally added by particle dispensers in most cases. In this chapter, fine particles will be referred to simply as "particles."

As for experimental observations, particles of micron sizes scatter laser beams, and their orbits are easily followed, for example, by ccd cameras. In addition to the importance of (often not welcome) effects of the existence of particles in reacting plasmas applied in semiconductor processes, the relative easiness of observation might be one of the motives of investigation.

Particles immersed in plasmas obtain negative charges of large magnitudes: the thermal velocity of electrons is much larger than that of ions. Even though the interaction between particles are screened by surrounding plasma (composed of electrons and ions), particles are sometimes mutually strongly coupled, and many phenomena in strongly coupled Coulomb or Coulomb-like systems are expected to be observed.

From the viewpoint of statistical physics, systems of particles resemble the Coulomb system with the charges of definite (negative) sign, the screening of charges by surrounding plasma being the most important difference from the pure Coulomb systems. The mutual interaction between particles is approximately given by the Yukawa or the screened Coulomb potential. The one-component plasma (OCP) is a typical model of Coulomb systems composed of charges of definite sign where the inert charges of opposite sign (the background charge) neutralize the particle charges. In fine particle plasmas, particle charges are screened and, at the same time, neutralized by the surrounding plasma.

Compared with electrons and ions, particles have macroscopic masses, and the gravity on the ground has significant influence on their behavior. In order to observe their generic properties, the experiments in the environment of microgravity have been and are performed on the International Space Station (ISS) as PK-3Plus and PK-4 Projects [6, 7].

The main topics in the statistical physics of fine particle plasmas may be the following:

- Charging of fine particles
- Interaction between particles and their microscopic structures
- Behavior of electrostatic potential under the existence of particles
- Effect of gravity
- Simple models of fine particle clouds under microgravity and usual gravity
- Effects of strong coupling

In what follows, we will discuss them as much as possible within a limited space.

2. Typical values of parameters

Here we show some typical values of important parameters related to fine particle plasmas discussed in this chapter. They are summarized in **Table 1**.

Since we have weakly ionized plasmas, the density of neutral atoms is much larger than that of electrons, ions, or particles. The neutral atom density n_n at the pressure p_n and the temperature T_n

$$n_n \sim 2.081 \cdot 10^{14} (p_n[\mathrm{Pa}]) \left(\frac{0.03}{k_B T_n[\mathrm{eV}]} \right) \ \mathrm{cm}^{-3} \tag{1}$$

is to be kept in mind in considering phenomena in fine particle plasmas. It is sometimes necessary to take the effects of collisions between neutral atoms and electrons and ions into account: the collision cross section σ gives the collision mean free path $\ell = 1/n_n\sigma$ and corresponding collision frequency. Typical values of cross sections and the mean free path in our range of energy are

$$\sigma_{electron-neutral} \sim 2 \cdot 10^{-16} \ \mathrm{cm}^2 \quad (\mathrm{e} - \mathrm{Ar}), \tag{2}$$

Quantities	Notation	Values
Electron density	n_e	$10^8 - 10^9\ \text{cm}^{-3}$
Ion density	n_i	$10^8 - 10^9\ \text{cm}^{-3}$
Electron temperature	T_e	$(1-5)\ \text{eV}/k_B$
Ion temperature	T_i	300 K
Fine particle size	r_p	1 μm
Fine particle density	n_p	$10^5\ \text{cm}^{-3}$
Fine particle temperature	T_p	300 K
Neutral gas pressure	p_n	$10 - 10^2$ Pa
Neutral gas temperature	T_n	300 K

Table 1.
Typical values of parameters of fine particle plasmas.

$$\sigma_{ion-neutral} \sim 1 \cdot 10^{-14}\ \text{cm}^2 \qquad (\text{Ar}^+ - \text{Ar}), \tag{3}$$

$$\ell \sim \frac{1}{(p_n[Pa])}\text{cm} \qquad (\text{Ar}^+ - \text{Ar}). \tag{4}$$

The neutral gas is considered to be at room temperature. The temperature of electrons T_e is often measured or reasonably extrapolated from the discharge conditions. As for the ion temperature T_i, on the other hand, we usually do not have measured values and implicitly assume it to be around room temperature. There exist experiments where the velocity of the ion motion suggests higher temperatures. Most of experimental results, however, seem to be compatible with the above assumption. It may be also justified by frequent collision with neutral atoms. The temperature of particles T_p can be measured by monitoring their motion, and the latter sometimes indicates rather high temperatures of even eVs. We here assume, however, that particles are also at room temperature, attributing apparent high temperatures to some instability.

As will be shown in Section 1.4, charges on particles are screened by the plasma of electrons and ions with characteristic screening or Debye lengths: the electron Debye lengths λ_e and ion Debye lengths λ_i are, respectively

$$\lambda_e \equiv \left(\frac{\varepsilon_0 k_B T_e}{n_e e^2}\right)^{1/2} \sim 7.434 \cdot 10^{-2} (k_B T_e[\text{eV}])^{1/2} \left(\frac{10^8}{n_e[\text{cm}^{-3}]}\right)^{1/2} \text{cm}, \tag{5}$$

$$\lambda_i \equiv \left(\frac{\varepsilon_0 k_B T_i}{n_i e^2}\right)^{1/2} \sim 1.288 \cdot 10^{-2} \left(\frac{k_B T_i[\text{eV}]}{0.03}\right)^{1/2} \left(\frac{10^8}{n_i[\text{cm}^{-3}]}\right)^{1/2} \text{cm}. \tag{6}$$

Since $T_e \gg T_i$, we have for the total screening length $\lambda = 1/k_D$ as

$$\lambda \equiv \left(\frac{1}{1/\lambda_e^2 + 1/\lambda_i^2}\right)^{1/2} \sim \lambda_i \sim 1.29 \cdot 10^2 \left(\frac{k_B T_i[\text{eV}]}{0.03}\right)^{1/2} \left(\frac{10^8}{n_i[\text{cm}^{-3}]}\right)^{1/2} \mu\text{m}. \tag{7}$$

The thermal velocity of electrons $v_{th,e}$ and that of ions $v_{th,i}$ for Ar^+ (Ar) are given, respectively, by

$$v_{th,e} \equiv \left(\frac{k_B T_e}{m_e}\right)^{1/2} \sim 4.194 \cdot 10^7 (k_B T_e[\text{eV}])^{1/2} \text{ cm/s}, \tag{8}$$

$$v_{th,i} \equiv \left(\frac{k_B T_i}{m_{Ar}}\right)^{1/2} \sim 2.692 \cdot 10^4 (k_B T_i[\text{eV}])^{1/2} \text{ cm/s}. \tag{9}$$

3. Charging of fine particles in plasmas

The charge on a particle $-Qe$ is one of the key parameters of our system. Since $v_{th,e} \gg v_{th,i}$, an object immersed in the plasma obtains negative charges. The resultant magnitude of the charge is determined so as to have balanced currents of electrons and ions onto the surface of particles. From the capacitance of a spherical capacitor of the radius r_p, $4\pi\varepsilon_0 r_p$, and the potential difference $k_B T_e/e$, we have a rough estimate as $Qe \sim 4\pi\varepsilon_0 r_p (k_B T_e/e)$. For micron-sized particles in the plasma with the electron temperature of the order of eV, the charge number Q easily has the magnitudes of $10^2 - 10^3$.

From the analysis of electron and ion currents, we have

$$Q = z \frac{k_B T_e}{e^2/4\pi\varepsilon_0 r_p} = z \times 6.94 \cdot 10^2 \ (r_p[\mu\text{m}]) \ (k_B T_e[\text{eV}]), \tag{10}$$

where z is determined by

$$\exp(-z) = \frac{n_i}{n_e}\left(\frac{m_e}{m_{Ari}}\frac{T}{T_e}\right)^{1/2}\left[1 + \frac{T_e}{T_i}z + 0.1\left(\frac{T_e}{T_i}\right)^2 z^2 \frac{\lambda}{\ell_i}\right]. \tag{11}$$

Here the first and second terms in [ ] on the right-hand side are given by the orbit-motion-limited (OML) theory [8, 9], and the third term is the effect of ion-neutral collisions, ℓ_i being the ion mean free path [10]. Typically this equation gives $z \sim 0.5$.

4. Mutual interaction and one-body potential in fine particle system

In the system composed of the electron-ion plasma and particles, we are mainly interested in the latter and take statistical average with respect to variables related to the former. Such a treatment of the uniform system was given previously [11, 12]. For systems of particles in plasmas considered here, however, the effects of nonuniformity are of essential importance. The corresponding inhomogeneous system has been analyzed, and mutual interactions and the one-body potential have been obtained [13]. In the process of statistical average, the importance of the charge neutrality of the whole system is pointed out. The main results are summarized below.

Since the thermal velocity of particles is much smaller than that of electronsor ions, we adopt the adiabatic approximation and regard configurations of particles as static. Under the conditions of fixed volume and fixed temperature, the work necessary to change the configuration of particles (the effective interaction) is given by the change in the Helmholtz free energy of the electron-ion system [14]. The effective interaction energy U_{ex} for particles is thus written as

$$U_{ex} = F_{id}^{(e)} + F_{id}^{(i)} + \left[\frac{1}{2}\int d\mathbf{r}\rho(\mathbf{r})\Psi(\mathbf{r}) - U_s\right]. \tag{12}$$

Here, $F_{id}^{(e)} + F_{id}^{(i)}$ is the Helmholtz free energy of the electron-ion plasma

$$F_{id}^{(e,i)} = k_B T_{e,i} \int d\boldsymbol{r} n_{e,i}(\boldsymbol{r}) \left(\ln \left[n_{e,i}(\boldsymbol{r}) \Lambda_{e,i}^3 \right] - 1 \right), \tag{13}$$

Λ_e and Λ_i being the thermal de Broglie lengths, $\rho(\boldsymbol{r})$ the charge density

$$\rho(\boldsymbol{r}) = \rho_p(\boldsymbol{r}) + \rho_{bg}(\boldsymbol{r}), \quad \rho_p(\boldsymbol{r}) = -Qe \sum_i \delta(\boldsymbol{r} - \boldsymbol{r}_i), \quad \rho_{bg}(\boldsymbol{r}) = en_i(\boldsymbol{r}) - en_e(\boldsymbol{r}), \tag{14}$$

and $\Psi(\boldsymbol{r})$ the electrostatic potential. (We subtract the self-energy $U_s = (1/2)\sum_{i=j}^{N}(-Qe)^2/4\pi\varepsilon_0 r_{ij}$ formally included in the integral.) We apply the ideal gas value to $F_{id}^{(e)} + F_{id}^{(i)}$ assuming weakly coupled electron-ion plasma.

The (microscopic) particle charge density $\rho_p(\boldsymbol{r})$ fluctuates around the average $\overline{\rho_p(\boldsymbol{r})}$ as $\rho_p = \overline{\rho}_p + \delta\rho_p$ and induces the density fluctuations of electrons and ions. The potential also fluctuates as $\Psi = \overline{\Psi} + \delta\Psi$. We assume that electrons and ions respond to $\delta\Psi$ as (+ for e and − for i)

$$\delta n_{e,i}(\boldsymbol{r}) \sim \overline{n_{e,i}}(\boldsymbol{r}) \left[\exp\left(\pm \frac{e\delta\Psi(\boldsymbol{r})}{k_B T_{e,i}} \right) - 1 \right] \sim \pm \overline{n_{e,i}}(\boldsymbol{r}) \frac{e\delta\Psi(\boldsymbol{r})}{k_B T_{e,i}} \tag{15}$$

and have

$$\delta\Psi(\boldsymbol{r}) \sim \int d\boldsymbol{r}' u(\boldsymbol{r}, \boldsymbol{r}') \delta\rho_p(\boldsymbol{r}'), \tag{16}$$

where

$$u(\boldsymbol{r}, \boldsymbol{r}') = \frac{\exp\left(-k_D^+ |\boldsymbol{r} - \boldsymbol{r}'|\right)}{4\pi\varepsilon_0 |\boldsymbol{r} - \boldsymbol{r}'|}, \qquad k_D^+ = k_D[(\boldsymbol{r} + \boldsymbol{r}')/2], \tag{17}$$

with the screening parameter evaluated at the midpoint $(\boldsymbol{r} + \boldsymbol{r}')/2$ noting the position dependence of electron and ion densities.

We expand $F_{id}^{(e)} + F_{id}^{(i)}$ with respect to fluctuations to the second order as

$$\sim F_{id,0} + \frac{1}{2} \int d\boldsymbol{r} \left[k_B T_e \frac{\delta n_e^2(\boldsymbol{r})}{\overline{n_e}(\boldsymbol{r})} + k_B T_i \frac{\delta n_i^2(\boldsymbol{r})}{\overline{n_i}(\boldsymbol{r})} \right] = F_{id,0} - \frac{1}{2} \int d\boldsymbol{r} \delta\rho_{bg}(\boldsymbol{r}) \delta\Psi(\boldsymbol{r}), \tag{18}$$

where $\delta\rho_{bg}(\boldsymbol{r}) = e\delta n_i(\boldsymbol{r}) - e\delta n_e(\boldsymbol{r})$ and

$$F_{id,0} = k_B T_e \int d\boldsymbol{r} \overline{n_e}(\boldsymbol{r}) \left[\ln \left[\overline{n_e}(\boldsymbol{r}) \Lambda_e^3 \right] - 1 \right] + k_B T_i \int d\boldsymbol{r} \overline{n_i}(\boldsymbol{r}) \left[\ln \left[\overline{n_i}(\boldsymbol{r}) \Lambda_i^3 \right] - 1 \right]. \tag{19}$$

The electrostatic energy is written as

$$\frac{1}{2} \int d\boldsymbol{r} \rho \Psi - U_s = \frac{1}{2} \int d\boldsymbol{r} \left[\overline{\rho_p} + \overline{\rho_{bg}} + \rho_{ext} \right] \overline{\Psi} + \frac{1}{2} \int d\boldsymbol{r} \left[\delta\rho_p + \delta\rho_{bg} \right] \delta\Psi - U_s. \tag{20}$$

From (18) and (20), we have finally

$$U_{ex} = F_{id,0} + \frac{1}{2}\int dr\left[\overline{\rho_p}(\boldsymbol{r}) + \overline{\rho_{bg}}(\boldsymbol{r}) + \rho_{ext}(\boldsymbol{r})\right]\overline{\Psi}(\boldsymbol{r}) + \left[\frac{1}{2}\int dr\delta\rho_p(\boldsymbol{r})\delta\Psi(\boldsymbol{r}) - U_s\right]. \tag{21}$$

The last term of (21) is rewritten as

$$\begin{aligned}\frac{1}{2}\int\int drdr'u(\boldsymbol{r},\boldsymbol{r}')\left[\rho_p(\boldsymbol{r}) - \overline{\rho_p}(\boldsymbol{r})\right]\left[\rho_p(\boldsymbol{r}') - \overline{\rho_p}(\boldsymbol{r}')\right] - U_s = \frac{(Qe)^2}{2}\sum_{i,j=1}^{N} u(\boldsymbol{r}_i,\boldsymbol{r}_j)\\ -U_s - (-Qe)\sum_{i=1}^{N}\int dr'u(\boldsymbol{r}_i,\boldsymbol{r}')\overline{\rho_p}(\boldsymbol{r}') + \frac{1}{2}\int\int drdr'u(\boldsymbol{r},\boldsymbol{r}')\overline{\rho_p}(\boldsymbol{r})\overline{\rho_p}(\boldsymbol{r}').\end{aligned} \tag{22}$$

The first two terms on the right-hand side of (22) reduce to the mutual Yukawa repulsion and the free energy stored in the sheath:

$$\frac{1}{2}\sum_{i,j=1}^{N}(Qe)^2u(\boldsymbol{r}_i,\boldsymbol{r}_j) - U_s = \frac{(Qe)^2}{2}\sum_{i\neq j}^{N}u(\boldsymbol{r}_i,\boldsymbol{r}_j) - \frac{1}{2}\sum_{i=1}^{N}\frac{(Qe)^2k_D(\boldsymbol{r}_i)}{4\pi\varepsilon_0} \tag{23}$$

and the Helmholtz free energy is finally given by

$$\begin{aligned}U_{ex} = F_{id,0} &+ \frac{1}{2}\int dr\left[\overline{\rho_p}(\boldsymbol{r}) + \overline{\rho_{bg}}(\boldsymbol{r}) + \rho_{ext}(\boldsymbol{r})\right]\overline{\Psi}(\boldsymbol{r})\\ &+ \left[\frac{1}{2}\sum_{i\neq j}^{N}(Qe)^2u(\boldsymbol{r}_i,\boldsymbol{r}_j) + \sum_{i=1}^{N}(-Qe)\int dr'u(\boldsymbol{r}_i,\boldsymbol{r}')\left[-\overline{\rho_p}(\boldsymbol{r}')\right]\right]\\ &+ \frac{1}{2}\int\int drdr'u(\boldsymbol{r},\boldsymbol{r}')\overline{\rho_p}(\boldsymbol{r})\overline{\rho_p}(\boldsymbol{r}') - \frac{1}{2}\sum_{i=1}^{N}\frac{(Qe)^2k_D(\boldsymbol{r}_i)}{4\pi\varepsilon_0}.\end{aligned} \tag{24}$$

The averages $\overline{\rho_p}(\boldsymbol{r}), \overline{\rho_{bg}}(\boldsymbol{r})$, and $\overline{\Psi}(\boldsymbol{r})$ are determined so as to be consistent with the plasma generation and loss and the ambipolar diffusion in the system (and also with the external potential, if any). Explicitly configuration-dependent terms in (24),

$$\frac{1}{2}\sum_{i\neq j}^{N}(Qe)^2u(\boldsymbol{r}_i,\boldsymbol{r}_j) + \sum_{i=1}^{N}(-Qe)\int dr'\left[-\overline{\rho_p}(\boldsymbol{r}')\right]u(\boldsymbol{r}_i,\boldsymbol{r}') - \frac{1}{2}\sum_{i=1}^{N}\frac{(Qe)^2k_D(\boldsymbol{r}_i)}{4\pi\varepsilon_0}, \tag{25}$$

describe the Helmholtz free energy for a given configuration of particles $\{\boldsymbol{r}_i\}_{i=1,\dots N}$. The first term gives the mutual Yukawa repulsion between fine particles. The integral in the second term

$$\int dr'\left[-\overline{\rho_p}(\boldsymbol{r}')\right]u(\boldsymbol{r}_i,\boldsymbol{r}') = \int dr'\frac{\left[-\overline{\rho_p}(\boldsymbol{r}')\right]}{4\pi\varepsilon_0|\boldsymbol{r}_i - \boldsymbol{r}'|}\exp\left(-k_D^{+}|\boldsymbol{r}_i - \boldsymbol{r}'|\right), \tag{26}$$

can be regarded as the Yukawa potential at $\boldsymbol{r}_i$ due to $\left[-\overline{\rho_p}(\boldsymbol{r}')\right]$, the (imaginary) charge density, which exactly cancels the average particle charge density $\overline{\rho_p}(\boldsymbol{r}')$. We may call $\left[-\overline{\rho_p}(\boldsymbol{r}')\right]$ the *shadow* to $\left[\overline{\rho_p}(\boldsymbol{r}')\right]$. The charge density of the shadow has the sign opposite to that of particles, and the potential due to the shadow is attractive for particles. Thus we may summarize the result as *particles are mutually interacting via the Yukawa repulsion and, at the same time, confined by the attractive potential due to the shadow charge density* $\left[-\overline{\rho_p}(\mathbf{r}')\right]$.

The last term in (24) and (25), the free energy stored in the sheath around each particle, works as a one-body potential for particles: this is called the polarization force [15].

5. Description by drift-diffusion equations

In Section 4, we have derived the effective interaction and the one-body *shadow* potential for particles. The behavior of average densities can be analyzed on the basis of the drift-diffusion equations [16, 17]. Here we derive average densities of electrons, ions, and particles and determine the one-body potential for particles [18–21] which enables microscopic simulations of fine particle distribution.

For densities $n_{e,i,p}$ and flux densities $\mathbf{\Gamma}_{e,i,p}$, the equations of continuity are written in the form

$$\frac{\partial n_e}{\partial t} + \nabla \cdot \mathbf{\Gamma}_e = \frac{\delta n_e}{\delta t}, \qquad \frac{\partial n_i}{\partial t} + \nabla \cdot \mathbf{\Gamma}_i = \frac{\delta n_i}{\delta t}, \qquad \frac{\partial n_p}{\partial t} + \nabla \cdot \mathbf{\Gamma}_p = 0, \tag{27}$$

where $\delta n_e/\delta t = \delta n_i/\delta t$ is the contribution of the plasma generation/loss. Flux densities are given by diffusion coefficients $D_{e,i,p}$ and mobilities $\mu_{e,i,p}$ as

$$\begin{aligned} \mathbf{\Gamma}_e &= -D_e \nabla n_e - n_e \mu_e \left(\frac{\mathbf{F}_e}{-e}\right), \quad \mathbf{\Gamma}_i = -D_i \nabla n_i + n_i \mu_i \left(\frac{\mathbf{F}_i}{e}\right), \\ \mathbf{\Gamma}_p &= -D_p \nabla n_p - n_p \mu_p \left(\frac{\mathbf{F}_p}{-Qe}\right). \end{aligned} \tag{28}$$

Here $\mathbf{F}_{e,i,p}$ are forces for an electron, an ion, and a particle, respectively. We assume Einstein relations between diffusion coefficients and mobilities

$$\frac{D_e}{\mu_e} = \frac{k_B T_e}{e}, \qquad \frac{D_i}{\mu_i} = \frac{k_B T_i}{e}, \qquad \frac{D_p}{\mu_p} = \frac{k_B T_p}{Qe} \tag{29}$$

with the temperature of each component, $T_{e,i,p}$. Solutions of (27) are characterized by the ambipolar diffusion length R_a. Forces $\mathbf{F}_{e,i,p}$ are due to the electric field, and, for $\mathbf{F}_p$, we also take the gravity $m_p\mathbf{g}$ and the ion drag force [22] into account: the reaction of the latter is to be included for ions but with small effect. We consider the stationary state.

5.1 Microgravity

We first neglect the gravity or consider the case under microgravity [18]. Examples of solutions of drift-diffusion equations are shown in **Figure 1** where we assume the cylindrical symmetry expecting applications to discharges in cylindrical tube. The main results are:

a. The charge neutrality is largely enhanced in fine particle clouds.

b. The ion distribution compensates the negative charge of particles.

c. The electron distribution is not sensitive to the existence of particles.

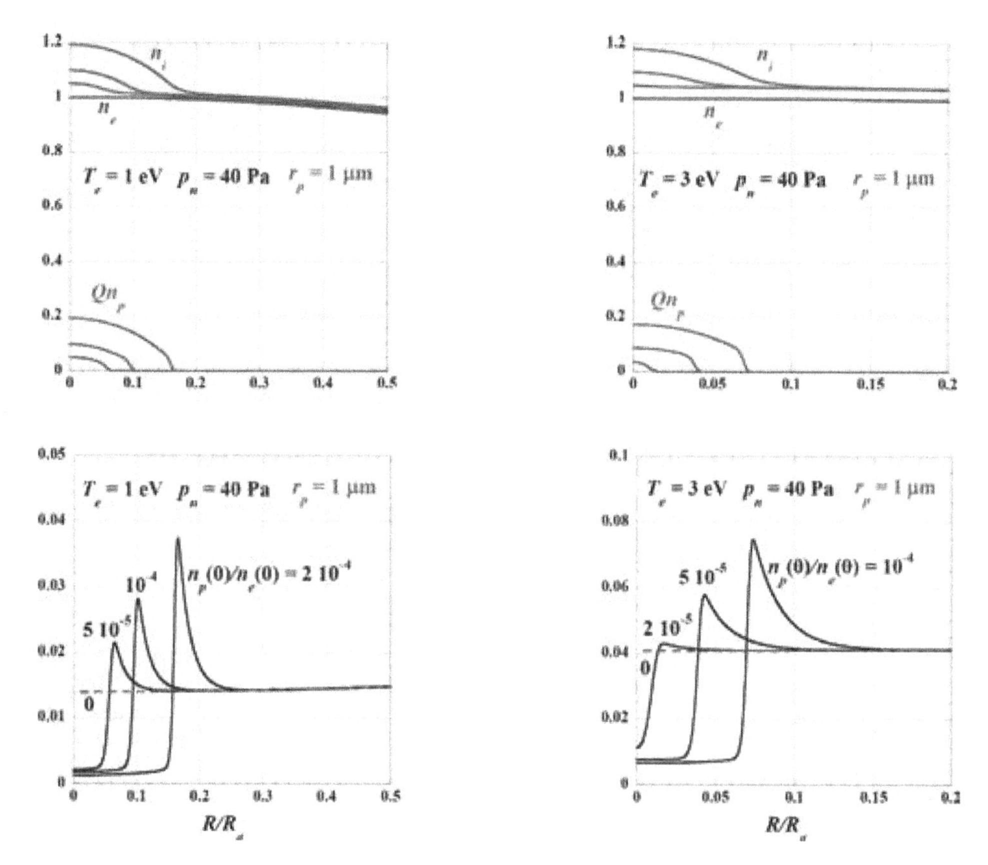

Figure 1.
Examples of solutions of drift-diffusion equations under microgravity. On the left panel, from top, densities and the net charge density both normalized by $n_e(0)$ at $T_e = 1eV$. On the right panel, those at $T_e = 3eV$. Three cases are shown at each electron temperature [18].

The enhancement of the charge neutrality is shown also analytically [23]. We obtain the condition for the formation of voids around the center: voids are formed when the electron density is high and the neutral gas pressure is low and the radius of particles is large. The critical density is given approximately by [18]

$$n_e^{critical}(0) \sim 4 \times \frac{(p_n[\text{Pa}])^{1.56}}{(r_p[\mu\text{m}])^{0.42}(T_e[\text{eV}])^{0.24}} 10^6 \ [\text{cm}^{-3}]. \tag{30}$$

5.2 Gravity

In the case under gravity, we have the same equations, but the effect of gravity is to be taken into account for the force on particles [20, 21]. The distributions do not have symmetries even in cylindrical discharges. Here we assume the one-dimensional structures: the assumption is applicable to the case under gravity at least at the central part of the distributions. Examples of solutions are shown in **Figure 2** [20, 21]. We also observe the large enhancement of the charge neutrality as in the case of microgravity. The enhancement is derived also analytically [24].

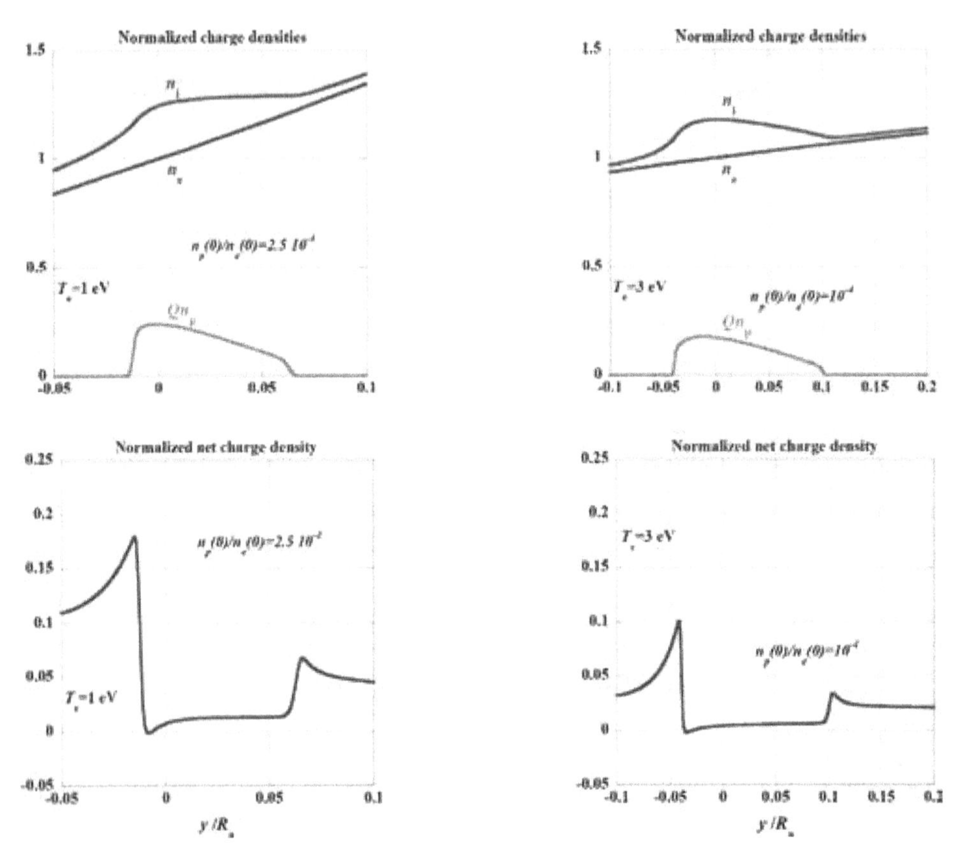

Figure 2.
Examples of solutions of drift-diffusion equations under gravity [21]. On the left panel, from top, densities and the net charge density both normalized by $n_e(0)$ at $T_e = 1$ eV. On the right panel, those at $T_e = 3$ eV. The origin $y = 0$ is (arbitrarily) taken in particle clouds. The slope of n_e and n_i outside of clouds corresponds to the electric field supporting particles against gravity.

6. Enhancement of charge neutrality in fine particle clouds: application to models

In clouds of particles, we have largely enhanced charge neutrality as in **Figures 1** and **2**, shown by numerical solutions of the drift-diffusion equations and analytically [18, 20, 21, 24]. This observation enables us to construct simple models of particle clouds in plasmas. The charge neutrality in the domain of fine particles' existence was noticed previously [25–27] but first established as one of the most important characteristics in particle clouds by our analyses.

6.1 Microgravity

In cylindrical fine particle plasmas under microgravity, particles gather around the center where the potential is maximum. Based on the enhanced charge neutrality in particle clouds, we are able to derive a relation between the radii of particle cloud and of the discharge tube from the boundary conditions of the potential. As shown in **Figure 3**, the cloud radius increases with the tube radius, and there exists a minimum of tube radius for the existence of the central cloud [23].

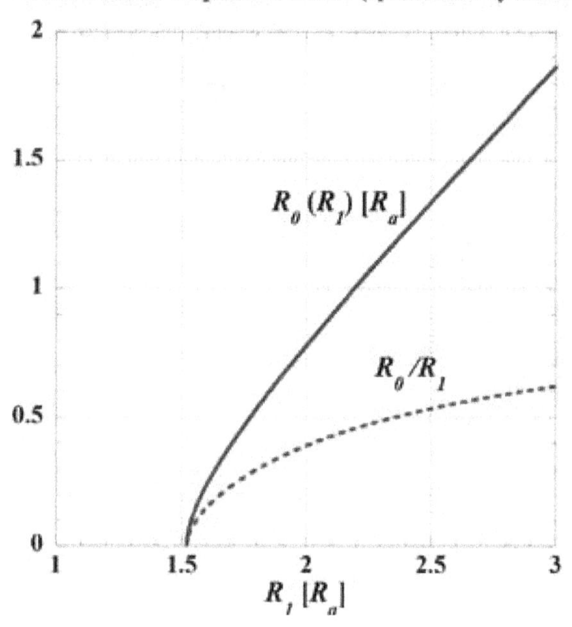

Figure 3.
Cloud radius R_0 and ratio R_0/R_1 as functions of plasma radius R_1, both radii in units of ambipolar diffusion length R_a [23].

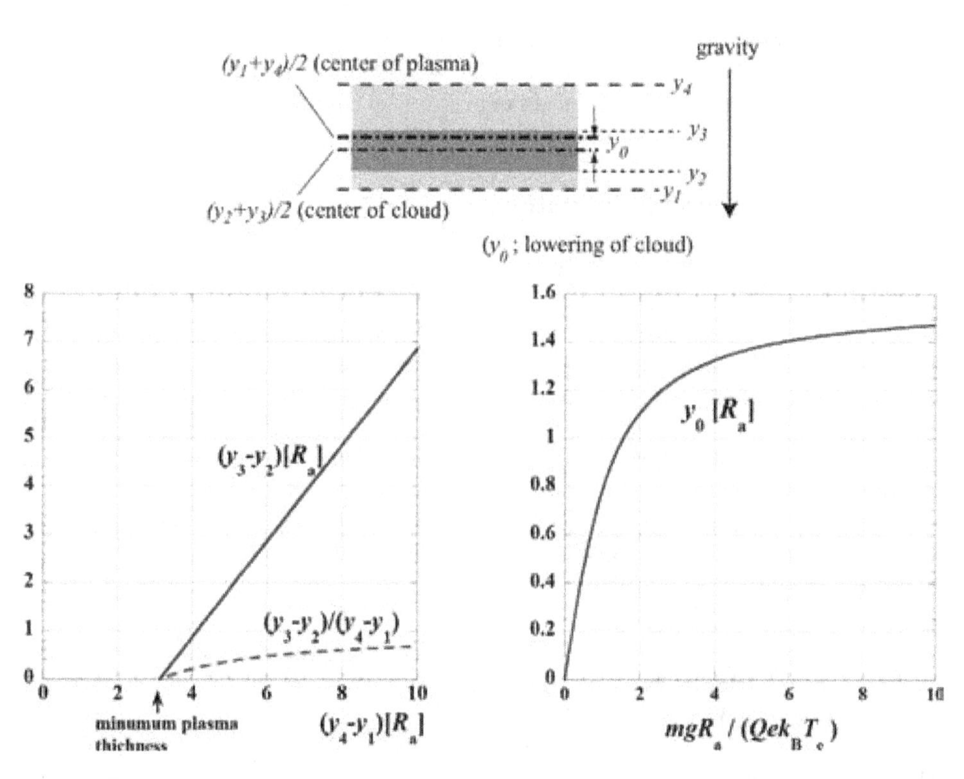

Figure 4.
Particle cloud (dark gray) in plasma (light gray) under gravity (upper panel), cloud thickness (lower left), and lowering of cloud center (lower right) in units of R_a [24].

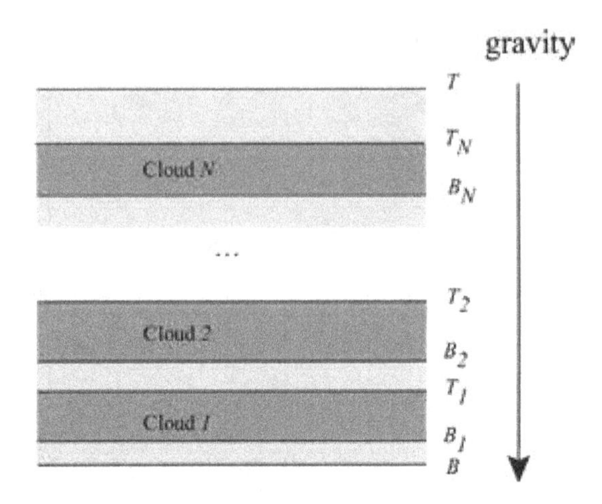

Figure 5.
Stratified particle clouds under gravity. Clouds (between T_i and B_i) of different species are separated by the difference in $\tan^{-1}\left(\frac{m_i}{Q_i}\frac{gR_a}{k_BT_e}\right)$ *[28].*

6.2 Gravity

In the case under gravity, we have a structure shown in the upper panel of **Figure 4**. In this case, the thickness of particle cloud is related to that of whole plasma as shown in the lower left panel, and the center of the cloud is located under the center of the whole plasma as in the lower right panel [24].

The result under gravity is extended to the case where we have multiple species of particles as shown in **Figure 5**. We have stratified clouds in the order of the charge/mass ratio, and the spacings are expressed by simple formulae [28].

7. Example of numerical simulations

For microscopic structures of Coulomb or Coulomb-like systems, there have been developed various theoretical methods including the ones applicable to inhomogeneous cases. Once the interaction and the one-body potentials for particles are established, however, numerical simulations of particle orbits may be the most useful. Their charges being screened by surrounding electron-ion plasma, particles mutually interact via the Yukawa repulsion. Since their charges are of the same (negative) sign, the system naturally tends to expand if we have no confinement.

In the case of infinite uniform system, one usually confines particles implicitly by imposing the periodic boundary conditions (with fixed volume). The number of particles being also fixed, we have effective background of smeared-out Yukawa particles with the opposite charge.

The necessity of confinement becomes clearer when one simulates finite and/or inhomogeneous systems. We here emphasize that, as shown in Section 4, particles are actually confined by the *shadow* potential corresponding to the average density of particles which, in principle, is to be determined self-consistently with the distribution of particles. It is also to be noted that, in addition to the *shadow* potential, there may exist additional external one-body potential, depending on experimental setups. In most cases, however, some kinds of parabolic confining potential or the *shadow* potential for the expected average distribution have been assumed, and its self-consistency has not been rigorously achieved.

7.1 Microgravity

Under microgravity, we may assume simple symmetries of the system coming from the geometry of the apparatus. The cylindrical symmetry is one such common example.

From the result of the drift-diffusion analyses, we expect the formation of clouds of fine particles around the symmetry axis. In the cloud, we have almost flat electrostatic potential due to enhanced charge neutrality in clouds. Microscopic structure of particles in the cloud can be obtained by numerical simulations.

Before drift-diffusion analyses, microscopic structures have been analyzed by assuming the *shadow* potential corresponding to the uniform distribution of particles [29]. This shadow potential is not exact but almost identical to the real one. The assumed uniform average distribution and the shadow potential are shown in **Figure 6**.

At low temperatures, particles take shell structures as shown in **Figure 7**, and these structures are expressed by simple interpolation formulae [29]. One may almost justify *aposteriori* the *shadow* potential corresponding to flat average distribution which was assumed in advance of the analysis given in Section 4.

7.2 Gravity

Examples of the potential under gravity are given by drift-diffusion analyses shown in **Figure 2** [20, 21]: potentials (not shown in figures) correspond to the almost charge-neutral density distribution in particle clouds. Microscopic distribution of particles has been simulated by assuming the *shadow* potential given by the drift-diffusion analysis as shown in **Figure 8** [20]. In this case, parameters approximately correspond to the case shown in the left part of **Figure 2**, and particles are organized into two horizontal layers.

Though the *shadow* potential has to be determined self-consistently, there may be the case where particles are vertically confined by a given potential: with the

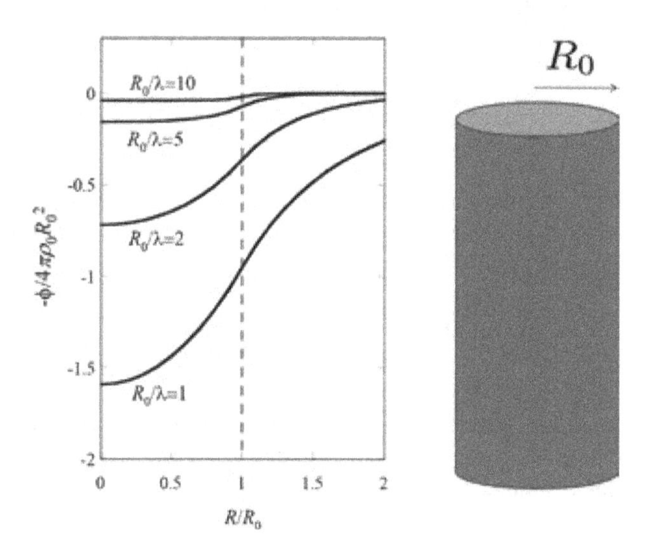

Figure 6.
Assumed shadow potential for simulations (left) and corresponding to uniform cylindrical distribution (right) [29].

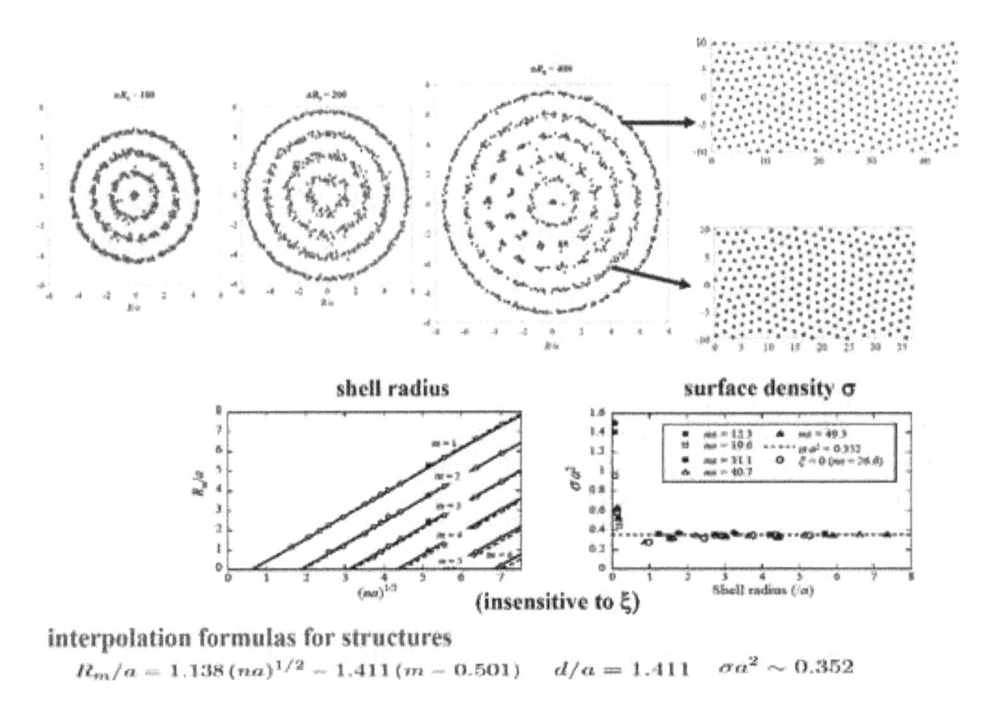

Figure 7.
*Shell structures under microgravity in the shadow potential of **Figure 6**. Particles are ordered into triangular lattice with defects of almost equal density in each shell, and shell radius is expressed by simple interpolation formula [29].*

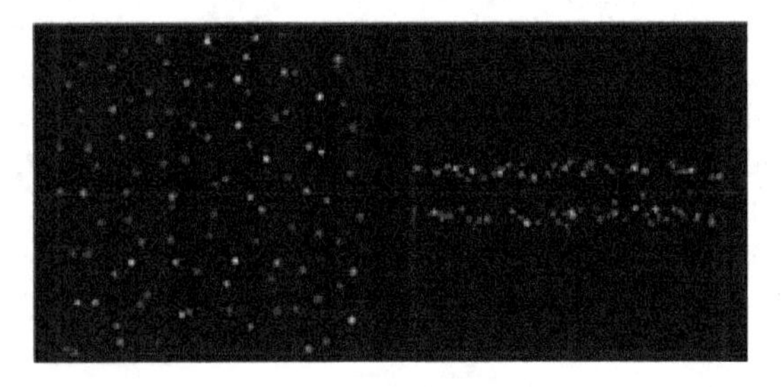

Figure 8.
*Horizontal (left) and vertical (right) distributions of particles in a cloud under downward gravity (with colors for identification) [20]. Parameters correspond to the left panel of **Figure 2** [20, 21].*

effect of the gravity, we may assume a one-dimensional parabolic confining potential. In this case, the structure of particle cloud at low temperatures is determined by the relative strength of the confinement and the mutual repulsion:

$$v(z) = \frac{1}{2}kz^2 \quad vs. \quad \frac{(Qe)^2}{4\pi\varepsilon_0 r}\exp(-r/\lambda) \tag{31}$$

At low temperatures, particles form horizontal layers, and the number of layers is shown in a phase diagram in **Figure 9** [30]. Parameters characterizing strengths of confinement η and screening ξ are defined, respectively, by

$$\eta = \frac{k}{\left[(Qe)^2/\varepsilon_0\right]N_S^{3/2}} = \frac{kN_S^{-1}}{(Qe)^2/\varepsilon_0 N_S^{-1/2}} = \frac{confinement}{repulsion}, \tag{32}$$

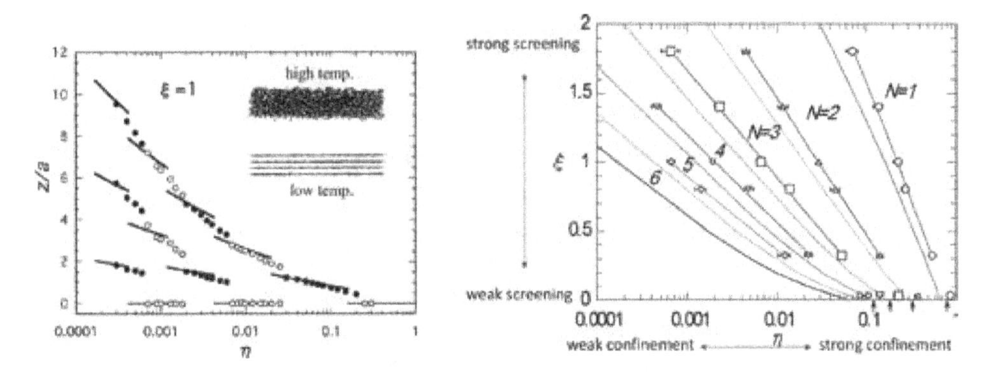

Figure 9.
Yukawa system confined by a parabolic potential. Layered structure (left) and phase diagram for number of layers (right). Dots are numerical simulation and lines are results of shell model [30].

$$\xi = \frac{1}{(\pi N_S)^{1/2}\lambda} = \frac{(\pi N_S)^{-1/2}}{\lambda} = \frac{\textit{mean distance}}{\textit{screening length}}. \tag{33}$$

Note that the mean distance $a = (\pi N_S)^{-1/2}$ defined by the total areal density N_S is used for characterization.

Results of simulations are reproduced by the shell model to a good accuracy (see Appendix).

8. Aspects as strongly coupled system

The system of particles with the Yukawa mutual interaction

$$\frac{(Qe)^2}{4\pi\varepsilon_0 r} \exp(-r/\lambda) \tag{34}$$

has long been explored as a model which interpolates systems with the short-ranged and long-ranged interactions [31]. Since particles of the same charge repel each other, the existence of the uniform *shadow* potential is (often implicitly) assumed. There are three phases at thermal equilibrium: fluid, bcc lattice, and fcc lattice. The parameters Γ and ξ defined, respectively, by

$$\Gamma = \frac{1}{k_B T}\frac{(Qe)^2}{4\pi\varepsilon_0 a}, \qquad \xi = \frac{a}{\lambda}, \tag{35}$$

a being the mean distance defined by the particle density $(4\pi a^3/3)n_p = 1$, characterize the system, and the phase diagram takes the form shown in **Figure 10** (ξ differs from the one defined by (33)). In computer simulations with a fixed number of particles and periodic boundary conditions with fixed shape or the volume of the fundamental cell, the existence of the *shadow* potential coming from the uniform average particle density is automatically built in.

When the coupling parameter Γ is sufficiently large and the screening is not so strong, the pressure obtained by these analyses of the Yukawa system takes on negative values, and the isothermal compressibility even diverges. Negative pressures may sound somewhat strange for this system with repulsion, but it is to be noted that the pressure obtained by these analyses of the Yukawa system is

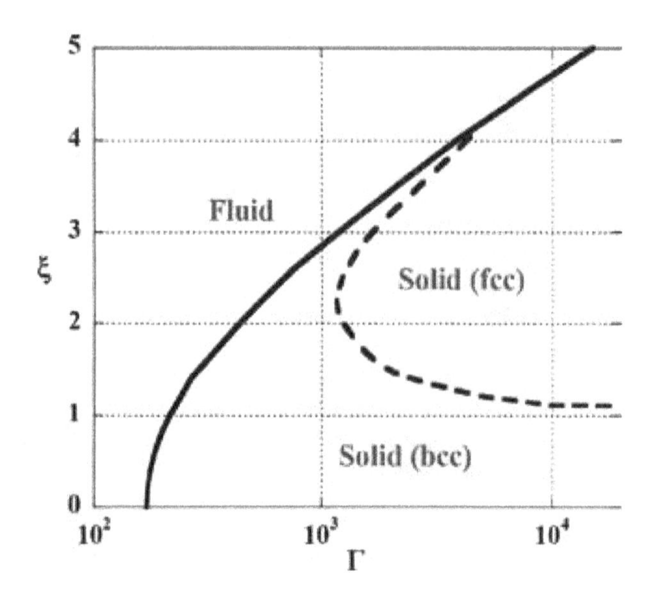

Figure 10.
Phase diagram of Yukawa system.

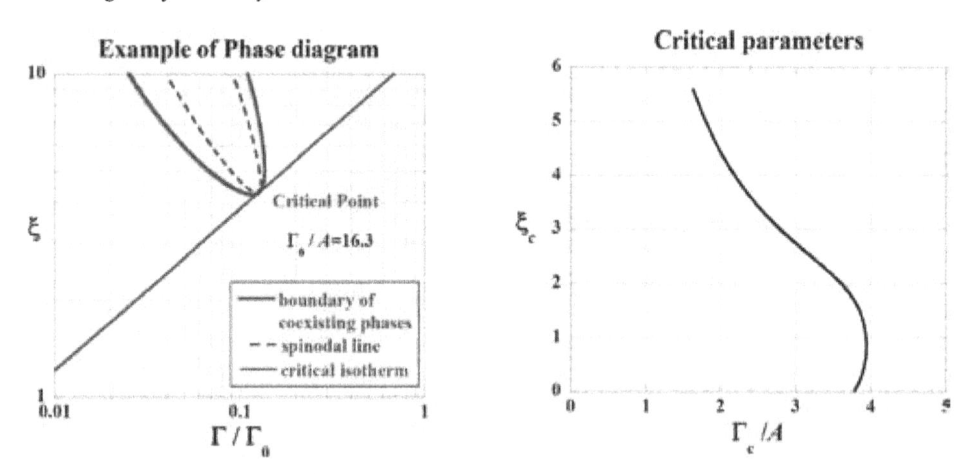

Figure 11.
Example of phase diagram of fine particle plasma [33].

assuming the deformation of the *shadow* potential without any cost of energy. In reality, we have the surrounding plasma which has positive pressure of sufficient magnitude, and these negative pressures usually cannot be observed.

In fine particle plasmas, the magnitude of particle charge can become very large, and we have a possibility of negative pressures and divergent isothermal compressibility of the whole system including the surrounding plasma [32]. As physical phenomena, the latter is very interesting, and the corresponding conditions have been predicted [33]. An example of the phase diagram is shown in the left panel of **Figure 11**. Here A is the ratio of the ideal gas pressure of the surrounding plasma (electrons and ions) to the ideal gas pressure of particles and is usually much larger than unity, and Γ_0 is defined similarly to Γ with a replaced by the radius of particle. The locus of the critical point in (Γ, ξ)-plane is shown in the right panel of **Figure 11**.

Along these considerations, the enhancement of long wavelength fluctuations of density has been expected in the PK-3Plus experiments on ISS, but, since the nonuniformity of the system (particle cloud) gives similar behavior of the fluctuation spectrum, the results have not been conclusive.

9. Conclusions

In this chapter, we have discussed important aspects of fine particle systems immersed in fine particle (dusty) plasmas from the viewpoint of basic physics. We have introduced them as an example of many body system, which allows direct observations of orbits of constituent particles. Mutual interactions and one-body potential for particles are derived on the basis of statistical mechanics, and the enhancement of charge neutrality in fine particle clouds is emphasized as a basis for constructing models of structures. Due to large magnitudes of fine particle charges, these systems have a possibility to manifest some interesting but not yet observed phenomena related to the strongly coupled Coulomb or Coulomb-like systems. They are also discussed with concrete examples.

Fine particles show various interesting dynamic behaviors which are not included in this chapter of limited space. From a theoretical point of view, we would like to stress the importance of understanding the structure of the electrostatic potential and distributions of electrons and ions which are usually not directly observable but are playing a fundamental role to determine the visible behavior of particles.

Acknowledgements

The author would like to thank organizers/participants of series of Workshop on Fine Particle Plasmas supported by the National Institute for Fusion Science (NIFS), Japan, for useful information exchange. For discussions, he is also indebted to Dr. K. Takahashi, Dr. S. Adachi, and the members of the previous ISAS/JAXA Working Group of dusty plasma microgravity experiments (supported by Japan Aerospace Exploration Agency). In relation to the PK-3Plus project, thanks are due to members of the Max Planck Institute for Extraterrestrial Physics (MPE) and Joint Institute for High Temperatures (JIHT, RAS), especially to Professor G. Morfill, Dr. H. Thomas, and Professor V. E. Fortov, including the late Dr. Vladimir Molotkov.

Appendix: Shell model for Coulomb and Coulomb-like systems

When the temperature is sufficiently low, particles are organized into some kind of lattice structures. In the case of uniform system, it is not difficult to determine the structure with the lowest energy. When the symmetry of the system is restricted by the geometry of boundary conditions, however, it is not so easy to determine the lowest energy state theoretically. For Coulomb or Coulomb-like systems, the shell model which has been first proposed for the structure analysis of ions in traps [34] is known to reproduce low-temperature structures to a good accuracy [35].

This model consists of three steps:

1. Assume that particles are organized into a collection of two-dimensional systems (generally not-planar sheets) which are in accordance with the symmetry of the system.

2. Calculate the energy taking the cohesive energy in sheets into account.

3. Minimize the energy of the system with respect to all parameters, the number of sheets, number densities in the sheets, and positions of sheets.

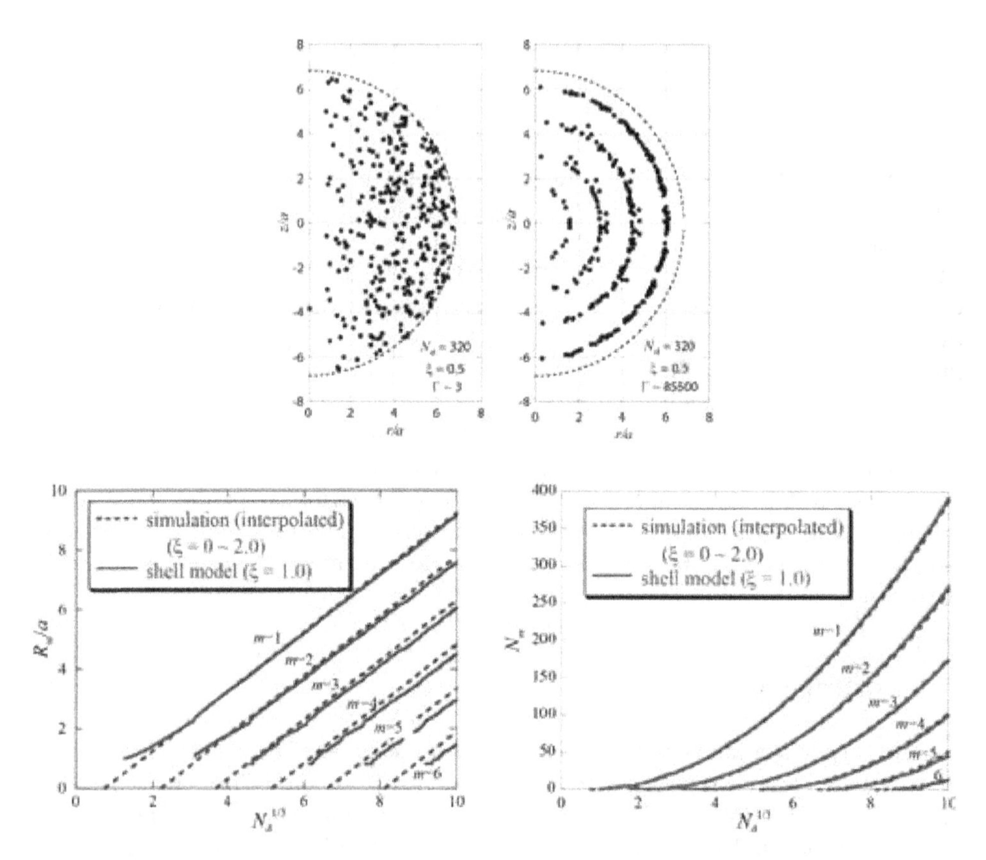

Figure 12.
Example of shell formation (upper panel), radii of shells (lower left), and number of particles on each shell (lower right). In lower panel, simulation results are expressed by interpolated expression [37].

In the above steps, the consideration of the two-dimensional cohesive energy is essential: without its contribution (which have a negative contribution of magnitudes increasing with the density in sheets), we have infinite number of sheets with infinitely low density at the ground state. We can reproduce numerical simulation or real experiments in the cases of Coulomb system in a cylindrically symmetric confining potential [34], Yukawa system in a one-dimensional confining potential, [30] and Yukawa system in a spherically symmetric confinement (Coulomb ball) [36–38].

An example of three-dimensional confinement by the *shadow* potential corresponding to uniform average distribution is shown in **Figure 12** [36, 37]. The case of four shells is in agreement with the experiment [38], and simulation results are reproduced by the shell model to a good accuracy [37].

Author details

Hiroo Totsuji[1,2]

1 Okayama University (Professor Emeritus), Okayama, Japan

2 Saginomiya, Nakanoku, Tokyo, Japan

*Address all correspondence to: totsuji-09@t.okadai.jp

References

[1] Shukla PK, Mamun AA. Introduction to Dusty Plasma Physics. London: Institute of Physics Publishing; 2002

[2] Fortov VE, Ivlev AV, Khrapak SA, Khrapak AG, Morfill GE. Complex (dusty) plasmas: Current status, open issues, perspectives. Physics Reports. 2005;**421**:1

[3] Tsytovich VN, Morfill GE, Vladimirov SV, Thomas H. Elementary Physics of Complex Plasmas. Berlin, Heidelberg: Springer; 2008

[4] Morfill GE, Ivlev AV. Complex plasmas: An interdisciplinary research field. Reviews of Modern Physics. 2009; **81**:1353

[5] Fortov VE, Morfill GE, editors. Complex and Dusty Plasmas-From Laboratory to Space. Boca Raton, London, New York: CRC Press/Taylor and Francis Group; 2010

[6] Fortov V, Morfill G, Petrov O, Thoma M, Usachev A, Höfner H, et al. The project 'Plasmakristall-4' (PK-4)—a new stage in investigations of dusty plasmas under microgravity conditions: first results and future plans. Plasma Physics and Controlled Fusion. 2005;**47**: B537

[7] Thomas HM, Morfill GE, Fortov VE, Ivlev AV, Molotkov VI, Lipaev AM, et al. Complex plasma laboratory PK-3 plus on the international space station. New Journal of Physics. 2008;**10**:033036

[8] Allen JE. Probe theory - the orbital motion approach. Physica Scripta. 1992; **45**:497

[9] Goree J. Charging of particles in a plasma. Plasma Sources Science and Technology. 1994;**3**:400

[10] Khrapak SA, Ratynskaia SV, Zobnin AV, Usachev AD, Yaroshenko VV, Thoma MH, et al. Particle charge in the bulk of gas discharges. Physical Review E. 2005;**72**: 016406

[11] Hamaguchi S, Farouki RT. Thermodynamics of strongly-coupled Yukawa systems near the one-component-plasma limit. I. Derivation of the excess energy. The Journal of Chemical Physics. 1994;**101**:9876

[12] Rosenfeld Y. Adiabatic pair potential for charged particulates in plasmas and electrolytes. Physical Review E. 1994;**49**:4425

[13] Totsuji H. Modeling fine-particle (dusty) plasmas and charge-stabilized colloidal suspensions as inhomogeneous yukawa systems. Journal of the Physical Society of Japan. 2015;**84**:064501

[14] Landau LD, Lifshitz EM. Statistical Physics, Part 1. 3rd ed. Oxford: Pergamon Press; 1980. Sect. 20

[15] Hamaguchi S, Farouki RT. Polarization force on a charged particulate in a nonuniform plasma. Physical Review E. 1994;**49**:4430

[16] Land V, Goedheer WJ. Effect of large-angle scattering, ion flow speed and ion-neutral collisions on dust transport under microgravity conditions. New Journal of Physics. 2006;**8**(8)

[17] Goedheer WJ, Land V. Simulation of dust voids in complex plasmas. Plasma Phys. Control. Fusion. 2008;**50**:124022

[18] Totsuji H. Distribution of electrons, ions, and fine (dust) particles in cylindrical fine particle (dusty) plasmas: drift-diffusion analysis. Plasma Physics and Controlled Fusion. 2016;**58**:045010

[19] A part of preliminary result has been given inTotsuji H. Behavior of dust

particles in cylindrical discharges: Structure formation, mixture and void, effect of gravity. Journal of Plasma Physics. 2014;**80**(part 6):843

[20] Totsuji H. Strongly coupled fine particle clouds in fine particle (dusty) plasmas. Contributions to Plasma Physics. 2017;**57**:463

[21] Totsuji H. Static behavior of fine particle clouds in fine particle (dusty) plasmas under gravity. Journal of Physics Communications. 2018;**2**: 025023

[22] Khrapak SA, Ivlev AV, Morfill GE, Thomas HM. Ion drag force in complex plasmas. Physical Review E. 2002;**66**: 046414

[23] Totsuji H. Simple model for fine particle (dust) clouds in plasmas. Physics Letters A. 2016;**380**:1442

[24] Totsuji H. Charge neutrality of fine particle (dusty) plasmas and fine particle cloud under gravity. Physics Letters A. 2017;**381**:903

[25] Polyakov DN, Shumova VV, Vasilyak LM, Fortov VE. Influence of dust particles on glow discharge. Physica Scripta. 2010;**82**:055501

[26] Polyakov DN, Shumova VV, Vasilyak LM, Fortov VE. Study of glow discharge positive column with cloud of disperse particles. Physics Letters A. 2011;**375**:3300

[27] Sukhinin GI, Fedoseev AV, Antipov SN, Petrov OF, Fortov VE. Dust particle radial confinement in a dc glow discharge. Physical Review E. 2013;**87**:013101

[28] Totsuji H. One-dimensional structure of multi-component fine particle (dust) clouds under gravity. Physics Letters A. 2019;**383**:2065

[29] Totsuji H, Totsuji C. Structures of yukawa and coulomb particles in cylinders: Simulations for fine particle plasmas and colloidal suspensions. Physical Review E. 2011;**84**:015401

[30] Totsuji H, Kishimoto T, Totsuji C, Tsuruta K. Structure of confined yukawa system (dusty plasma). Physical Review Letters. 1997;**78**:3113

[31] For example,Robbins MO, Kremer K, Grest GS. Phase diagram and dynamics of Yukawa systems. Journal of Chemical Physics. 1988;**88**:3286 and references therein

[32] Totsuji H. Thermodynamic instability and critical fluctuations in dusty plasmas modelled as yukawa OCP. Journal of Physics A: Mathematical and General. 2006;**39**:4565

[33] Totsuji H. Thermodynamics of strongly coupled repulsive yukawa particles in ambient neutralizing plasma: Thermodynamic instability and the possibility of observation in fine particle plasmas. Physics of Plasmas. 2008;**15**: 072111

[34] Totsuji H, Barrat J-L. Structure of a nonneutral classical plasma in a magnetic field. Physical Review Letters. 1988;**60**:2484

[35] Dubin DHE, O'Neil TM. Trapped nonneutral plasmas, liquids, and crystals (the thermal equilibrium states). Reviews of Modern Physics. 1999;**71**:87

[36] Totsuji H, Totsuji C, Ogawa T, Tsuruta K. Ordering of dust particles in dusty plasmas under microgravity. Physical Review E. 2005;**71**:045401(R)

[37] Totsuji H, Ogawa T, Totsuji C, Tsuruta K. Structure of spherical yukawa clusters: A model for dust particles in dusty plasmas in an isotropic environment. Physical Review E. 2005; **72**:036406

[38] Arp O, Block D, Piel A, Melzer A. Dust coulomb balls: Three-dimensional plasma crystals. Physical Review Letters. 2004;**93**:165004

9

Dynamic Behavior of Dust Particles in Plasmas

Yoshifumi Saitou and Osamu Ishihara

Abstract

Experimentally observed dynamic behavior, such as a particle circulation under magnetic field, a bow shock formation in an upper stream of an obstacle, etc., will be reviewed. Dust particles confined in a cylindrical glass tube show a dynamic circulation when strong magnetic field is applied from the bottom of the tube using a permanent magnet. The circulation consists of two kinds of motions: one is a toroidal rotation around the tube axis, and the other is a poloidal rotation. Dust particles are blown upward from near the bottom of the tube against the gravity neighborhood of the tube axis. A two-dimensional supersonic flow of dust particles forms a bow shock in front of a needlelike-shaped obstacle when the flow crosses the obstacle. The slower flow passes the obstacle as a laminar flow. A streamline-shaped void where dust particles are not observed is formed around the obstacle.

Keywords: dust flow, dust fluid, bow shock, dynamic circulation, storm in a glass tube

1. Introduction

The natural world is filled with fluids. Fluids present various phenomena such as waves, oscillations, vortices, etc. Scales of such phenomena vary widely. The bow shock formed near the heliopause is in the astrophysical scale, while the Great Red Spot of Jupiter is in the planetary scale, a tornado is in the earth's atmosphere scale, and a swirling tea in a teacup is in the tabletop scale.

It is often observed that collective behavior of individual particles can be regarded as a fluid. A complex plasma, defined as a plasma in which microparticles are embedded in the background of electrons, ions, and neutral particles, provides one of the examples of the case.

In 1986, Ikezi theoretically predicted existence of a crystalized structure with small particles contained in a plasma [1]. It was in 1994 that Hayashi et al., Thomas et al., and Chu et al. separately found in their experiments that charged dust particles formed the crystalized structure in plasmas [2–4]. Since then, research on dusty plasmas has been actively conducted [5–41]. Looking back on the past, Galilei discovered in 1610 that Saturn had "ears." It was found later that the "ears" were a ring or rings by Huygens, Cassini, etc. [42]. Further later in 1856, Maxwell considered the stability of Saturn's ring and concluded that the stable Saturn's ring must consist of independent particles [43]. The interplanetary space is a plasma state dominantly filled by protons brought by the solar wind. Planetary rings like the Saturn's ring is one of the examples that ubiquitously exist in the universe.

Research of complex plasmas including dust particles is unique in a sense that we can chase the motion of individual dust particles by the naked eye using the visible laser light on site without time delay.

In this chapter, experimentally observed dynamic behaviors, such as a circulation of dust particles under magnetic field and a bow shock formation in an upper stream of an obstacle, will be reviewed. A two-dimensional supersonic flow of dust particles forms a bow shock in front of a needlelike-shaped obstacle when the flow passes the obstacle. The slower flow passes the obstacle as a laminar flow. A streamline-shaped void where dust particles are absent is formed around the obstacle. On the other hand, dust particles confined in a cylindrical glass tube show a three-dimensional dynamic circulation when strong enough magnetic field is applied from the bottom of the tube. The circulation consists of two kinds of motions: one is a toroidal rotation around the tube axis, and the other is a poloidal rotation. Dust particles are blown upward from near the bottom of the tube against the gravity around the tube axis.

2. Bow shock formation in two-dimensional dust flow

A NASA's Spitzer Space Telescope observed a shock structure formed in front of the speedster star known as Kappa Cassiopeia in 1994 [44]. The shock is formed near the boundary between a stellar wind and interstellar medium. Another example of a shock wave can be seen around a boundary between a planetary magnetosphere and a stellar wind. These shock waves are similar to a shock excited in front of a bow of a ship cruising fast a water surface and are called a bow shock.

The bow shock is also observable in a supersonic flow of charged dust particles in a complex plasma. In this section, we will look back our experimental work on the bow shock formation [45]. Charged dust particles levitate at height where the gravity and the sheath electrostatic force acting on each particle are balanced in an experimental device on the ground. Therefore, monosized dust particles distribute and flow in an almost two-dimensional plane. An obstacle is placed in the middle of the dust flow just like the star or the planet in the solar wind or the ship on the ocean. The obstacle is a thin needlelike conducting wire and forms a potential barrier against the dust flow. The bow shock is formed when the dust flow interacts with the potential barrier.

2.1 Experimental setup

The schematic of the experimental device Yokohama Complex Plasma Experiment (YCOPEX) is shown in **Figure 1** [46]. Detailed description on the device and experimental setup can be seen in Ref. [45]. The device consists of a glass chamber and a flat metal plate. The size of the metal plate is 800 mm in length (x direction = the main flow direction) and 120 mm in width (y direction). The device is equipped with an up-and-down gate which is electrically controlled from outside. The up-and-down gate separates the plate into two regions: the reservoir of dust particles and the experimental region. A needlelike conducting wire is placed in the experimental region and is used as an obstacle. The potential of the obstacle is floating against the plasma potential here.

The argon gas pressure is 3.6 Pa. To avoid the drag by neutral particles or by ions [45–49], the vacuum pump and the gas feeding are stopped when the pressure reached the set value. Plasma is generated with an rf discharge of 5 W (13.56 MHz). The measured plasma parameters are $n_e \sim 5 \times 10^{14}$ m^{-3}, $T_e \sim 5$ eV. The plasma potential is ~ -30 V.

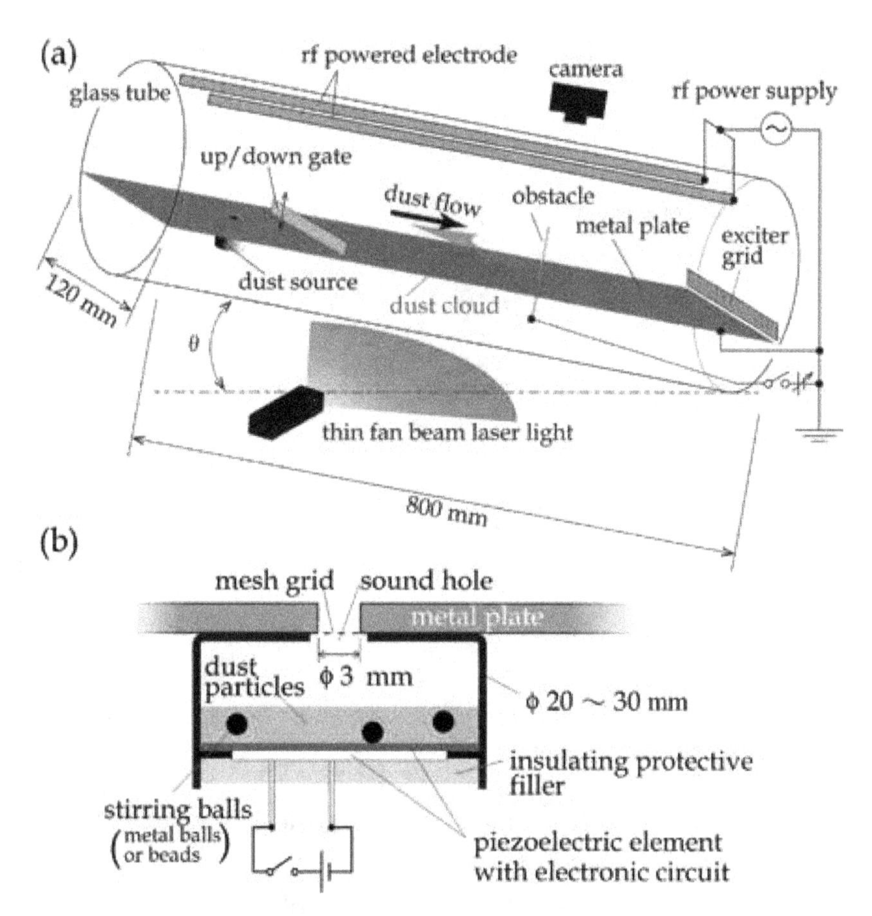

Figure 1.
Schematic drawings of the experimental glass chamber: a YCOPEX device (a) and a piezoelectric buzzer as the dust source (b).

Each dust particle is an Au-coated silica sphere of $a = 5\,\mu\text{m}$ in diameter and $m_d = 1.68 \times 10^{-13}$ kg in mass. The particles are charged to $Q = Z_d e = -(4.4 \pm 0.5) \times 10^4 e$, where e is the elementary electric charge [49]. The particles levitate near the boundary between the plasma and the sheath, whose height is approximately 8 mm above the metal plate. The dust particles are irradiated with two thin fan laser lights from the radial directions. Mie-scattered laser light from the particles is observed and recorded with a camera placed outside the device.

In the initial state, the dust particles are accumulated in a cylindrical piezoelectric buzzer which is placed under the metal plate and acts as a dust source. A part of the accumulated dust particles is hopped into the plasma by energizing the buzzer. The dust particles are stored in the reservoir region above the metal plate when the up-and-down gate is in condition to the up position. By tilting the entire device at angle θ and lowering the gate, the stocked dust particles begin moving and form the almost two-dimensional flow. The flow velocity is controlled by changing angle θ. The velocity reaches a terminal velocity before the particles arrive near the obstacle.

2.2 Wave modes observed in a complex plasma

It is known that there are extremely low-frequency longitudinal wave modes in complex plasmas. Typically, one is the dust acoustic (DA) mode, and the other is the dust lattice (DL) mode. The n-dimensional DA wave velocity, C_{DA}^{nD}, and the DL velocity, C_{DL}, are given by

$$C_d = u(Z_d, m_d) f(\kappa), \tag{1}$$

where $C_d = C_{DA}^{nD}$ or C_{DL}, and

$$u(Z_d, m_d) = \sqrt{\frac{Z_d^2 e^2}{\varepsilon_0 m_d \lambda_{Di}}} \tag{2}$$

with ε_0 the permittivity of free space, λ_{Di}, the ion Debye length. The function $f(\kappa)$ is given by

$$f(\kappa) = \begin{cases} \dfrac{1}{\kappa^{3/2}} & \text{for } C_{DA}^{3D} \\ \dfrac{1}{\sqrt{2\pi\kappa^2}} & \text{for } C_{DA}^{2D} \\ \sqrt{\dfrac{4\pi\kappa}{(\kappa^2 + 2\kappa + 2)\exp(-\kappa)}} & \text{for } C_{DL}\ (\kappa \gg 1) \end{cases}, \tag{3}$$

where $\kappa = d/\lambda_{Di}$ with d as the interparticle distance [16, 19, 20]. The distance is given by

$$d = \begin{cases} \dfrac{1}{\left(n_d^{3D}\right)^{1/3}} & \text{for } 3-\text{D dust distribution} \\ \dfrac{1}{\sqrt{\pi n_d^{2D}}} & \text{for } 2-\text{D dust distribution} \end{cases}, \tag{4}$$

where n_d^{3D} and n_d^{2D} are three- and two-dimensional dust densities, respectively.

The velocity of a wave excited in the dusty plasma, C_d, is measured using the time-of-flight method at $\theta = 0$ degree. The velocity of dust acoustic modes coincides well with the velocity of the dust lattice mode around $\kappa = 3 - 6$. The three-dimensional dust acoustic mode with velocity C_{DA}^{3D} is likely the candidate for the observed mode of the wave although the strict mode identification is still to be determined.

2.3 Bow shock formation

The particles flow from the reservoir region to the obstacle by changing the tilting angle θ. The void which has a streamline-like shape can be seen in the hatched area of **Figure 2**. The void is an area where dust particles are absent. The dust flow near the leading edge of the void is decelerated. The trajectories of dust particles are deflected toward the $\pm y$ direction in front of the void.

The flow velocity, v_f, in the upstream area has a constant value which is mainly determined by a balance of the gravitational force controlled by angle θ and the neutral drag force. The flow is almost uniform, and there is no prominent structure in the upstream area when v_f is small. When v_f increases, an arcuate structure where the intensity of the scattered laser light is enhanced is formed in front of the leading edge of the void. For further increase of v_f, a curvature of the arc becomes larger. The tail of the void is extended with increasing values of v_f.

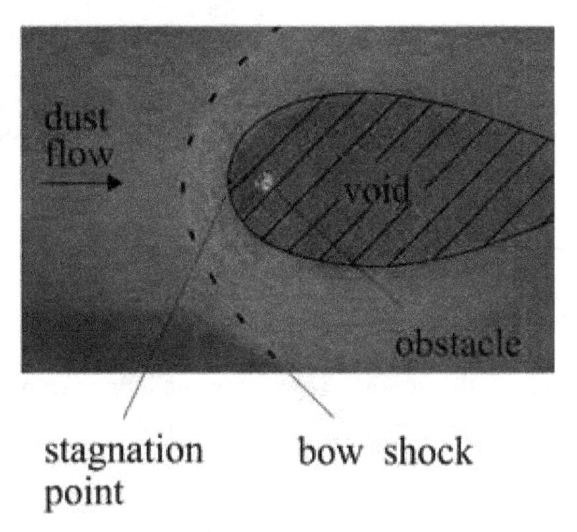

Figure 2.
Typical example of the formed bow shock. Modified from ***Figure 3(b)*** *of Ref. [45].*

The arcuate structure is the bow shock. The value of v_f is required to exceed Mach number 1 when the bow shock is formed, where the Mach number is defined as the ratio of the flow velocity to the dust acoustic velocity. The experimentally measured velocity is 71 m/s (Mach number $M = 1$). It is found that the arcuate structure is distinctive when the flow is supersonic. In addition, there exists a deceleration region between the leading edge of the arcuate structure and the void as shown in **Figure 2**, that is, there is a region where the flow velocity is reduced in order to keep the flux constant around the obstacle. The presence of such a deceleration region, a subsonic flow region, between the wave front and the stagnation point is one of the defining features of the bow shock [50].

The density ratio n_{dp}/n_{d0} is shown as a function of the Mach number, where n_{dp} is the density in front of the stagnation point and n_{d0} is the density of the upstream area. It is known that a polytropic hydrodynamic model provides criterion for the shock wave formation [50]:

$$\frac{n_{dp}}{n_{d0}} = \begin{cases} \left(1+\frac{\gamma-1}{2}M^2\right)^{1/(\gamma-1)} & (M<1) \\ \left(\frac{\gamma+1}{2}\right)^{(\gamma+1)/(\gamma-1)} \dfrac{M^2}{1+\dfrac{\gamma-1}{2}M^2}\left(\gamma-\dfrac{\gamma-1}{2M^2}\right)^{-1/(\gamma-1)} & (M\geq 1) \end{cases}, \quad (5)$$

where γ is the polytropic index.

2.4 Bow shock by a simulation

A molecular-dynamics simulation code is carried out to examine the bow shock formation. The density ratio n_{dp}/n_{d0} and its spatial distribution are calculated. The simulation result on the density ratio corresponds to the numerical result of Eq. (5) with $\gamma = 2.2$. The experimental result on the density ratio seems to correspond to the case of $\gamma = 5/3$ (= the specific heat ratio of monoatomic gas) ~ 2.2 though there is a small deviation.

The polytropic index found in the simulation and experimental observation may result from the fact that the significant amount of internal energy of the polytropic fluid, which consists of charged dust particles, may be stored in the background plasma. The value of the complex plasma polytropic index indicates that the present complex plasma is far from isothermal ($\gamma = 1$).

As for the spatial density distribution, the density contour plot shows the arcuate structure, and its curvature increases with increasing Mach number as seen in the experiment.

2.5 Bow shock formation in two-dimensional flow

Under the polytropic process which is a quasi-static process, $p/n^{\gamma} = const.$ and $T/n^{\gamma-1} = const.$, are held with the polytropic index γ, where p is the pressure, T is the temperature, and n is the density. The polytropic index means

$$\gamma = \begin{cases} 0 & \text{isobaric process} \\ 1 & \text{isothermal process} \\ \kappa_h & \text{isentropic process} \\ \infty & \text{isochoric process} \end{cases}, \tag{6}$$

where κ_h is the ratio of specific heat. The experimentally obtained polytropic index lies between $5/3$ and 2.2. The value $5/3$ is equivalent to the ratio of specific heat of ideal monoatomic gas. The bow shock forms under the almost adiabatic process. The value around 2 is suggested for the investigation on the solar wind [51, 52]. In addition, the dust flow consists of a collection of dust particles with finite size. It is hard to regard the fluid component as ideal. Hence, the polytropic index deviates from $5/3$.

The bow shock formation is a nonisothermal process. The pressure ratio and the temperature ratio p_{dp}/p_{d0} and T_{dp}/T_{d0} dependence on the density ratio n_{dp}/n_{d0} are given by $p_{dp}/p_{d0} = (n_{dp}/n_{d0})^{\gamma}$ and $T_{dp}/T_{d0} = (n_{dp}/n_{d0})^{\gamma-1}$, where p_{dp} and T_{dp} are the pressure and temperature at the stagnation point and p_{d0} and T_{d0} are those at the upstream area, respectively. The results are shown in **Figure 3**. The bow shock is formed for $n_{dp}/n_{d0} > 1$. These results suggest that the polytropic index may be

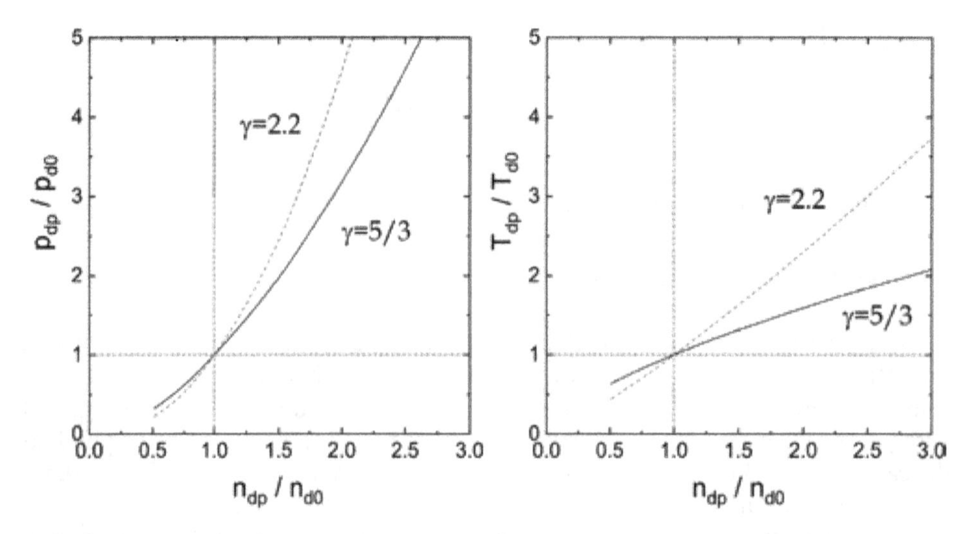

Figure 3.
Expected changes in the ratios of the pressure and the temperature by the bow shock formation $n_{dp}/n_{d0} > 1$ under the polytropic process.

determined by measuring the pressure ratio p_{dp}/p_{d0} or the temperature ratio T_{dp}/T_{d0}.

In this polytropic process, there is a small amount of heat exchange with the outside of the system. The first law of thermodynamics gives

$$dQ = dU + pdV = C_V dT + pdV, \tag{7}$$

where dQ is a differential heat added to the system, dU is the differential internal energy of the system, and C_V is the specific heat at constant volume. By integrating this equation from state 0 to state 1

$$Q_{01} = C_\gamma (T_1 - T_0), \tag{8}$$

where $C_\gamma = C_V(\gamma - \kappa)/(\gamma - 1)(>0)$ is the polytropic specific heat. As seen in **Figure 3**, $T_1 - T_0 = T_{dp} - T_{d0} > 0$, and as a result, $Q_{01} > 0$. It is expected that the heat Q_{01} is added to the system for the bow shock formation.

3. Dynamic circulation under magnetic field

You may watch a dynamic motion of tea leaves, set on the bottom of the teacup, by stirring the tea by a teaspoon. The tea leaves get close to the center and rise near the tea surface as illustrated in **Figure 4**. We can see a similar phenomenon in a complex plasma system. In this section, we will look back our experimental work on such a dynamic motion of dust particles in a complex plasma [53].

The observation of particle motion in the dynamic circulation similar to the motion of the tea leaves helps to understand the simple but profound nature of the ubiquitous vortex commonly encountered in nature.

3.1 Experimental setup

The experiment is performed in a cylindrical glass tube as shown in **Figure 5**. Detailed explanation on the experimental setup is given in Ref. [53]. The cylindrical coordinates (r, θ, z) are with the origin at the inner bottom of the tube, and the gravity is in the negative z direction.

The argon gas pressure is $p = 5 - 25$ Pa. A geometry of the gas supply and exhaust system is configured to avoid the neutral drag force acting on the dust

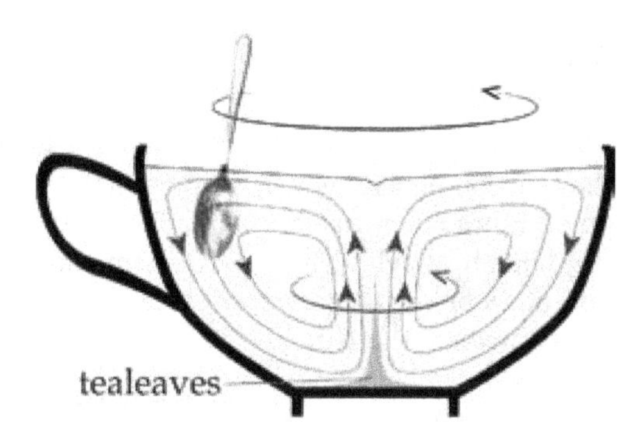

Figure 4.
Schematic of tea leaves in tea stirred in a teacup.

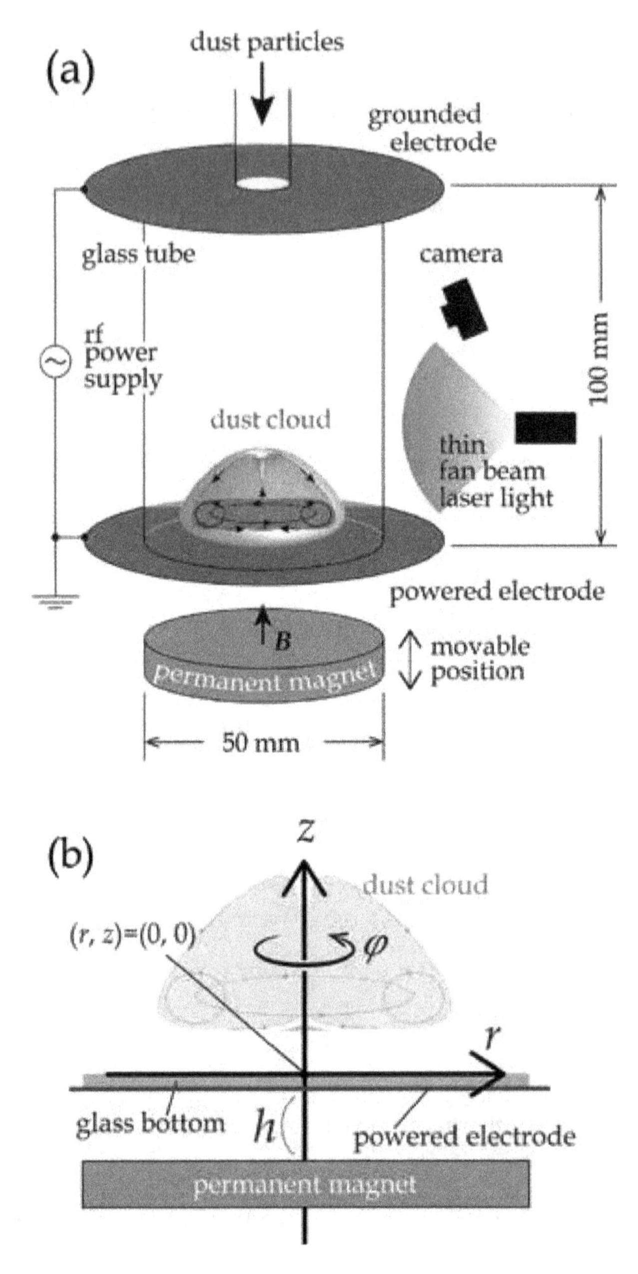

Figure 5.
*Schematic drawing of the experimental glass tube (a) and the coordinate system (b). The structure of the dust cloud can be seen in **Figures 7** and **8**.*

particles in the experimental region. The plasma is produced by an rf discharge of 20 W (13.56 MHz). The electron density is $\sim 10^{14}$ m^{-3}, the electron temperature is ~ 3 eV, and the ion temperature is estimated to be ~ 0.03 eV.

A magnetic field is applied by a cylindrical permanent magnet of 50 mm in diameter placed at a distance h below the powered electrode, and the magnetic field strength is controlled by adjusting the distance h by a jack. The strength of the magnetic field at $r = 0$ is given by

$$B(h, z) \approx 0.29 \left(\frac{50}{z + h + 50} \right)^3 \text{T}, \tag{9}$$

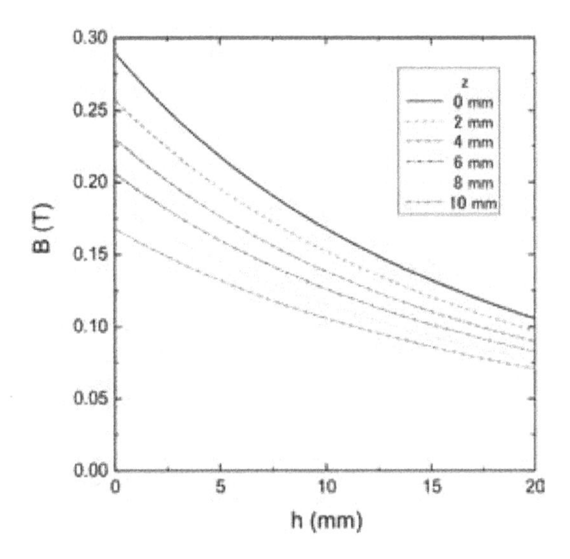

Figure 6.
Strength of the magnetic field at $r = 0$.

where h and z are measured in mm. The calculated $B(h, z)$ is shown in **Figure 6**.

Dust particles which are acrylic resin spheres of $a = 3$ μm in diameter and $m_d = 1.7 \times 10^{-14}$ kg in mass are supplied from a dust reservoir on the top of the glass tube. Each dust particle is charged in the plasma to $Q \sim -10^4 e$ [49]. The particles in the experimental region are irradiated with a thin fan laser light from the radial directions. The laser sheet can be rotated around the laser axis. The scattered laser light from the particles is observed and recorded with a camera placed outside of the tube.

3.2 Behavior of dust particles in a glass cylinder

Because of the cylindrical symmetry of the glass tube, the motion of dust particles is well observable by watching in a meridional (vertical) plane as shown in **Figures 5, 7,** and **8.**

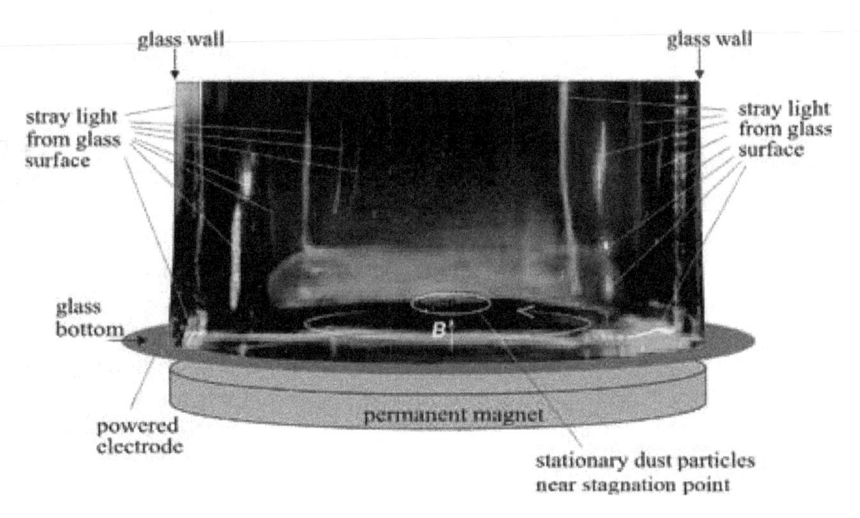

Figure 7.
Typical example of the formed structure. B = 1.5 kG. Modified from ***Figure 2****(c) of Ref. [53]).*

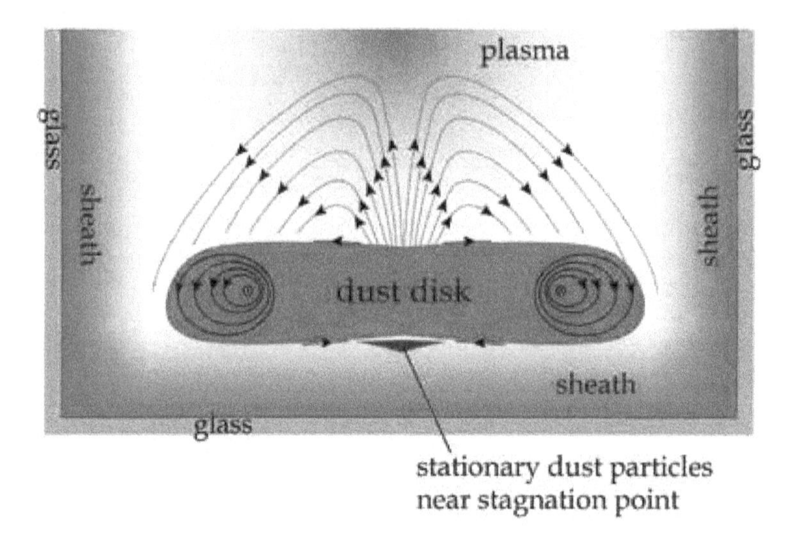

Figure 8.
Schematic of dust particle motion in the meridional plane. The arrows indicate the directions of dust particle motion.

When $B(h \gtrsim 137, z = 0) \leq 0.006$ T, the dust particles levitate a few mm above the glass bottom forming a thin disk of radius 20 mm with a dense group of particles at the rim of the disk near the outer wall. When B is increased, the stored dust particles near the wall moved inward to the center and formed a disk of uniformly distributed at $z > 0$. For $B(17, 0) = 0.12$ T with $p = 20$ Pa, where the electrons and ions are weakly magnetized and dust particles are rotating around the axis of the tube. The radial electric field is induced by the ambipolar diffusion and the vertically applied $\boldsymbol{B}$ field produces $\boldsymbol{E} \times \boldsymbol{B}$ drift motion of plasma particles, resulting in a solid body-like azimuthal motion of dust particles with angular velocity $10 - 20$ mm/s.

With a further increase of B and reached $B(12, 0) \approx 0.15$ T, the dust disk becomes the form as shown in **Figure** 7, i.e., the disk is thicker and its radius is smaller. A meridional plane reveals a spectacular movement of dust particles in the thick disk.

Typical trajectories of the spectacular particle motions are shown with arrows in **Figure 8**. There are two small poloidal rotations in the meridional plane near the edges within the thick disk, i.e., one is the clockwise rotation on the right-hand plane, and the other is the counterclockwise rotation on the left-hand plane. The particle motion in the disk as seen in a meridional plane is somewhat similar to the motion of tea leaves as shown in **Figure 4**.

The dust particles move upward against the gravity near $r = 0$. Especially, a part of the particles blows up and exceeds the disk thickness. Such ascending motion of dust particles is followed by radial movement toward the outer wall and then downward. The situation that the particles gush near the top looks like fireworks. After hitting close to the tube bottom, dust particles move inward along the tube bottom. At the same time, dust particles localized near the outer edge of the disk, which do not readily approach $r \approx 0$, form a local circulation in the meridional plane. While dust particles move in these closed circles in a meridional plane, the dust cloud rotates around the z axis and forms a toroidal rotation. As a result, dust particles form a helical motion around the z axis. A schematic illustration of the observed movement of dust particles is something similar as shown in **Figure 4**. In addition, there is a stagnation area around $r \approx 0$ near the tube bottom, where a group of dust particles is not involved in the dynamic meridional rotation as shown in **Figures 4** and 5.

3.3 MHD dust flow as a rotating fluid

The radial ambipolar diffusion is suppressed due to the electron magnetization, and the current starts to flow in an azimuthal direction in a magnetized plasma. The azimuthal electric field associated with the current density is given by

$$E_\theta = \frac{1}{|\omega_{ce}|\tau_{en}} \frac{\kappa T_e}{e} \frac{1}{n_e} \frac{\partial n_e}{\partial r} \tag{10}$$

In the present case, $E_\theta \approx 9\ \mathrm{V/m}$ because $|\omega_{ce}|\tau_{en} \approx 17$, $\kappa T_e/e \approx 3\ \mathrm{eV}$, and $(\partial n_e/n_e \partial r)^{-1} \approx 2 \times 10^{-2}$ m, where ω_{ce} is the electron cyclotron angular frequency, τ_{en} is the mean-free-time of the electron-neutral collision, and κ is the Boltzmann constant. This electric field will produce the azimuthal motion of ions with angular velocity $v_{\theta,\ ion} = eE_\theta/m_i\nu_{in} \approx 40$ m/s. Those ions circling around the tube axis will move dust particles resulting in a dust flow around the axis. Our observed maximum dust angular velocity is $v_{\theta,\,dust} \approx 0.02$ m/s, near the wall.

The rotating magnetohydrodynamics (MHD) fluid involving dust particles may be described by the Navier-Stokes equation with the continuity equation of an incompressible fluid of constant mass density:

$$\left(\frac{\partial}{\partial t} + \boldsymbol{v}\cdot\nabla\right)\boldsymbol{v} = -\frac{1}{m_d n_d}\nabla p + \nu\nabla^2\boldsymbol{v} + \frac{1}{m_d n_d}\boldsymbol{J}\times\boldsymbol{B} + \boldsymbol{f}, \tag{11}$$

$$\nabla\cdot\boldsymbol{v}=0, \tag{12}$$

where ν is a kinematic viscosity of the dust fluid and $\boldsymbol{f}\left(=\boldsymbol{f}_g + \boldsymbol{f}_d + \boldsymbol{f}_T + \cdots\right)$ is an external force. The gravitational force $\boldsymbol{f}_g$ is in the negative z direction, while the drag forces $\boldsymbol{f}_d$ by neutral particles, and ions are in the azimuthal direction. The thermophoretic force $\boldsymbol{f}_T$ and the other external forces are negligible in our experimental conditions [54].

We consider rotating fluid with constant angular frequency Ω far from the tube bottom $(z = 0)$. Eqs. (11) and (12) can be solved for steady, axisymmetric flow with $\boldsymbol{v} = (v_r, v_\theta, v_z)$ in cylindrical coordinates with boundary conditions $v_\theta = r\Omega$, $v_r = 0$ at $z = \infty$, and $v_r = v_\theta = v_z = 0$ at $z = 0$. We assume $\boldsymbol{J} = (0, J_\theta, 0)$ and $\boldsymbol{B} = (0, 0, B)$. The equilibrium condition requires

$$-r\Omega^2 = \frac{1}{m_d n_d}\left(\frac{\partial p}{\partial r} - J_\theta B\right), \tag{13}$$

indicating that the centrifugal force on a dust particle is balanced by the pressure gradient and the Lorentz force.

By introducing dimensionless parameters, $\overline{v}_r = v_r/r\Omega$, $\overline{v}_\theta = v_\theta/r\Omega$, $\overline{v}_z = v_z/\sqrt{\nu\Omega}$, $\overline{p} = p/\rho\nu\Omega$, and $\overline{g} = g/\sqrt{\nu\Omega^3}$ with $\overline{z} = z/\sqrt{\nu/\Omega}$. Eqs. (11) and (12) can be expressed by a set of three ordinary differential equations:

$$\begin{cases} \dfrac{d^2\overline{v}_r}{d\overline{z}^2} = 1 + \overline{v}_r^2 - \overline{v}_\theta^2 + \overline{v}_z \dfrac{d\overline{v}_r}{d\overline{z}} \\ \dfrac{d^2\overline{v}_\theta}{d\overline{z}^2} = 2\overline{v}_r\overline{v}_\theta + \dfrac{d\overline{v}_\theta}{d\overline{z}}\overline{v}_z \\ \dfrac{d\overline{v}_z}{d\overline{z}} = -2\overline{v}_r \end{cases}, \tag{14}$$

supplemented by the pressure gradient equation:

$$\frac{d\overline{p}}{d\overline{z}} = \frac{d^2\overline{v}_z}{d\overline{z}^2} - \overline{v}_z\frac{d\overline{v}_z}{d\overline{z}} - \overline{g}. \tag{15}$$

The set of equations is well studied as similarity solutions for the rotating fluid [55, 56]. The solution shows the presence of a stagnation point at $(r, z) = (0, 0)$ and the presence of a thin boundary layer near the bottom where the fluid moves inward. Our observation shows the boundary layer $3\sqrt{\nu/\Omega} \approx 5$ mm.

As Eqs. (11) and (12) show, dust particles drift in the azimuthal direction, and the centrifugal force on a particle is given by $f_C = m_d r\,(\alpha\Omega)^2$ with α, a constant less than unity. The centrifugal force is balanced by an inward drag force by neutral particles $f_{dn} = C_D \pi a^2 m_n n_n v_r^2/8$, where C_D is a drag coefficient, m_n is a neutral mass, n_n is a neutral density, and $v_r (= \beta r\Omega)$ is a representative radial velocity of dust particles with a constant $\beta < 1$. The balancing equation gives the equilibrium radius as

$$r = \frac{8}{3C_D}\frac{m_d n_d}{m_n n_n}\left(\frac{\alpha}{\beta}\right)^2 \frac{a}{2}. \tag{16}$$

Eq. (16) with $\alpha/\beta \approx 0.03$ gives an equilibrium radius of about 0.02 m, which agrees well with our experimental observation.

3.4 Storm in a glass tube

The mechanism of the meridional dust flow is understood in the following way. Initially dust particles are driven by the ion azimuthal motion caused by the radial plasma density gradient in the presence of a strong vertical magnetic field. While the MHD dust fluid forms a rotation around the tube axis, the angular velocity of dust particles near the tube bottom is reduced by the friction from the sheath plasma transition area. The friction reduces the centrifugal force. As a result, the pressure gradient force together with the Lorentz force which remains the same near the bottom generates a radial inward flow of dust particles. Because of the continuity, the radial inward motion will be compensated by an axial upward flow. Dust particles near the bottom ascend along the tube axis, where the sheath electric force pushes charged dust particles upward. When rising dust particles move outside of the sheath, the dust particles feel only the gravitational force. The ascending motion of dust particles near the axis is followed by the outward movement, and then the particles descend.

A circulation with an inward flow at the bottom has been known as a teacup phenomenon [57], also known as Einstein's tea leaves [58]. In 1926, Einstein explained that tea leaves gather in the center of the teacup when the tea is stirred as a result of a secondary, rim-to-center circulation caused by the fluid rubbing against the bottom of the cup. It is indeed observed in our complex plasma experiment that there were some levitated dust particles staying close to the bottom near the center.

4. Discussion: collective behavior of dust particles as a fluid

A complex plasma is a system consisting of electrons, ions, neutral gas particles, and dust particles. The dust particles are macroparticles of nanometers to micrometers in size. In our experiments, monodisperse dust particles of 3 or 5 or 5.6 μm in diameter were used. Behavior of dust particles can be regarded as MHD fluid if $l_{mfp} \ll L$, $\tau_p \gg \tau_d$, and the system keeps quasi-neutrality. Here l_{mfp} is the mean free path of dust particles, L is a representative scale length of a phenomenon, τ_p is time scale of the evolution of the phenomenon, and τ_d is the dust plasma period [59]. The quasi-neutrality is always kept. Typically, $l_{mfp} \sim 0.1$ mm, $L \sim 1$ cm, $\tau_p \sim 1$ s, and $\tau_d \sim 0.1$ s in our experiments. Hence, the dust cloud can be treated as an MHD fluid.

In water or air or other fluids, a tracer such as aluminum powder or smoke is often used for visualizing a motion of fluid elements. The tea leaves in a teacup are, of course, one of the examples of the tracer as well. This is an indirect observation of the motion because a different tracer has unique characteristics, e.g., a size or a specific weight. The uniqueness comes down to a variation in trackability of the tracer to the fluid element and affects the observation results. The various trackability may give a different result in a measurement. Schlieren imaging and shadowgraph are often used to visualize a flow, too. These methods observe a fluctuation of a density or a refractive index. The setting of the optical system, etc. requires high precision for these methods.

In contrast, in the dust fluid, it is possible to regard each dust particle as a fluid element itself. The particle can be visualized by illuminating using a visible laser light in experiments. The laser light suffers Mie scattering because the size of the dust particle ($\lesssim$10 μm) is usually larger than the wavelength of the visible laser light ($\sim$ several hundreds nm). The motion of the fluid element is directly visualized without being bothered about both the trackability and the optical precision. It is worth emphasizing that the visualization is achieved on the spot without time lag in experiment.

One of the applications of such a dust fluid is the new method to estimate the dust charge [60]. A dust particle has an electric charge Q in a plasma and levitates at a height where the electrostatic force due to the sheath electric field $\boldsymbol{E}$ and the gravitational force f_g are balancing, $Q\boldsymbol{E} = f_g$, on the ground. In experiments to measure the charge of an individual dust particle, it has been hard to separate Q and $\boldsymbol{E}$ independently. In addition, the conventional measurement methods require to change the experimental setup to measure Q and $\boldsymbol{E}$. However, by regarding the collection of dust particles as a fluid, it is possible to measure the resonant frequency of the dust fluid, i.e., the dust plasma frequency by externally applying the sinusoidal oscillation. The dust charge Q_A is calculated from the resonant frequency. The charge Q_A is an averaged charge for all dust particles present in the experimental region in this case.

In fluid dynamics, the Reynolds number is one of the important parameters. The Reynolds number is given by (inertial forces)/(viscous forces). The Reynolds number is also important in the dust flow. There are investigations relating to the widely changed Reynolds number or the viscosity of dust fluid by the simulation methods [61, 62]. However, it is hard to observe turbulence in our experiments on the dust flow, i.e., it is expected that the Reynolds number is rather small even when $M > 1$.

It is clear that collective behavior of dust particles can be described as a fluid globally. The fluid picture is held where the MHD conditions are satisfied. Intrinsically, however, the dust fluid is a group consisting of independent particles. Therefore, it is expected that the complex plasma includes unique features that is peculiar to a particle system, i.e., properties that are insufficient and difficult to be described by the MHD equation or the Navier–Stokes equation. Such a situation is possible where the MHD conditions do not hold locally. In fact, a few irregular particles are observed in quite rare case. For example, there is a dust particle whose orbit is irregular and different from the others in the way like the dust particle is reflected in a larger angle with faster speed by the obstacle in the bow shock experiment.

In addition, the following experiment may give another example. The schematic of the experimental device is shown in **Figure 9(a)** [63]. The dust cloud exists under an influence of an axisymmetric nonuniform magnetic field applied by a

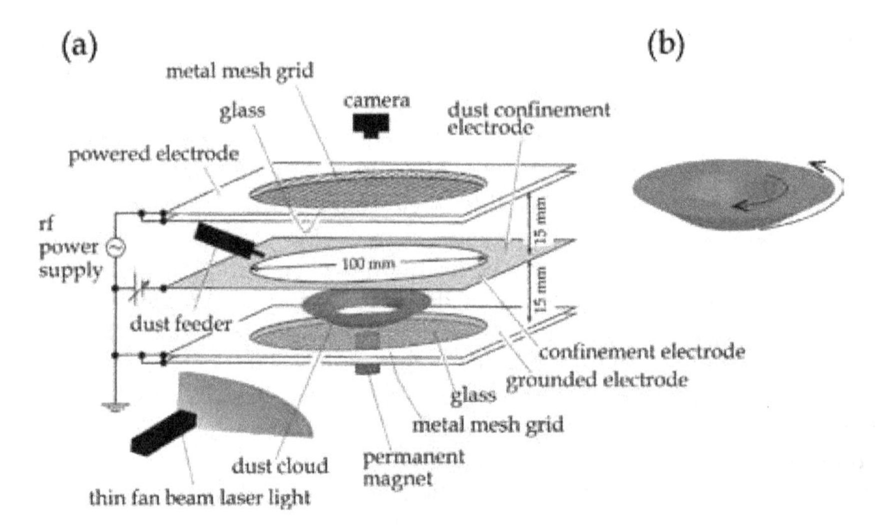

Figure 9.
Schematic of the experimental device (a) and the enhanced illustration of the dust cloud (b). The arrows in (b) are the directions that the particles at the edges rotate. The magnetic field is vertically upward.

small permanent magnet. The dust particles frequently collide with other particles surrounding it. Only the dust particles locate quite near the inner edge and the outer edge rotating along the edges as shown in **Figure 9(b)**. Excluding these edge regions, the collective motion of dust particles seems to be like a fluid. As described in the previous paragraph, the dust fluid has high viscosity. In that case, the rotating particles at the edges have to transfer their momenta to the neighboring particles and must drag the neighbors to their rotating directions. However, in the experiment, the particles do not drag their neighboring particles. It is considered that, at both edges, the dust particles behave as individual particles. Hence, the MHD conditions may be locally broken near the edges, and the particle motion there may suggest one of the particulate-like properties.

5. Summary and subsequent development

It is found in our experiments that the group of dust particles collectively behaves in a similar way to a fluid. In the fast flow of $M > 1$, the bow shock is formed in front of the obstacle. Under the strong magnetic field applied with the permanent magnet, the dust fluid shows a dynamic circulation.

In addition, there is the experimental result which may suggest a particle property of the group of dust particles. It is expected that the dusty plasma or the complex plasma bridges the different nature between continuum mechanics such as fluids and kinetics of particles.

Our results have inspired other researchers in wider fields beyond plasma physics [64–76]. The followings are examples. Tiwari et al. constructed two-dimensional generalized hydrodynamic model and discussed on turbulence in a strongly coupled plasma [64]. They reported that the turbulence was able to occur at a low Reynolds number if the Weissenberg number was high. Kähler et al. derived the ion susceptibility in a partially ionized plasma [65]. Tadsen et al. reported that the dust cloud confined in a magnetized plasma was diamagnetic [66]. Gibson et al. gave an improved understanding of magnetized electron behavior in a dipole magnetic field [67]. Laishram et al. investigated the dust vortex formation in a plasma [68].

Research on dusty plasmas has a strong influence to various areas of physics as seen above. Further progress will be expected.

Acknowledgements

We thank Prof. T. Kamimura and Prof. Y. Nakamura for their collaboration on the bow shock research. The work on the bow shock formation is supported by the Asian Office of Aerospace Research and Development under Grant No. AOARD 104158 and JSPS Grants-in-Aid for Scientific Research (A) under Grant No. 23244110. The work on the dynamic circulation is supported by the Asian Office of Aerospace Research and Development FA2386-12-1-4077 and JSPS Grants-in-Aid for Scientific Research (A) 23244110 and for Challenging Exploratory Research 24654188. One of the authors, Saitou, thanks Prof. Y. Hayashi for his support to the experiment using the small permanent magnet, and this work is partly supported by the Japan Society for the Promotion of Science (JSPS) Grant-in-Aid for Scientific Research (A) 24244094.

Author details

Yoshifumi Saitou[1*] and Osamu Ishihara[2]

1 School of Engineering, Utsunomiya University, Utsunomiya, Tochigi, Japan

2 Chubu University, Kasugai, Aichi, Japan

*Address all correspondence to: saitou@cc.utsunomiya-u.ac.jp

References

[1] Ikezi H. Coulomb solid of small particles in plasmas. Physics of Fluids. 1986;**29**:1764

[2] Hayashi Y, Tachibana K. Observation of coulomb-crystal formation from carbon particles grown. Japanese Journal of Applied Physics. 1994;**33**:L804

[3] Thomas H, Morfill GE, Demmel V, Goree J, Feuerbacher B, Möhlmann D. Plasma crystal: Coulomb crystallization in a dusty plasma. Physical Review Letters. 1994;**73**:652

[4] Chu JH, Lin I. Direct observation of Coulomb crystals and liquids in strongly coupled rf dusty plasmas. Physical Review Letters. 1994;**72**:4009

[5] Nosenko V, Goree J, Ma ZW, Piel A. Observation of shear-wave Mach cones in a 2D dusty-plasma crystal. Physical Review Letters. 2002;**88**:135001

[6] Samsonov D, Goree J, Thomas HM, Morfill GE. Mach cone shocks in a two-dimensional Yukawa solid using a complex plasma. Physical Review E. 2000;**61**:5557

[7] Havnes O, Li F, Melansø F, Aslaksen T, Hartquist TW, Morfill GE, et al. Diagnostic of dusty plasma conditions by the observation of Mach cones caused by dust acoustic waves. Journal of Vacuum Science and Technology A. 1996;**14**:525

[8] Melzer A, Nunomura S, Samsonov D, Ma ZW, Goree J. Laser-excited Mach cones in a dusty plasma crystal. Physical Review E. 2000;**62**:4162

[9] Gosh S. Shock wave in a two-dimensional dusty plasma crystal. Physics of Plasmas. 2009;**16**:103701

[10] Luo Q-Z, D'Angelo N, Merlino RL. Experimental study of shock formation in a dusty plasma. Physics of Plasmas. 1999;**6**:3455

[11] Nosenko V, Zhdanov S, Morfill GE. Supersonic dislocations observed in a plasma crystal. Physical Review Letters. 2007;**99**:025002

[12] Ness NF, Scearce CS, Seek JB. Initial results of the imp 1 magnetic field experiment. Journal of Geophysical Research. 1964;**69**:3531

[13] Morfill GE, Thomas HM, Konopka U, Rothermel H, Zuzic M, Ivlev A, et al. Condensed plasmas under microgravity. Physical Review Letters. 1999;**83**:1598

[14] Vladimirov SV, Tsytovich VN, Morfill GE. Stability of dust voids. Physics of Plasmas. 2005;**12**:052117

[15] Thomas E Jr, Avinash K, Merlino RL. Probe induced voids in a dusty plasma. Physics of Plasmas. 2004;**11**:1770

[16] Thompson CO, D'Angelo N, Merlino RL. The interaction of stationary and moving objects with dusty plasmas. Physics of Plasmas. 1999; **6**:1421

[17] Ishihara O. Complex plasma: Dusts in plasma. Journal of Physics D. 2007; **40**:R121

[18] Morfill GE, Ivlev A. Complex plasmas: An interdisciplinary research field. Reviews of Modern Physics. 2009; **81**:1353

[19] Kalman GJ, Hartmann P, Donkó Z, Rosenberg M. Two-dimensional Yukawa liquids: Correlation and dynamics. Physical Review Letters. 2004;**92**:065001

[20] Donkó Z, Kalman GJ, Hartmann P. Dynamical correlations and collective

excitations of Yukawa liquids. Journal of Physics, Condensed Matter. 2008;**20**: 413101

[21] Yeo LY, Friend JR, Arifin DR. Electric tempest in a teacup: The tea leaf analogy to microfluidic blood plasma separation. Applied Physics Letters. 2006;**89**:103516

[22] Arfin DR, Yeo LY, Fried JR. Microfluidic blood plasma separation via bulk electrohydrodynamic flows. Biomicrofluidics. 2007;**1**:014103

[23] Melzer A, Trottenberg T, Piel A. Experimental determination of the charge on dust particles forming Coulomb lattices. Physics Letters A. 1994;**191**:301

[24] Smith BA, Soderblom L, Batson R, Bridges P, Inge J, Masursky H, et al. A new look at the Saturn system: The Voyager 2 images. Science. 1982;**215**:504

[25] Ishihara O. Complex plasma research under extreme conditions. In: Mendonça JT, Resendes DP, Shukla PK, editors. Multifacets of Dusty Plasmas. AIP Conference Proceedings. Vol. 1041. Melville, NY: AIP; 2008. pp. 139-142

[26] Tsytovich VN, Morfill G, Vladimirov SV, Thomas H. Elementary Physics of Complex Plasmas. Heidelberg: Springer; 2008

[27] Piel A. Plasma Physics: An Introduction to Laboratory, Space, and Fusion Plasmas. Dordrecht: Springer; 2010

[28] Ishihara O. Low-dimensional structures in a complex cryogenic plasma. Plasma Physics and Controlled Fusion. 2012;**54**:124020

[29] Nebbat E, Annou R. On vortex dust structures in magnetized dusty plasmas. Physics of Plasmas. 2010;**17**:093702

[30] Tsytovich VN, Gusein-zade NG. Nonlinear screening of dust grains and structurization of dusty plasma. Plasma Physics Reports. 2013;**39**:515

[31] Kamimura T, Suga Y, Ishihara O. Configurations of Coulomb clusters in plasma. Physics of Plasmas. 2007;**14**: 123706

[32] Kamimura T, Ishihara O. Coulomb double helical structure. Physical Review E. 2012;**85**:016406

[33] Schwabe M, Rubin-Zuzic M, Zhdanov S, Ivlev AV, Thomas HM, Morfill GE. Formation of bubbles, blobs, and surface cusps in complex plasmas. Physical Review Letters. 2009;**102**: 255005

[34] Saitou Y, Ishihara O. Tempest in a glass tube: A helical vortex formation in a complex plasma. Journal of Plasma Physics. 2014;**80**:869

[35] Thomas E Jr, Merlino RL, Rosenberg M. Magnetized dusty plasmas: the next frontier for complex plasma research. Plasma Physics and Controlled Fusion. 2012;**54**:124034

[36] Konopka U, Samsonov D, Ivlev AV, Goree J, Steinberg V, Morfill GE. Rigid and differential plasma crystal rotation induced by magnetic fields. Physical Review E. 2000;**61**:1890

[37] Sato N, Uchida G, Kaneko T, Shimizu S, Iizuka S. Dynamics of fine particles in magnetized plasmas. Physics of Plasmas. 2001;**8**:1786

[38] Ishihara O, Kamimura T, Hirose KI, Sato N. Rotation of a two-dimensional Coulomb cluster in a magnetic field. Physical Review E. 2002;**66**:046406

[39] Ishihara O, Sato N. On the rotation of a dust particulate in an ion flow in a magnetic field. IEEE Transactions on Plasma Science. 2001;**29**:179

[40] Tsytovich VN, Sato N, Morfill GE. Note on the charging and spinning of

dust particles in complex plasmas in a strong magnetic field. New Journal of Physics. 2003;**5**:43

[41] Hutchinson H. Spin stability of asymmetrically charged plasma dust. New Journal of Physics. 2004;**6**:43

[42] Seal D. Jet Propulsion Laboratory [Internet]. 2019. Available from: https://web.archive.org/web/20090321071339/http://www2.jpl.nasa.gov/saturn/back. html

[43] Maxwell JC. On the stability of the motion of Saturn's rings: An essay, which obtained the Adams prize for the year 1856. London: University of Cambridge, Macmillan and Co.; 1859

[44] Spitzer Telescope [Internet]. 2019. Available from: https://www.nasa.gov/jpl/spitzer/bow-shock-wave-20140220

[45] Saitou Y, Nakamura Y, Kamimura T, Ishihara O. Bow shock formation in a complex plasma. Physical Review Letters. 2012;**108**:065004

[46] Nakamura Y, Ishihara O. A complex plasma device of large surface area. The Review of Scientific Instruments. 2008;**79**: 033504

[47] Epstein PS. On the Resistance experienced by spheres in their motion through gases. Physics Review. 1924; **23**:710

[48] Shukla PK, Mamun AA. Introduction to Dusty Plasma Physics. London: Institute of Physics Publishing Ltd.; 2002. Chap. 3, p. 70

[49] Nakamura Y, Ishihara O. Measurements of electric charge and screening length of microparticles in a plasma sheath. Physics of Plasmas. 2009;**16**:043704

[50] Landau LD, Lifshitz EM. Fluid Mechanics. Oxford: Butterworth-Heinemann; 2002. Chaps. 9, 12, and 13, pp. 313-360 and 435-483

[51] Newbury JA, Russel CT, Lindsay GM. Solar wind polytropic index in the vicinity of stream interactions. Geophysical Research Letters. 1997;**24**:1431

[52] Roussev II, Gombosi TI, Sokolov IV, Velli M, Manchester W IV, Dezeeuw DL, et al. A three-dimensional model of the solar wind incorporating solar magnetogram observations. The Astrophysical Journal. 2003;**595**:L57

[53] Saitou Y, Ishihara O. Dynamic circulation in a complex plasma. Physical Review Letters. 2013;**111**: 185003

[54] Ishihara O. Polygon structure of plasma crystals. Physics of Plasmas. 1998;**5**:357

[55] Greenspan HP. The Theory of Rotating Fluids. Cambridge, England: Cambridge University Press; 1968. Chap. 3, pp. 133-184

[56] Tritton DJ. Physical Fluid Dynamics, 2nd ed. Oxford: Clarendon Press; 2011. Chap. 16, pp. 215-242

[57] Maxworthy T. A Storm in a Teacup. Journal of Applied Mechanics. 1968;**35**: 836

[58] Einstein A. Ursache der Mäanderbildung der Flussläufe und des sogenannten Baerschen Gesetzes. Naturwissenschaften. 1926;**14**:223

[59] Landau LD, Lifshitz EM, Pitaevskii LP. Electrodynamics of Continuous Media. Oxford: Butterworth-Heinemann; 2000. Chap. 8, pp. 225-256

[60] Saitou Y. A simple method of dust charge estimation using an externally applied oscillating electric field. Physics of Plasmas. 2018;**25**:073701

[61] Saigo T, Hamaguchi S. Shear viscosity of strongly coupled Yukawa

systems. Physics of Plasmas. 2002;**9**: 1210

[62] Laishram M, Zhu P. Structural transition of vortices to nonlinear regimes in a dusty plasma. Physics of Plasmas. 2018;**25**:103701

[63] Saitou Y. Motions of dust particles in a complex plasma with an axisymmetric nonuniform magnetic field. Physics of Plasmas. 2016;**23**: 013709

[64] Tiwari SK, Dharodi VS, Das A, Patel BG. Turbulence in strongly coupled dusty plasmas using generalized hydrodynamic description. Physics of Plasmas. 2015;**22**:023710

[65] Kähler H, Joost JP, Ludwig P, Bonitz M. Streaming complex plasmas: Ion susceptibility for a partially ionized plasma in parallel electric and magnetic fields. Contributions to Plasma Physics. 2016;**56**:204

[66] Tadsen B, Greiner F, Piel A. Probing a dusty magnetized plasma with self-excited dust-density waves. Physical Review E. 2018;**97**:033203

[67] Gibson J, Coppins M. Theory of electron density in a collisionless plasma in the vicinity of a magnetic dipole. Physics of Plasmas. 2018;**25**:112103

[68] Laishram M, Sharma D, Kaw PK. Analytic structure of a drag-driven confined dust vortex flow in plasma. Physical Review E. 2015;**91**:063110

[69] Kaur M, Bose S, Chattopadhyay PK, Sharma D, Ghosh J. Observation of dust torus with poloidal rotation in direct current glow discharge plasma. Physics of Plasmas. 2015;**22**:033703

[70] Deka T, Bouruah A, Sharma SK, Bailing H. Observation of self-excited dust acoustic wave in dusty plasma with nanometer size dust grains. Physics of Plasmas. 2017;**24**:093706

[71] Meyer JK, Merlino R. Transient bow shock around a cylinder in a supersonic dusty plasma. Physics of Plasmas. 2013; **20**:074501

[72] Jaiswal S, Bandyopadhyay P, Sen A. Experimental observation of precursor solitons in a flowing complex plasma. Physical Review E. 2016;**93**:041201(R)

[73] Ludwig P, Miloch WJ, Kählert H, Bonitz M. On the wake structure in streaming complex plasmas. New Journal of Physics. 2012;**14**:053016

[74] Wilms J, Reichstein T, Piel A. Experimental observation of crystalline particle flows in toroidal dust clouds. Physics of Plasmas. 2015;**22**:063701

[75] Vladimirov SV, Ishihara O. Electromagnetic wave band structure due to surface plasmon resonances in a complex plasma. Physical Review E. 2016;**94**:013202

[76] Sakakibara N, Matsubayashi Y, Iyo T, Terashima K. Formation of pseudo-microgravity environment for dusty plasmas in supercritical carbon dioxide. Physics of Plasmas. 2018;**25**:010704

10

Turbulence Generation in Inhomogeneous Magnetized Plasma Pertaining to Damping Effects on Wave Propagation

Ravinder Goyal and R.P. Sharma

Abstract

The damping phenomenon is studied due to the collisions of ions and neutral particles and Landau approach on the turbulent spectra of kinetic Alfvén wave (KAW) in magnetized plasma which is inhomogeneous as well. The localization of waves is largely affected by inhomogeneities in plasma which are taken in transverse as well as parallel directions to the ambient magnetic field. There is significant effect of damping on the wave localization and turbulent spectra. Numerical solutions of the equations governing kinetic Alfvén waves in the linear regime give the importance of wave damping phenomena while retaining the effects of Landau (collisionless) damping and ion-neutral collisional damping. A comparative study of the two damping effects reveals that the Landau damping effect is more profound under similar plasma conditions.

Keywords: kinetic Alfvén wave, inhomogeneous plasma, ion-neutral collisional damping, Landau damping, laboratory plasma, turbulence

1. Introduction

As far as enormous space plasma phenomena like solar wind turbulence, acceleration of solar wind, heating of solar coronal loops, solar flares, etc. are concerned, kinetic Alfvén wave (KAW) has a vital role to play [1]. The dispersion characteristics of KAW make it distinguishable from its parent Alfvén wave [2]. When Alfvén wave propagation develops a large value for wave number perpendicular ($k_\perp$) to the ambient magnetic field, then it gives rise to kinetic Alfvén wave (KAW). This mode conversion may lead to transportation of large amount of electromagnetic energy across geomagnetic field lines [3]. In Tokamak/ITER plasma where KAW plays an important role, the anomalous diffusion coefficient which depends upon the turbulence level is inversely related to the confinement time. In the scenario where the electric fields are nonstationary, this wave plays a vital role in the phenomenon of particle acceleration [4]. The acceleration and heating [5] of plasma particles are dependent on perpendicular wave number and is attributed to KAW carrying finite perturbations parallel to electric field. Also, the plasma heating is caused by dissipation of turbulence and is essential to interpret the observations of most astronomical [6] and laboratory [7] systems. The turbulent behavior of KAW can be thought of as the probable mechanism to describe the magnetic fluctuations which are observed near earth's magnetopause [8]. The theoretical and experimental studies [9, 10] support the importance of KAW in solar wind turbulence energy cascade and particle heating. Earth's magnetopause is also known to have a vital role played by KAW in the mechanism of stochastic ion heating [11]. Nykyri et al. [12] observed the turbulent spectra in high altitude cusp and revealed that break in observed spectra might be due to damping of obliquely propagating KAW.

In general, both space as well as laboratory plasmas incorporate various kinds of inhomogeneities, few of these being magnetic field [13], temperature, and density perturbations. The coronal heating has been theoretically studied by Davila [14] by assuming Alfven absorption when Alfven velocity is dependent on background plasma density in transverse direction to ambient magnetic field. Also, the magnetic field fluctuations due to shear Alfvén waves have been observed experimentally [15]. The Alfven wave energy changes along radial direction have also been studied by Vincena et al. [16]. The present work is inspired from the experimental studies of formation of varying magnetic fields and electric fields by propagation of KAW in inhomogeneous plasmas [17].

A comparative study of experimental results with theoretical ones [18] fully support the launching of KAW in inhomogeneous plasma. There may be qualitative and quantitative deviation of ideal behavior [19] of KAW from that when inhomogeneities or nonlinearities are introduced due to kinetic theory. Therefore, it is always better to have understanding of magnetic fluctuations with an introduction of Landau damping, which is considered to be an important phenomenon as far as KAW propagation is concerned [20, 21]. Based on Landau fluid model, several theories [5, 22–24] have been proposed regarding the propagation of magnetohydrodynamic waves in collisionless plasma and Alfven filamentation process due to density channels. Taking into consideration the experimental observations by Houshmandyar and Scime [17], Sharma et al. [25] have studied the effect of Landau damping on KAW propagation in inhomogeneous magnetized plasma and have concluded that there is considerable damping effect on wave propagation due to phenomenologically incorporated Landau damping factor.

The damping refers to gradual decrease in wave intensity and wave amplitude. Not only collisionless damping (like Landau damping) but collisional damping also plays an important role in wave energy decay phenomenon and hence marks its role in Tokamak/ITER fusion plasma processing. The electron-neutral and ion-neutral damping is dominant in scrape-off region of these fusion plasma devices where energy transportation of wave takes place, and hence, neutral particles are the one among the important factors playing a vital role [26] in drawing the turbulence level. Also, the wave dispersion characteristics in space plasmas may get affected by the inclusion of neutral particles in the wave dynamics. Some of these characteristics include particle heating, spicules formation [27], solar wind turbulence and acceleration [28], and chromospheric energy balance [29]. As a consequence of the charged particles of plasma colliding with the neutral ones, the plasma waves may get absorbed by interstellar clouds, photosphere, and chromosphere [30–33]. As far as effect of neutral-dominated collisions on Alfvén waves is concerned, several research groups have worked in this direction and have given their fruitful theories in their respective domains. Based on the theory of chromospheric heating by Alfvén waves, the ion-neutral collisional damping has been studied analytically by De Pontieu et al. [29] in partially ionized chromosphere. Alfvénic turbulence has been considered as the base by Krishan and Gangadhara [34] to get intercoupling mechanism among three plasma species which are neutrals, ions, and electrons. The mass loading of ions by neutral particles has been used by Houshmandyar and

Scime [17] to experimentally study the collisional impact due to collisions between ion and neutral particles on the KAW propagation in plasma. Recently, Goyal and Sharma [35] have studied the effect of ion-neutral collisional damping on KAW turbulence.

The present research deals with a comparative study of the two damping phenomena discussed above viz. ion-neutral collisional damping and Landau (collisionless) damping on the propagation of KAW which have been studied separately by Sharma et al. [25] and Goyal and Sharma [35] in the light of inhomogeneous magnetized plasma. This comparison is very much important with respect to scrape-off regions of the fusion energy devices like Tokamak and ITER. The plasma inhomogeneity is taken into account by considering spatial (in x-z plane) inhomogeneities having scale lengths both in longitudinal and transverse directions to background magnetic field. Distinct model equations for KAW have been taken by considering damping factors corresponding to collisional and collisionless damping, and the results have been drawn, respectively, using two fluid simulation technique.

2. Equations governing the propagation of KAW

The governing equations of KAW come into their present form by taking into account the propagation of the wave in two dimensions (x-z surface), which imitates the propagation vector to be $\vec{k}_o = k_{ox}\hat{x} + k_{oz}\hat{z}$ where $k_{Ox}(k_{Oz})$ symbolizes the component of the wave vector in perpendicular (parallel) direction to ambient magnetic field $\vec{B}_0$ along z direction. Also, the plasma having background density n_0 is considered to be inhomogeneous in x as well as in z directions and hence is supposed to follow the density profile n_0 exp $\left(-x^2/L_x^2 - z^2/L_z^2\right)$, where L_x implies the inhomogeneity scale length perpendicular to ambient magnetic field $\vec{B}_0$ while L_z implies the same in parallel direction. Initially, the wave field profile is considered to follow hollow Gaussian behavior [36]. As the current numerical study is inspired from the experimental technique [17] where the plasma environment was developed experimentally to study the wave propagation, we have used the same laboratory parameters in our model which are as follows: $B_0 \approx 560$ G, plasma density $n_0 \approx 6 \times 10^{12}$ cm^{-3}, neutral particle density $n_n \approx 5 \times 10^{12}$ cm^{-3}, ion density $n_i = n_n/2 \approx 2.5 \times 10^{12}$ cm^{-3}, T_e = 81,200 K, $T_n \approx T_i$ = 2900 K. Using these experimental values, the other characteristic plasma parameters may be obtained like: $v_A \approx 2.5 \times 10^7$ cm/s, $c_s = 1.3 \times 10^6$ cm/s, $\beta \left(= c_s^2/v_A^2\right) \approx 0.002$, $v_{te} \approx 1.1 \times 10^8$ cm/s, $v_{ti} \approx 2.445 \times 10^5$ cm/s, $\omega_{ci} = 2.68 \times 10^6$ rad/s, λ_e = 0.1 cm, $\rho_i \approx 0.09$ cm, $\rho_s \approx 0.49$ cm, $k_{0x} \approx 0.85$ cm^{-1}, and $k_{0z} \approx 0.16$ cm^{-1} for $\omega/\omega_{ci} = f/f_{ci} = 0.9$.

Using the well-established methods [26–28], Maxwell's equations and Faraday's law, the two dynamical equations for finite amplitude KAW incorporating the effects of ion-neutral collisional damping [35] and Landau damping [25] can, respectively, be written as follows:

$$\frac{\partial^2 B_y}{\partial t^2} + 2\gamma_n \frac{\partial B_y}{\partial t} = \lambda_e^2 \frac{\partial^4 B_y}{\partial x^2 \partial t^2} - \left(v_{te}^2 \lambda_e^2 + v_A^2 \rho_i^2\right) \frac{\partial^4 B_y}{\partial x^2 \partial z^2} + v_A^2 \exp\left(x^2/L_x^2 + z^2/L_z^2\right) \frac{\partial^2 B_y}{\partial z^2} + \frac{v_A^2}{\omega_{ci}^2}\left(1 + \frac{2(n_n v_{ti}/n_i v_{tn})}{1 + \omega^2/\nu_{ni}^2}\right) \frac{\partial^4 B_y}{\partial z^2 \partial t^2} \tag{1}$$

and

$$\frac{\partial^2 B_y}{\partial t^2} + 2\gamma_L \frac{\partial B_y}{\partial t} = \left(\frac{\partial^2}{\partial t^2} + 2\gamma_L \frac{\partial}{\partial t}\right)\lambda_e^2 \frac{\partial^2 B_y}{\partial x^2} - v_A^2 \rho_s^2 \frac{\partial^4 B_y}{\partial z^2 \partial x^2} + \frac{v_A^2}{\omega_{ci}^2}\left(\omega_{ci}^2 + \frac{\partial^2}{\partial t^2}\right) \exp\left[\frac{x^2}{L_x^2} + \frac{z^2}{L_z^2}\right] \frac{\partial^2 B_y}{\partial z^2} \quad (2)$$

where γ_n is damping factor representing ion-neutral collisions and γ_L is phenomenologically incorporated Landau damping factor obtained using the dispersion relation used by Lysak and Lotko [20] and Hasegawa and Chen [21]. The perturbations being very small in comparison to other terms in the above equations; their spatial derivatives may not be considered. Now, in order to study the damping mechanism, wave localization phenomenon is very much important which is done by using an envelope solution [36] into Eqs. (1) and (2) to obtain a set of normalised equations [25, 35] which look mathematically similar to nonlinear Schrödinger (NLS) equation. So, on solving these equations individually by following the standard algorithm which has been developed in order to solve NLS equation, one can infer about the wave localization, energy dissipation, and turbulent cascading of wave energy. Post examining the accuracy for NLS invariants up to 10^{-5}, these normalized equations are accommodated keeping in view the modifications being done in order to solve NLS equation. By considering hollow Gaussian profile (initial simulation condition) before the start of simulation process (at $t = 0$), these equations are solved numerically. The hollow Gaussian profile at $t = 0$ is given by [35]:

$$B(x,z,0) = a_0\left(x^2/r_{10}^2 + z^2/r_{20}^2\right) \exp\left(-x^2/r_{10}^2 - z^2/r_{20}^2\right) \quad (3)$$

Here, a_0 and r_{10}, respectively, dictate the amplitude of KAW and the scale size of wave field in perpendicular direction to the ambient magnetic field B_0 before start of the simulation and r_{20} is the same in parallel direction. $r_{10}(r_{20})$ may be termed as the system length in 2-D geometry. These system lengths while spreading in 2-D geometry expand even beyond the stature of hollow Gaussian profile in transverse as well as longitudinal directions to B_0. Also, as $k_x(k_z)$ is the normalized propagation constant in transverse (longitudinal) direction to B_0, one can infer $r_{10} = k_x^{-1}$ and $r_{20} = k_z^{-1}$. The regulation of KAW evolution is done by initially taking the system constants to be $a_0 = 0.02$ and $k_x = k_z = 0.2$ where a_0 is computed by the use of experimental parameters [17] and the wave amplitude is normalized by ambient magnetic field B_0. Sharma and Singh [37] have established a distinct algorithm in order to numerically solve Eqs. (1) and (2). This algorithm is framed in such a way that it relies upon a pseudo spectral mechanism for spatial integrals having recurring lengths $l_x = r_{10}$, $l_z = r_{20}$, and a grid having squared dimensions (grid points) given by $2^6 \times 2^6$. And, the temporal evolution is studied by the use of the predictor corrector mechanism besides finite difference method where the evolution advances with a step size $\Delta t \approx 10^{-4}$.

3. Outcomes and results

The damping effects on the wave propagation are plainly detectable by having a view of **Figure 1** where the time-dependent variations in normalized magnetic field intensity for specific values of parallel and transverse inhomogeneity scales are shown. The figure depicts the formation of localized structures or normalized

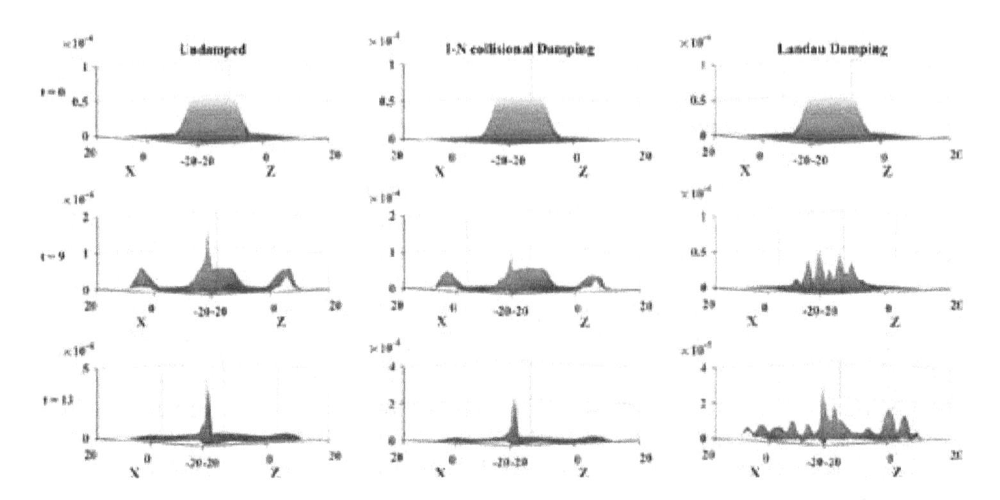

Figure 1.
KAW having variations in intensity profile of magnetic field (normalized) with the direction of ambient magnetic field (z) and transverse direction (x) at t = 0, t = 9, and t = 13 with and without damping. At initial time, the presumed magnetic field profile, i.e., hollow Gaussian is followed. But as there is progress in time, the respective damping effects come into play which is depicted from the decrease in amplitude of the wave and hence the intensity. At times greater than t = 0, there is an enhancement in the peak intensity of respective curves for undamped, ion-neutral collision-dominated damping, and Landau damping cases, but on comparison, one can easily state that the Landau damping peaks are the lowest for a particular time. There is formation of chaotic structures in case of Landau damping as is clear from smaller intensity peaks. Hence, one can infer that Landau damping is most profound on KAW propagation in inhomogeneous magnetized plasma.

magnetic field intensity peaks due to density inhomogeneity in plasma. The wave gets focused and defocused due to these inhomogeneities. At a fixed time, the normalized magnetic field intensity has spatial spreads (in x-z plane), in almost similar manner in undamped as well as damped (ion-neutral collisions) cases but its peak values differ for different cases viz. undamped, ion-neutral collisions, and Landau damping. As expected, the field intensity has lower peak values for damped cases (ion-neutral collisions and Landau damping). A comparative picture of ion-neutral collisional damping and Landau damping reveals that the Landau damping effect seems to be more dominating due to smaller magnetic field intensity peaks but it can also be seen that the spatial spread is also more in Landau damping case. We can make an inference that due to Landau damping effect, the spatial energy spread in smaller peaks is more pronounced than that in ion-neutral damping case where the intensity peaks resemble almost the ones in undamped case. Also, the damping phenomena get more pronounced as we scale up the time domain.

The damping effect may also be observed on the temporal variation of magnetic field in **Figure 2** for a specific spatial location (x, z). It can be clearly observed from **Figure 2** that at the initial times, all the fluctuations are of high amplitude (magnetic field) but with the passage of time, the damping effects come into play as fluctuations die down for both ion-neutral collisional as well as Landau damping effect. This behavior is indicated from **Figure 1** as well. Comparatively, Landau damping effect is more pronounced than the collisional effect. The Fourier transform of undamped as well as damped magnetic field fluctuations as given in **Figure 2** results into the magnetic power spectra of the respective cases, which are indicated as the $|B_\omega|^2$ varying with frequency ω in **Figure 3**. These spectra have been obtained using a magnetic field fluctuations' time window ~10 μs and indicate a clear difference due to damping effects for some specified values of inhomogeneity scale lengths. Once again, it is depicted from **Figure 3** that the Landau damping effect is more pronounced than ion-neutral collisional effect on the propagation of KAW which is indicative from the initial spectral structures.

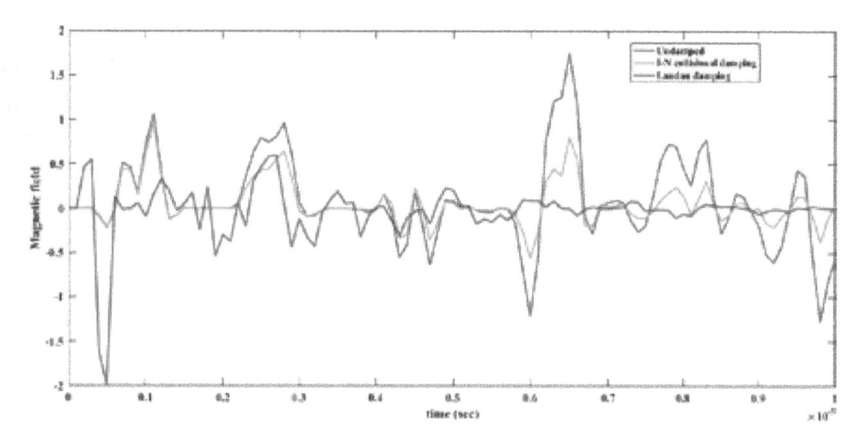

Figure 2.
The time scales of wave magnetic field fluctuations showing the effects of ion-neutral collisional damping and Landau damping at a particular location (x, z). As the time advances, the Landau damping is more dominant over ion-neutral collisional damping, which is indicated from the dying down of fluctuations in case of Landau damping.

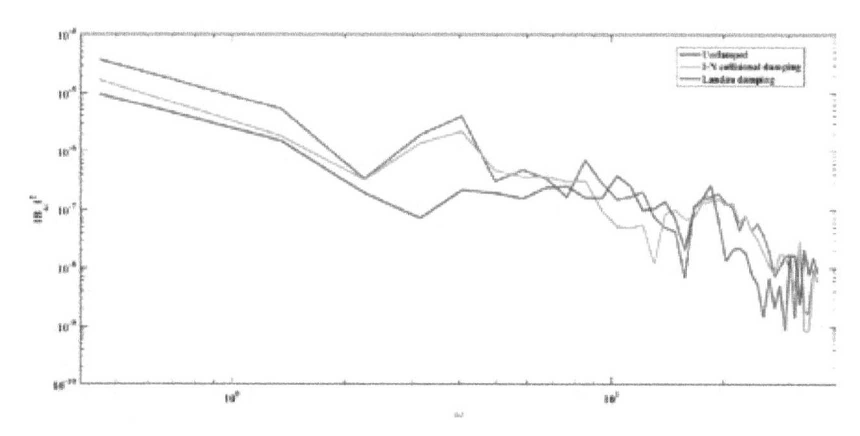

Figure 3.
Taking the Fourier transform of the data obtained in ***Figure 2*** *depicts the power spectra of the respective data for undamped, ion-neutral collisional-dominated and Landau-dominated wave propagation. The power spectra are shown to be varying* $|B_w|^2$ *versus* ω *and can clearly indicate the dominance of Landau damping shown as least on energy scale.*

Also, it is clearly observed from **Figure 3** that on inclusion of damping effects, the spectral behavior (almost) retains its trend in the initial frequency regimes which indicates the linear propagation of KAW. But at the later stage (higher frequencies), the turbulent effects come into play and KAW propagation ceases to follow linear trend (undamped oscillations). Or this behavior might also be due to some unidentified nonlinearity or due to drawback of the numerical model used. The alteration in wave localization occurs due to the involvement of damping effects as is observed in **Figure 1**. The rates at which the spectral components are associated with spatially localized filamentary structures might emerge with time at enunciated rates. So, at a specific spatial location, the wave magnetic field is supposed to have fluctuation behavior with time, as is indicated in **Figure 2**.

In the present study, we deal only with the comparison of Landau damping and collisional damping on kinetic Alfvén waves in which plasma is considered to be inhomogeneous and magnetized. The purpose of the study is to look for the effect of these damping effects on wave propagation. In case of ion-neutral collisions, the inclusion of dust particles could have further affect the damping phenomena. As the size of the dust particles is much larger than the neutral atoms, their collisional cross section would be much larger and hence the collisional damping would have been enhanced. Also, the frequency of collisions would be increased by incorporating dust particles into the plasma, their negative charges may also put adverse effect on the wave propagation and hence, damping phenomenon. In this way, the dust particles' role is unavoidable when we talk about collisional effects on plasma, but in the present study, we have concentrated only on ion-neutral collisions. The effect of dust particles will be studied in our future work.

4. Conclusion

A comparative study about relative profoundness of damping effects has been done when KAW propagates in two-dimensional inhomogeneous magnetized plasma. The study is based on the numerical simulation technique that has been employed to solve the dynamical equations using two-fluid approach. These equations serve as the numerical representations of KAW propagation when ion-neutral collisional damping and Landau damping effects are separately incorporated into the KAW dynamics. In collisionless plasma, the prominent dissipation phenomenon, i.e., Landau damping is due to electrons unlike neutral particle-dominated plasma where the dissipation is primarily by ion-neutral collisions. Likewise, the neutral atoms colliding with other plasma species may cause dissipation of interplanetary magnetic field. Also, in scrape-off region of Tokamak/ITER, the major cause of particle energization is due to ion-neutral collisions because of presence of neutral atoms in plasma. But it is longitudinal component of electric field that causes particle heating in collisionless plasma because KAW follows this field due to its large transverse wavenumber and hence small wavelength. Due to dependency of Landau damping on wavenumber, it affects the localization amplitude, and hence, the localization of KAW happens due to wave packets which have varying wave numbers. The formation of these wave packets is attributed to the inclusion of inhomogeneities in KAW dynamics. It is clear from the figures that there is appreciable effect of damping phenomena on magnetic field fluctuations as well as on intensity profiles and the energy spread over a wide frequency range can be interpreted from the turbulence analysis. The collisional dynamics which considers ions and neutral particles to be at same temperatures has clear effect on wave dissipation but it is the Landau damping which shows relatively weighty effect as indicated by the wave amplitudes and the resultant frequency spectra. Furthermore, there is no nonlinearity included in KAW dynamics. Some interesting features of the dynamics are expected to be observed when temperature difference between ion and neutral particle dynamics are considered and/or when the nonlinear dynamics are included.

Acknowledgements

This work is partially supported by DST (India) and ISRO (India) under RESPOND program.

Author details

Ravinder Goyal* and R.P. Sharma
Centre for Energy Studies, Indian Institute of Technology, Delhi, India

*Address all correspondence to: ravig.iitd@gmail.com

References

[1] Chaston CC, Phan TD, Bonnell JW, Mozer FS, Auna M, Goldstein ML, et al. Drift-Kinetic Alfvén waves observed near a reconnection X line in the Earth's magnetopause. Physical Review Letters. 2005;**95**:065002

[2] Hasegawa A, Mima K. Exact solitary Alfvén wave. Physical Review Letters. 1976;**37**:690

[3] Chaston CC, Wilber M, Mozer FS, Fujimoto M, Goldstein ML, Acuna M, et al. Mode conversion and anomalous transport in Kelvin-Helmholtz vortices and kinetic Alfvén waves at the Earth's magnetopause. Physical Review Letters. 2007;**99**:175004

[4] Goertz CK. Kinetic Alfvén waves on auroral field lines. Planetary and Space Science. 1984;**32**:1387

[5] Hasegawa A, Chen L. Kinetic process of plasma heating due to Alfvén wave excitation. Physical Review Letters. 1975;**35**:370

[6] Salem CS, Howes GG, Sundkvist D, Bale SD, Chaston CC, Chen CHK, et al. Identification of kinetic Alfvén wave turbulence in the solar wind. Astrophysical Journal Letters. 2012;**745**: L9

[7] Carter TA, Maggs JE. Modifications of turbulence and turbulent transport associated with a bias-induced confinement transition in the large plasma device. Physics of Plasmas. 2009;**16**:012304

[8] Chaston C, Bonnell J, McFadden JP, Carlson CW, Cully C, Le CO, et al. Turbulent heating and cross-field transport near the magnetopause from THEMIS. Geophysical Research Letters. 2008;**35**:L17S08

[9] Sahraoui F, Goldstein ML, Robert P, Khotyaintsev YV. Evidence of a cascade and dissipation of solar-wind turbulence at the electron gyroscale. Physical Review Letters. 2009;**102**:231102

[10] Howes GG, Dorland W, Cowley SC, Hammett GW, Quataert E, Schekochihin AA, et al. Kinetic simulations of magnetized turbulence in astrophysical plasmas. Physical Review Letters. 2008;**100**:065004

[11] Johnson JR, Cheng CZ. Stochastic ion heating at the magnetopause due to kinetic Alfvén waves. Geophysical Research Letters. 2001;**28**:4421

[12] Nykyri K, Grison B, Cargill PJ, Lavraud B, Lucek E, Dandouras I, et al. Origin of the turbulent spectra in the high-altitude cusp: Cluster spacecraft observations. Annales de Geophysique. 2006;**24**:1057

[13] Chen HH, Liu CS. Solitons in nonuniform media. Physical Review Letters. 1976;**37**:693

[14] Davila JM. Heating of the solar corona by the resonant absorption of Alfven waves. The Astrophysical Journal. 1987;**317**:514

[15] Gekelman W, Vincena S, Leneman D, Maggs J. Laboratory experiments on shear Alfvén waves and their relationship to space plasmas. Journal of Geophysical Research. 1997; **102**:7225

[16] Vincena S, Gekelman W, Maggs J. Shear Alfvén wave perpendicular propagation from the kinetic to the inertial regime. Physical Review Letters. 2004;**93**:105003

[17] Houshmandyar S, Scime E. Ducted kinetic Alfvén waves in plasma with steep density gradients. Physics of Plasmas. 2011;**18**:112111

[18] Scime EE, Keiter PA, Balkey MM, Kline JL, Sun X, Keesee AM, et al. The

hot helicon experiment: HELIX. Journal of Plasma Physics. 2015;**28**:125

[19] Chen L, Zonca F. Gyrokinetic theory of parametric decays of kinetic Alfvén waves. Europhysics Letters. 2011;**96**:35001

[20] Lysak RL, Lotko W. On the kinetic dispersion relation for shear Alfvén waves. Journal of Geophysical Research. 1996;**101**:5085

[21] Hasegawa A, Chen L. Kinetic processes in plasma heating by resonant mode conversion of Alfvén wave. Physics of Fluids. 1976;**19**:1924

[22] Borgogno D, Hellinger P, Passot T, Sulem PL, Trávníček PM. Alfvén wave filamentation and dispersive phase mixing in a high-density channel: Landau fluid and hybrid simulations. Nonlinear Processes in Geophysics. 2009;**16**:275

[23] Passot T, Sulem PL. Long-Alfvén-wave trains in collisionless plasmas. II. A Landau-fluid approach. Physics of Plasmas. 2003;**10**:3906

[24] Passot T, Sulem PL. Filamentation instability of long Alfvén waves in warm collisionless plasmas. Physics of Plasmas. 2003;**10**:3914

[25] Sharma RP, Goyal R, Gaur N, Scime EE. Linear kinetic Alfvén waves in inhomogeneous plasma: Effects of Landau damping. Europhysics Letters. 2016;**113**:25001

[26] Houshmandyar S, Scime E. Enhanced neutral depletion in a static helium helicon discharge. Plasma Sources Science and Technology. 2012; **21**:035008

[27] Haerendel G. Weakly damped Alfvén waves as drivers of solar chromospheric spicules. Nature (London). 1992;**360**:241

[28] Allen LA, Habbal SR, Qiu HY. Thermal coupling of protons and neutral hydrogen in the fast solar wind. Journal of Geophysical Research. 1998;**103**:6551

[29] De Pontieu B, Martens PCH, Hudson S. Electromagnetic field equations for a moving medium with Hall conductivity. The Astrophysical Journal. 2001;**558**:859

[30] Piddington JH. Electromagnetic field equations for a moving medium with Hall conductivity. MNRAS. 1954; **114**:638

[31] Piddington JH. The motion of ionized gas in combined magnetic, electric and mechanical fields of force. MNRAS. 1954;**114**:651

[32] Lehnert B. Plasma physics on cosmical and laboratory scale. Nuovo Cimento. 1959;**13**(Suppl):59

[33] Osterbrock DE. The heating of the solar chromosphere, plages, and corona by magnetohydrodynamic waves. The Astrophysical Journal. 1961;**134**:347

[34] Krishan V, Gangadhara RT. Mean-field dynamo in partially ionized plasmas – I. Monthly Notices of the Royal Astronomical Society. 2008;**385**: 849

[35] Goyal R, Sharma RP. Effect of ion-neutral collisions on the evolution of kinetic Alfvén waves in plasmas. Plasma Physics and Controlled Fusion. 2018;**60**: 035002

[36] Goyal R, Sharma RP, Scime EE. Time dependent evolution of linear kinetic Alfvén waves in inhomogeneous plasma. Physics of Plasmas. 2015;**22**: 022101

[37] Sharma RP, Singh HD. Density cavities associated with inertial Alfvén waves in the auroral plasma. Journal of Geophysical Research. 2009;**114**: A03109

11

Microgravity Experiments Using Parabolic Flights for Dusty Plasmas

Kazuo Takahashi

Abstract

The chapter is dedicated to descriptions of the microgravity experiments, which were done by using parabolic flights and for analyzing behaviors of dust particles in plasmas, i.e., dusty plasmas under microgravity. There are projects of the microgravity experiments of the dusty plasmas using the International Space Stations, where time for microgravity is long and has a scale of hours. Conversely, it is significant to find out phenomena of the dusty plasmas in the short time scale of a few 10 s including the transition from gravity to microgravity, which is performed by parabolic flights of aircrafts on the earth. Methodology and results of the experiments are shown here for further investigations of the dusty plasmas in future.

Keywords: dusty plasma, coulomb crystal, microgravity experiment, parabolic flight

1. Introduction

Dust particles in plasmas have been investigated under microgravity for this quarter century, since they attract interest of physicists working on the dusty plasmas which are plasmas containing them and the Coulomb crystals formed by them resulting from charging and strongly coupled interactions [1–4]. Experiments of the dusty plasmas under microgravity were initiated by a Russian team accessing to a space station of the Mir in the late 1990s. The team was a pioneer for the experiments and started to collaborate with a German team [5]. After collaboration between the teams, the experiments under microgravity came to be done by other facilities, drop towers in Germany and Japan [6], sounding rockets used by a German team and the International Space Station (ISS) intensively promoted by the Russian-German joint project of the PKE-Nefedov and PK-3 plus [7, 8]. The project on the ISS of PK-4 has been in progress since 2014 by an international team organized by the European Space Agency [9–11]. In this history, microgravity experiments by using aircrafts with parabolic flights have played an important role in feasibility studies for missions on the ISS. Opportunities of the parabolic flights for the dusty plasmas have been mainly provided in campaigns from the ESA and the Japan Aerospace Exploration Agency (JAXA). A Japanese team was engaging in the experiments with apparatuses similar to PK-3 plus and PK-4 to contribute preliminary results of these projects. Here the results obtained by the team are introduced as knowledge for experiments in the future. It costs a lot and takes so long time to prepare and perform the microgravity experiments by the aircrafts. Pieces of information in the chapter help successors to reduce money and time and get much more scientific merit in future microgravity experiments.

2. Microgravity experiments by parabolic flights

2.1 Background

The Japanese team experienced the missions of PK-3 plus, where the team aimed at demonstrating a critical phenomenon in dusty plasmas [12–14]. The PK-3 plus had a configuration of parallel-plate electrodes for rf (13.56 MHz) discharges. The team did microgravity experiments by aircrafts using the chamber the same as the PK-3 plus in Japan. This opportunity allowed the team to proceed to the next parabolic flight campaign for the next project of PK-4. The team built the apparatus of PK-4J whose glass tube for discharges was scaled down from the original one to load a rack system on the aircrafts for parabolic flights in Japan. The PK-4J had two phases of the microgravity experiments in Japan. The Phase 1 was dedicated to sophisticate the apparatus and acquire wide experience for the experiments using the parabolic flights. In the other, Phase 2, the behaviors of the dust particles were observed in cylindrical discharges [15], which had been preliminarily studied in numerical simulations [16, 17]. In parallel with the microgravity experiments, measuring parameters such as ion density and electron temperature in plasmas had been developed to reach comprehensive analyses of phenomena observed under microgravity [18–21].

2.2 Parabolic flights

In Japan, the parabolic flights are carried out by a company, Diamond Air Service, Inc. It has two aircrafts (**Figure 1**). One is smaller than the other and equipped with two racks. The larger one has four racks. Everything for the experiments is integrated into the rack before loading on the aircraft. The whole rack is inspected that all components inside are tightly fixed and checked before flights not to electromagnetically interfere with the aircrafts during operation. The aircraft makes a condition free from gravity with a flight pattern of parabola in an airspace above the Sea of Japan or the Pacific Ocean. It takes 1 h to obtain a set of parabolas not more than 15 times a day. **Figure 2(a)** shows a graph of an altitude as a function of time corresponding to the flight pattern of the aircraft. Angles of elevation and

Figure 1.
Aircrafts for parabolic flights operated by DAS. There are two types of the aircrafts. One is MU-300 from Mitsubishi heavy industries loading two racks (left), and the other is G-II from Grumman aircraft engineering Corp. For four racks (right).

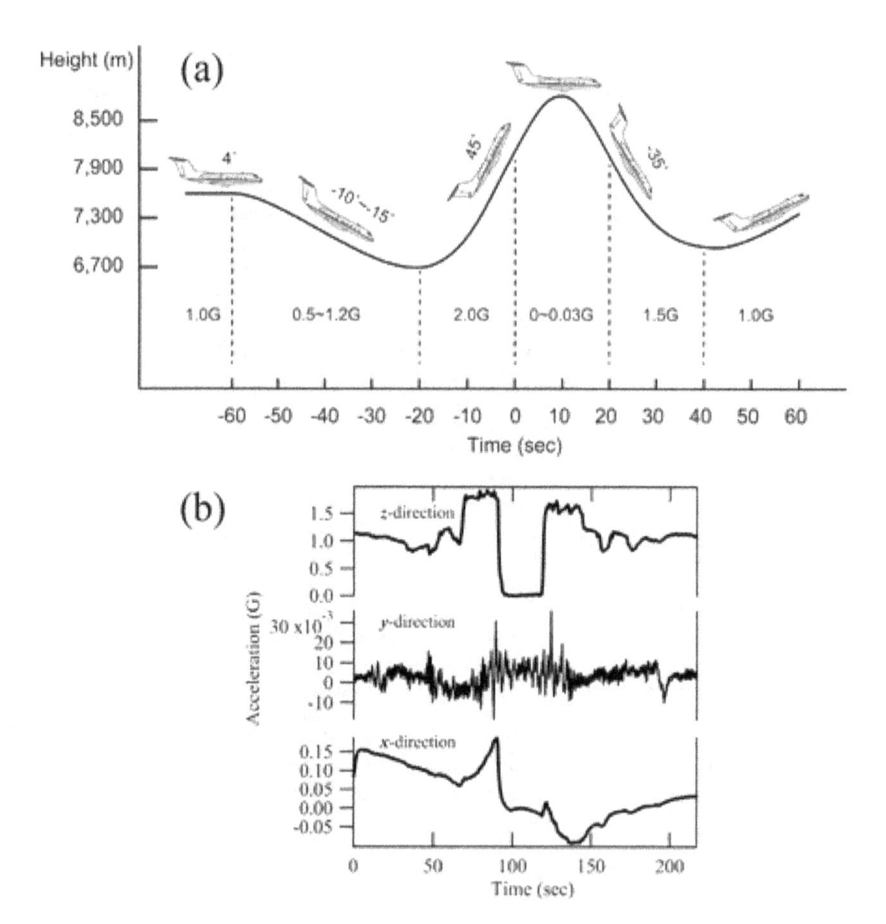

Figure 2.
(a) A flight pattern of the aircraft to obtain a condition under microgravity shown by an altitude of the aircraft as a function of time. The figure shows an angle of elevation and the acceleration parallel to gravity in a unit of G which can be felt by everything onboard. Here the 1 G is equivalent to the acceleration of gravity on the earth, i.e., 9.8 m/s². One can feel free from gravity during around 20 s in a parabola. (b) The time variations of accelerations onboard the aircraft measured in a parabolic flight. The x, y, and z are corresponding to the directions parallel to the traveling direction of the aircraft, to its wing, and to the gravity, respectively.

accelerations to be feel onboard are also indicated on the graph. The 1 G is corresponding to the acceleration of gravity on the earth, i.e., 9.8 m/s^2. To make a parabola, the aircraft rapidly accelerates and gains speed while getting the nose down between −60 and −20 s. After that, it elevates steeply while getting the nose up between −20 and 0 s and reaches the release point of an elevation of 45°, where a pilot closes the throttle. The aircraft continues flying by its inertia and makes the parabola between 0 and 20 s, where gravity is well balanced with the inertia. This is why everything onboard does not feel gravity. One can typically have a time of 20 s under microgravity during the aircraft moving up and dropping down in around 1000 m. After the parabola, the aircraft recovers speed and its altitude. Before and after the parabola, a load force normally acts on the aircraft by the acceleration less than 2 G.

Figure 2(b) is a graph of accelerations onboard the aircraft measured in a parabolic flight. The accelerations were detected by the directions of x, y, and z parallel to the traveling direction of the aircraft, to its wing, and to the gravity, respectively. The acceleration in the z-direction, i.e., parallel to gravity, dropped down to 0 G (μG) after an entry to parabola with 1.8 G and kept a constant stable for around 20 s. In the y-direction along to the wing, the acceleration was detected to be less than 4×10^{-2} G and had no effect on the experiments. However, in the x-direction parallel to the traveling direction, the acceleration was increased when the throttle was closed at the entry. It caused a problem in a feasibility study of Phase 1 as described later.

2.3 Methodology and design of apparatus

PK-4 was supposed to observe dust particles in a cylindrical discharge generated inside a glass tube. The Japanese team had built the apparatus of PK-4J based on dimensions and functions of PK-4 since 2011 (**Figure 3(a)**). The glass tube was an π-shape chamber consisting of a main tube with an inner diameter of 30 mm and branch tubes welded onto the main. The glass tube was fixed on an optical bench as the branch tubes being perpendicular to the floor on the aircraft. A pair of electrodes was placed in the branch tubes. Plasmas were generated by a rectangular pulse voltage applied between the electrodes. The voltage was varying between 650 and 750 V peak-to-peak at 1 kHz and supplied to the electrodes out of phase. This meant that polarity between the electrodes alternated at the frequency which did not allow the dust particles to follow an electric field by the voltage. Plasma frequencies of electron and ion are much higher than the frequency of the voltage, so that they are running between the electrodes back and forth. A current around 2.0 mA was observed in the discharge. The Ar gas was fed from a port on either of the branch tubes at a flow rate of four SCCM (SCCM denotes cubic centimeter per minute at the standard conditions) and pumped down from another port closed to the feeding port. Therefore the gas flow was negligible in the main tube like a stagnant pool of the gas. It was aimed at trapping the dust particles inside the main tube without perturbation by the gas flow. The dust particles of melamine-formaldehyde were injected by a dust dispenser mounted on the branch tube. Their diameter was 2.55 μm. The amount of the dust particles was regulated by a pulse voltage driving the dispenser.

The first apparatus was used for a feasibility study of Phase 1 and tested in microgravity experiments of parabolic flights between 2011 and 2012. The size of the main tube, whose length was 550 mm, was identical to the original one of PK-4

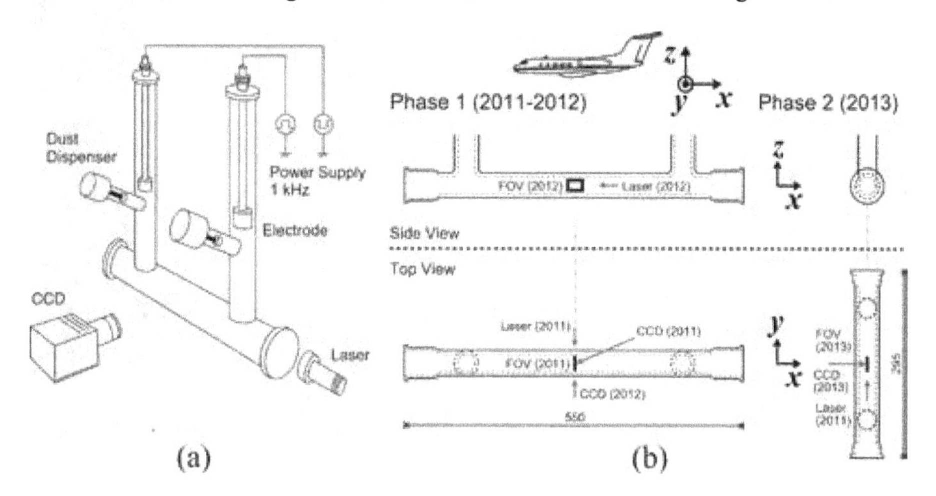

Figure 3.
Schematic of the apparatus of PK-4J. (a) It was developed to be loaded on a rack system of the aircraft operated by Diamond Air Service, Inc. the π-shape tube enabled to observe the dust particles in a cylindrical discharge. (b) Design of the glass tube was progressed from phase 1 to phase 2. In phase 1, the main tube had been placed parallel to the traveling direction. However, it was rotated to be perpendicular to the direction to avoid the effect of inertia by acceleration along the traveling direction of the aircraft just before entry to a parabola. In order to keep the main tube inside the rack, its length was reduced. The field of view (FOV) of the CCD camera was a cross section perpendicular to the axis of the main tube in phase 1 and changed to a plane along to the axis in phase 2.

(**Figure 3(b)**). The main tube was set along to the traveling direction of the aircraft (x-direction). The dust particles were illuminated on the cross section perpendicular to the axis of the main tube by a laser of 532 nm in wavelength, which corresponded to a field of view (FOV) observed diagonally by a charge-coupled device (CCD) camera.

The first campaign of parabolic flights in Phase 1 made it clear that the dust particles were levitated around a plasma-sheath boundary near the bottom of the main tube under gravity and moved up to the center of the tube under microgravity (**Figure 4(a)**). The cloud of the dust particles of μG seemed to elongate along the axis of the tube as a shape of a cylinder since a disk was observed on the FOV. In the next campaign, the FOV was set on the plane parallel to the axis. **Figure 4(b)** shows that the shape of the cloud was a cylinder elongating along the axis and spatial distribution of the dust clouds moved up from around the plasma-sheath boundary at 7 mm to the axial center. Under microgravity, the dust particles seem to form a bundle of thread. Furthermore, it found that the dust particles were forced to be moved by acceleration along the traveling direction. Before the entry of a parabola, they went toward the back of the aircraft. They came to flow to the front of the aircraft after the entry. This was the problem to be solved in order to precisely observe arrangements of the dust particles as mentioned above. Therefore, it was decided that the main tube was placed crossing with the traveling direction of the aircraft in Phase 2 (**Figure 3(b)**). The length of the main tube was reduced to be 420 mm to put it in a frame of the rack on the aircraft.

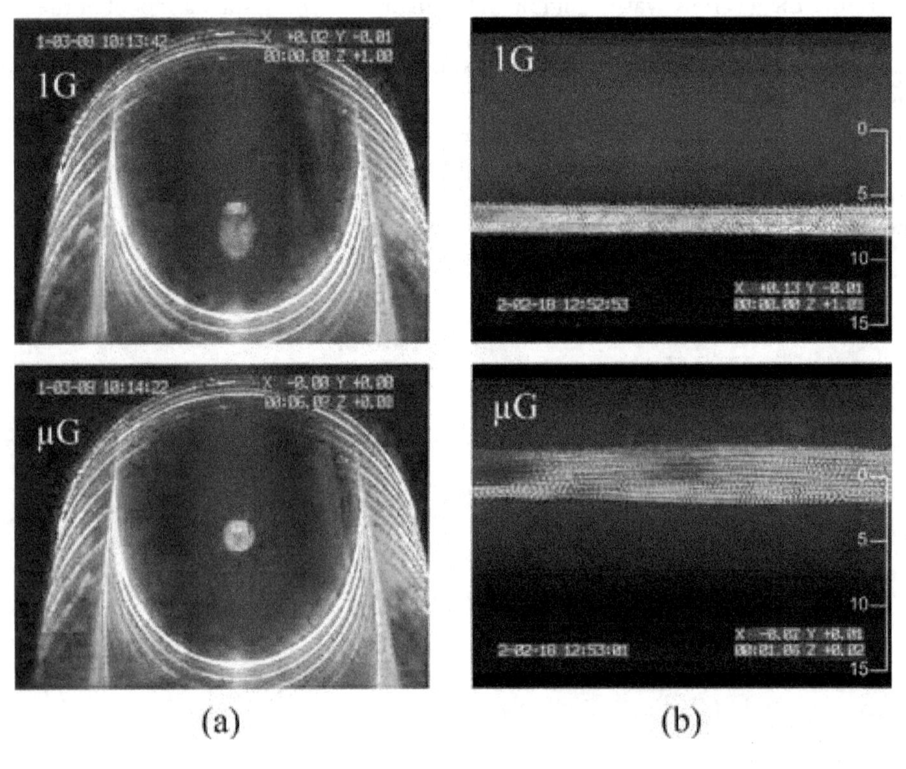

Figure 4.
Images of the clouds of the dust particles obtained in microgravity experiments of phase 1. (a) The cloud was observed on a cross section perpendicular to the axis of the main tube. The cloud stayed around the plasma-sheath boundary near the bottom of the tube under gravity (1G). Under microgravity (μG), it moved up and made a disk. (b) The scale bar shows distance from the axis. The 15 mm corresponds to the inner wall of the glass tube. The dust particles were levitated around the plasma-sheath boundary near the bottom of the tube (at 7 mm from the axis) (1G). The cloud elongated along the axis. The dust particles were distributed around the axis under microgravity (μG).

In Phase 2, a laser of 660 nm in wavelength was used to illuminate the dust particles. Its light was fine-shaped as the thickness (FWHM of intensity) of 50 μm by optics to make slice images of the cloud. The laser and CCD camera were mounted on a translation stage. The stage moved to make a scan in the direction along the traveling direction (x-axis). The CCD of a resolution of 480 × 640 pixels accumulated images at 200 fps while scanning at the speed of 6.5 mm/s. The field of view was 4.3 × 5.8 mm^2.

3. Dusty plasmas under microgravity

Figure 5 shows spatial distributions of the dust particles observed in Phase 2, where the peak-to-peak voltage and gas pressure were set at 700 V and 33 Pa, respectively [15]. The axes of x, y, and z correspond to the direction in traveling, that of the wing, and that perpendicular to the floor of the aircraft, respectively. Two cylinders of the clouds for the cases under gravity (1G) and microgravity (μG) are shown the **Figure 5(a)**. Both of them elongate along the y-axis and seem to be tapered. In the main tube, striations appeared in the discharge. The cloud became fat in swollen parts of brighter glow. In a level flight, i.e., under gravity, the main body of the cloud was placed below the axis, and the dust particles were distributed in $-4.7 \leq z < -2.0$ mm. The cloud of 1G consists of a shell-like structure of three layers clearly appearing in the bottom part as indicated in other experimental and theoretical studies [17, 22, 23]. The outermost shell along the arrow in **Figure 5(a)** was unfolded on a plane of the cylinder surface (**Figure 5(b)**). Coordinates of the dust particles are plotted on the plane as parameters of a circumference of the cylinder (L) and y. The dust particles arranged to make a close packed structure in 2-dimension by a Coulomb repulsive force between them. This arrangement formed the layered structure stacking in the radial direction. Another work proved that the

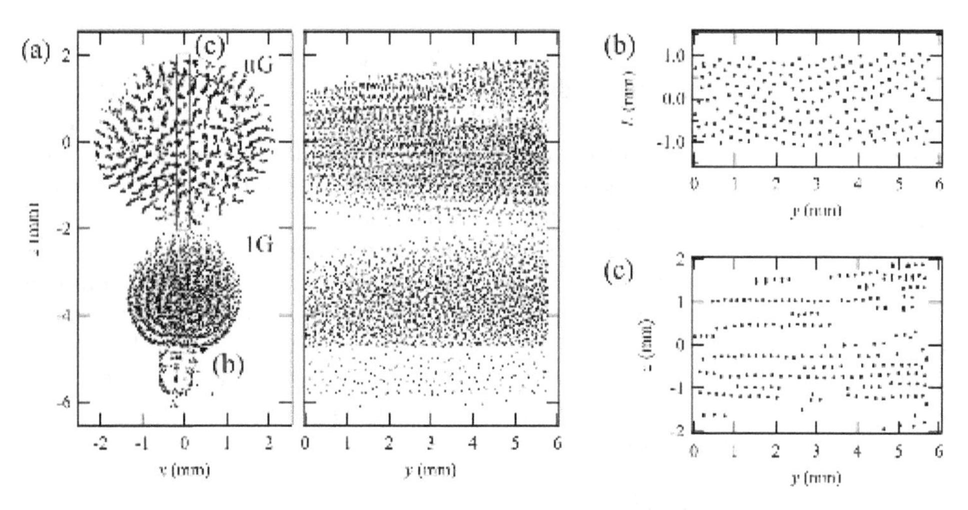

Figure 5.
Spatial distributions of the dust particles observed in phase 2 (the graph was reproduced by the data from the previous paper [15]). Each dot corresponds to each coordinate of a dust particle. (a) The clouds were shown for the conditions under gravity (1G) and microgravity (μG). The x, y, and z-axes correspond to the directions in traveling, of the wing and perpendicular to the floor of the aircraft, respectively. The left graph is a cross-sectional view shown as projection along the y-axis ($0 \leq y < 6$ mm). The right one is a projection along to the x-axis ($-2 \leq x \leq 2$ mm). (b) An axis of L is defined as a circumference of the outermost shell along an arrow in (a). An arrangement of the dust particles is shown on a L-y plane. (c) The coordinates of the dust particles trimmed by the rectangle shown in the figure (a) are plotted on a y-z plane as a projection along to the x-direction.

structure was the face-centered orthorhombic (FCO) lattice [24]. The FCO structure tends to appear rather than isotropic structures such as body-centered and face-centered cubics under the conditions that a stress works in one direction [25–27]. Here the stress means gravity.

Under microgravity (μG), the cloud moved up and the dust particles were distributed around the axis. The cylinder of the cloud got thicker than that under gravity. Its axis was exactly identical to that of the main tube. **Figure 5(c)** shows coordinates of the dust particles in a region trimmed by a rectangle (indicated by **Figure 5(a)**) as a projection along the x-axis. The FCO structure was never found under microgravity. The dust particles formed an assembly of linear chains elongating along the y-axis, i.e., the main tube [28, 29]. The electric field, whose direction is alternatively switched at 1 kHz between the electrodes, makes an ion stream along its direction. The ion stream causes the wake potential around the dust particles which makes them interact by an attractive force in addition to the Coulomb repulsive force [30–32]. Linear chains of the dust particles had normally been observed near the electrodes in rf discharges which accelerate ions in sheath [33–35]. In the discharge of the main tube, the ion stream has two components. One is from the ions going back and forth between electrodes, and the other is from those flowing toward walls by diffusion. At the axis, the ions going through the discharge are much more than those flowing toward the walls. Therefore, the dust particles form the chains around the axis by the wake potential, when they move up from the bottom to the center of the discharge under microgravity. Regarding the wake potential, its characteristics were made clear in several experiments under gravity [34, 36, 37]. Further, microscopic dynamics for causing the wake potential, e.g., visualization of the wake potential, will be expected to be analyzed in experiments under microgravity in addition to a calculation with a classical manner [38].

4. Concluding remarks

An apparatus of dusty plasmas was developed for observing dust particles of cylindrical discharges in a glass tube under microgravity. It was built step by step while testing its functions and observing the dust particles on board an aircraft. In order to analyze the arrangements of the dust particles, positioning the glass tube and field of views were considered in experiments of parabolic flights. It was significant for building the apparatus to suppress an effect of acceleration in the traveling direction of the aircraft caused around entry of a parabola.

In the experiments, coordinates of the dust particles were recorded in conditions under gravity as well as microgravity. They were located near the plasma-sheath boundary below the axis of the glass tube and found to form staking layers in a bottom part of clouds under gravity. The layers were not an isotropic three-dimensional structure such as body-centered or face-centered cubic. The FCO lattice appeared in the cloud, which seemed to be deformed by a stress in one direction originated in gravity. Switching the condition from gravity to microgravity, at first, location of the dust particles was changed around the axis. The dust particles distributing around the axis drastically changed their arrangement from the FCO structure to an assembly of linear chains. The chains were possibly formed by an attractive force from a wake potential. The wake potential was promised to be caused by streaming of the ion which was going back and forth between electrodes and driven by electric field alternative at 1 kHz. The dust particles switched dominant interaction potential from Coulomb repulsive under gravity to wake under microgravity. This was unexpected in simulations removing a term of gravity and an example that the microgravity condition possibly revealed a phenomenon hidden under the influence of gravity.

In microgravity experiments by parabolic flights, it is meaningful to have transition state from gravity to microgravity as well as to have microgravity only. Furthermore acceleration parallel to gravity controlled in operation of the aircraft by pilots is fascinating in precisely analyzing responses of physical phenomena to gravity. This is something that cannot be done on the ISS. The mechanism in formation of the linear chains was clearly understood in observation of the dust particles moving in the transition state where the acceleration was gradually changed. There are likely advantages to use the parabolic flights in diverse fields other than microgravity science.

The behavior of the dust particles, which are visible in an invisible ensemble of plasmas consisting of electrons and ions, clearly reflects phenomena of physics in the plasmas and makes them easily understood. Indeed a phenomenon of invisible wave might be visualized and comprehended by the dust particles. There are so much dust in the universe, pollutants in the atmosphere, cosmic dust in the interstellar, the regolith on the moon, etc. They attract much interest for investigating the origin of the universe from the point of view of natural science. Particulate matter is widely used and produced in industries. The dust particles charged in the plasmas are available for seeking ways of application and new ideas in technology. The microgravity experiments of dusty plasmas are promising to open new ideas in future science and technology. The know-how introduced in the chapter will be hopefully useful for the future.

Acknowledgements

The author would like to thank Dr. Satoshi Adachi of JAXA for the technical support in experiments. Microgravity experiments were performed as an activity of the scientific working group supported by JAXA with Dr. Hiroo Totsuji (Professor Emeritus of Okayama University), Dr. Yasuaki Hayashi (Professor Emeritus of Kyoto Institute of Technology), and former students of the author's laboratory, Manami Tonouchi and Tomo Ide. The author obtained a chance to join experimental projects on the ISS when he belonged to a scientific team at Max-Planck-Institut für extraterrestrishe Physik between 2005 and 2006 in Germany. He appreciates broad-mindedness of the then supervisor of the team, Professor Gregor E. Morfill, who supported his activities in Japan.

Author details

Kazuo Takahashi
Faculty of Electrical Engineering and Electronics, Kyoto Institute of Technology, Kyoto, Japan

*Address all correspondence to: takahash@kit.jp

References

[1] Hayashi Y, Tachibana K. Observation of Coulomb-crystal formation from carbon particles grown in a methane plasma. Japanese Journal of Applied Physics. 1994;**33**:L804-L806

[2] Thomas H, Morfill GE, Demmel V. Plasma crystal: Coulomb crystallization in a dusty plasma. Physical Review Letters. 1994;**73**:652-655

[3] Chu JH, I L. Direct observation of Coulomb crystals and liquids in strongly coupled rf dusty plasmas. Physical Review Letters. 1994;**72**:4009-4012

[4] Melzer A, Trottenberg T, Piel A. Experimental determination of the charge on dust particles forming Coulomb lattices. Physics Letters A. 1994;**191**:301-308

[5] Morfill GE, Thomas HM, Konopka U, Rothermel H, Zuzic M, Ivlev A, et al. Condensed plasmas under microgravity. Physical Review Letters. 1999;**83**: 1598-1601

[6] Hayashi Y. Radial ordering of fine particles in plasma under microgravity condition. Japanese Journal of Applied Physics. 2005;**44**:1436-1440

[7] Nefedov AP, Morfill GE, Fortov VE, Thomas HM, Rothermel H, Hagl T, et al. PKE-Nefedov: Plasma crystal experiments on the international Space Station. New Journal of Physics. 2003;**5**:33

[8] Thomas HM, Morfill GE, Fortov VE, Ivlev AV, Molotkov VI, Lipaev AM, et al. Complex plasma laboratory PK-3 plus on the International Space Station. New Journal of Physics. 2008;**10**:033036

[9] Usachev A, Zobnin A, Petrov O, Fortov V, Thoma M, Kretschmer M, et al. The project "Plasmakristall-4" (PK-4) is a dusty plasma experiment in a combined dc/rf(i) discharge plasma under microgravity conditions. Czechoslovak Journal of Physics. 2004; **54**:C639-C647

[10] Fortov V, Morfill G, Petrov O, Thoma M, Usachev A, Hoefner H, et al. The project 'Plasmakristall-4' (PK-4)—A new stage in investigations of dusty plasmas under microgravity conditions: First results and future plans. Plasma Physics and Controlled Fusion. 2005;**47**: B537-B549

[11] Pustylnik MY, Fink MA, Nosenko V, Antonova T, Hagl T, Thomas HM, et al. Plasmakristall-4: New complex (dusty) plasma laboratory on board the International Space Station. Review of Scientific Instruments. 2016;**87**:093505

[12] Totsuji H. Thermodynamics of strongly coupled repulsive Yukawa particles in ambient neutralizing plasma: Thermodynamic instability and the possibility of observation in fine particle plasmas. Physics of Plasmas. 2008;**15**: 072111

[13] Totsuji H. Thermodynamic instability and the critical point of fine particle (dusty) plasmas: Enhancement of density fluctuations and experimental conditions for observation. Journal of Physics A: Mathematical and Theoretical. 2009;**42**:214022

[14] Totsuji H. Possible observation of critical phenomena in fine particle (dusty, complex) plasmas. Microgravity Science and Technology. 2011;**23**: 159-167

[15] Takahashi K, Tonouchi M, Adachi S, Totsuji H. Study of cylindrical dusty plasmas in PK-4J; experiments. International Journal of Microgravity Science and Application. 2014;**31**:62-65

[16] Totsuji H, Totsuji C. Structures of Yukawa and Coulomb particles in cylinders: Simulations for fine particle

plasmas and colloidal suspensions. Physical Review E. 2011;**84**:015401(R)

[17] Totsuji H, Totsuji C, Takahashi K, Adachi S. Study of cylindrical dusty plasmas in PK-4J; Theory and simulations. International Journal of Microgravity Science and Application. 2014;**31**:55-61

[18] Takahashi K, Hayashi Y, Adachi S. Measurement of electron density in complex plasmas of the PK-3 plus apparatus on the international Space Station. Journal of Applied Physics. 2011;**110**:013307

[19] Takahashi K, Thomas HM, Molotkov VI, Morfill GE, Adachi S. Estimation of plasma parameters in dusty plasmas for microgravity experiments. International Journal of Microgravity Science and Application. 2015;**32**:320409

[20] Takahashi K, Adachi S, Totsuji H. Measurement of ion density and electron temperature by double-probe method to study critical phenomena in dusty plasmas. JAXA Research and Development Report. 2015;**14-012E**:7-11

[21] Takahashi K, Lin J, Hénault M, Boufendi L. Measurements of ion density and electron temperature around voids in dusty plasmas. IEEE Transaction on Plasma Science. 2018;**46**: 704-708

[22] Mitic S, Klumov BA, Konopka U, Thoma MH, Morfill GE. Structural properties of complex plasmas in a homogeneous dc discharge. Physical Review Letters. 2008;**101**:125002

[23] Totsuji H. Behavior of dust particles in cylindrical discharges: Structure formation, mixture and void, effect of gravity. Journal of Plasma Physics. 2014; **80**:843-848

[24] Takahashi K, Totsuji H. Structure of Coulomb crystals in cylindrical discharge plasmas under gravity and microgravity. IEEE Transaction on Plasma Sciences. 2019;**47**:4213-4218

[25] Fujii Y, Hase K, Hamaya N, Ohishi Y, Onodera A. Pressure-induced face-centered-cubic phase of monatomic metallic iodine. Physical Review Letters. 1987;**58**:796-799

[26] Stoeva SI, Prasad BLV, Uma S, Stoimenov PK, Zaikovski V, Sorensen CM, et al. Face-centered cubic and hexagonal closed-packed nanocrystal superlattices of gold nanoparticles prepared by different methods. Journal of Physical Chemistry B. 2003;**107**:7441-7448

[27] Hayashi Y. Structure of a three-dimensional Coulomb crystal in a fine-particle plasma. Physical Review Letters. 1999;**83**:4764-4767

[28] Ivlev AV, Thoma MH, Räth C, Joyce G, Morfill GE. Complex plasmas in external fields: The role of non-Hamiltonian interactions. Physical Review Letters. 2011;**106**:155001

[29] Dietz C, Kretschmer M, Steinmüller B, Thoma MH. Recent microgravity experiments with complex direct current plasmas. Contributions to Plasma Physics. 2018;**58**:21-29

[30] Nambu M, Vladimirov SV, Shukla PK. Attractive forces between charged particulates in plasmas. Physical Letters A. 1995;**203**:40-42

[31] Vladimirov SV, Nambu M. Attraction of charged particulates in plasmas with finite flows. Physical Review E. 1995;**52**:R2172-R2174

[32] Ishihara O, Vladimirov SV. Wake potential of a dust grain in a plasma with ion flow. Physics of Plasmas. 1997;**4**: 69-74

[33] Takahashi K, Oishi T, Shimomai K, Hayashi Y, Nishino S. Simple hexagonal

Coulomb crystal near a deformed plasma sheath boundary in a dusty plasma. Japanese Journal of Applied Physics. 1998;**37**:6609-6614

[34] Takahashi K, Oishi T, Shimomai K, Hayashi Y, Nishino S. Analyses of attractive forces between particles in Coulomb crystal of dusty plasmas by optical manipulations. Physical Review E. 1998;**58**:7805-7811

[35] Ivlev AV, Morfill GE, Thomas HM, Räth C, Joyce G, Huber P, et al. First observation of electrorheological plasmas. Physical Review Letters. 2008; **100**:095003

[36] Melzer A, Schweigert VA, Piel A. Transition from attractive to repulsive forces between dust molecules in a plasma sheath. Physical Review Letters. 1999;**83**:3194-3197

[37] Chen M, Dropmann M, Zhang B, Matthews LS, Hyde TW. Ion-wake field inside a glass box. Physical Review Letters. 2016;**94**:033201

[38] Melandsø F, Goree J. Polarized supersonic plasma flow simulation for charged bodies such as dust particles and spacecraft. Physical Review E. 1995; **52**:5312-5326

12

Kinetic Equations of Granular Media

Viktor Gerasimenko

Abstract

Approaches to the rigorous derivation of a priori kinetic equations, namely, the Enskog-type and Boltzmann-type kinetic equations, describing granular media from the dynamics of inelastically colliding particles are reviewed. We also consider the problem of potential possibilities inherent in describing the evolution of the states of a system of many hard spheres with inelastic collisions by means of a one-particle distribution function.

Keywords: granular media, inelastic collision, Boltzmann equation, Enskog equation

1. Introduction

It is well known that the properties of granular media (sand, powders, cements, seeds, etc.) have been extensively studied, in the last decades, by means of experiments, computer simulations, and analytical methods, and a huge amount of physical literature on this topic has been published (for pointers to physical literature, see in [1–6]).

Granular media are systems of many particles that attract considerable interest not only because of their numerous applications but also as systems whose collective behavior differs from the statistical behavior of ordinary media, i.e., typical macroscopic properties of media, for example, gases. In particular, the most spectacular effects include with the phenomena of collapse or cooling effect at the kinetic scale or clustering at the hydrodynamical scale, spontaneous loss of homogeneity, modification of Fourier's law and non-Maxwellian equilibrium kinetic distributions [1–3].

In modern works [4–6], it is assumed that the microscopic dynamics of granular media is dissipative, and it is described by a system of many hard spheres with inelastic collisions. The purpose of this chapter is to review some advances in the mathematical understanding of kinetic equations of systems with inelastic collisions.

As is known [7], the collective behavior of many-particle systems can be effectively described by means of a one-particle distribution function governed by the kinetic equation derived from underlying dynamics in a suitable scaling limit. At present the considerable advance is observed in a problem of the rigorous derivation of the Boltzmann kinetic equation for a system of hard spheres in the Boltzmann–Grad scaling limit [7–10]. At the same time, many recent papers [5, 11] (and see references therein) consider the Boltzmann-type and the Enskog-type kinetic equations for inelastically interacting hard spheres, modelling the behavior of granular gases, as the original evolution equations and the rigorous derivation of such kinetic equations remain still an open problem [12, 13].

Hereinafter, an approach will be formulated, which makes it possible to rigorously justify the kinetic equations previously introduced a priori for the description of granular media, namely, the Enskog-type and Boltzmann-type kinetic equations. In addition, we will consider the problem of potential possibilities inherent in describing the evolution of the states of a system of many hard spheres with inelastic collisions by means of a one-particle distribution function.

2. Dynamics of hard spheres with inelastic collisions

As mentioned above, the microscopic dynamics of granular media is described by a system of many hard spheres with inelastic collisions. We consider a system of a non-fixed, i.e., arbitrary, but finite average number of identical particles of a unit mass with the diameter $\sigma > 0$, interacting as hard spheres with inelastic collisions. Every particle is characterized by the phase coordinates: $(q_i, p_i) \equiv x_i \in \mathbb{R}^3 \times \mathbb{R}^3, i \geq 1$.

Let C_γ be the space of sequences $b = (b_0, b_1, \ldots, b_n, \ldots)$ of bounded continuous functions $b_n \in C_n$ defined on the phase space of n hard spheres that are symmetric with respect to the permutations of the arguments $x_1, \ldots, x_n$, equal to zero on the set of forbidden configurations $\mathbb{W}_n \doteq \{(q_1, \ldots, q_n) \in \mathbb{R}^{3n} \| q_i - q_j| < \sigma$ for at least one pair $(i,j) : i \neq j \in (1, \ldots, n)\}$ and equipped with the norm: $\|b\|_{C_\gamma} = \max_{n \geq 0} \frac{\gamma^n}{n!} \|b_n\|_{C_n} = \max_{n \geq 0} \frac{\gamma^n}{n!} \sup_{x_1, \ldots, x_n} |b_n(x_1, \ldots, x_n)|$. We denote the set of continuously differentiable functions with compact supports by $C_{n,0} \subset C_n$.

We introduce the semigroup of operators $S_n(t), t \geq 0$, that describes dynamics of n hard spheres. It is defined by means of the phase trajectories of a hard sphere system with inelastic collisions almost everywhere on the phase space $\mathbb{R}^{3n} \times (\mathbb{R}^{3n} \backslash \mathbb{W}_n)$, namely, outside the set $\mathbb{M}_n^0$ of the zero Lebesgue measure, as follows [14]:

$$(S_n(t)b_n)(x_1, \ldots, x_n) \equiv S_n(t, 1, \ldots, n)b_n(x_1, \ldots, x_n) \doteq \begin{cases} b_n(X_1(t), \ldots, X_n(t)), & \text{if } (x_1, \ldots, x_n) \in (\mathbb{R}^{3n} \backslash \mathbb{W}_n) \times \mathbb{R}^{3n}, \\ 0, & \text{if } (q_1, \ldots, q_n) \in \mathbb{W}_n, \end{cases} \tag{1}$$

where the function $X_i(t) \equiv X_i(t, x_1, \ldots, x_n)$ is a phase trajectory of ith particle constructed in [7] and the set $\mathbb{M}_n^0$ consists from phase space points-specified initial data $x_1, \ldots, x_n$ that generate multiple collisions during the evolution.

On the space C_n one-parameter mapping (1) is a bounded $*$-weak continuous semigroup of operators, and $\|S_n(t)\|_{C_n} < 1$.

The infinitesimal generator $\mathcal{L}_n$ of the semigroup of operators (1) is defined in the sense of a $*$-weak convergence of the space C_n, and it has the structure $\mathcal{L}_n = \sum_{j=1}^n \mathcal{L}(j) + \sum_{j_1<j_2=1}^n \mathcal{L}_{\mathrm{int}}(j_1, j_2)$, , and the operators $\mathcal{L}(j)$ and $\mathcal{L}_{\mathrm{int}}(j_1, j_2)$ are defined by formulas:

$$\mathcal{L}(j) \doteq \left\langle p_j, \frac{\partial}{\partial q_j} \right\rangle, \tag{2}$$

and

$$\mathcal{L}_{\mathrm{int}}(j_1,j_2)b_s \doteq \sigma^2\int_{\mathbb{S}^2_+} d\eta\langle\eta,(p_{j_1}-p_{j_2})\rangle(b_s(x_1,\dots,x^*_{j_1},\dots,x^*_{j_2},\dots,x_s)- b_s(x_1,\dots,x_s))\delta(q_{j_1}-q_{j_2}+\sigma\eta), \tag{3}$$

respectively. In (2) and (3) the following notations are used: $x^*_j \equiv (q_j,p^*_j)$, the symbol $\langle\cdot,\cdot\rangle$ means a scalar product, δ is the Dirac measure, $\mathbb{S}^2_+ \doteq \{\eta\in\mathbb{R}^3 \,|\, |\eta|=1, \langle\eta,(p_{j_1}-p_{j_2})\rangle\geq 0\}$, and the post-collision momenta are determined by the expressions:

$$\begin{aligned} p^*_{j_1} &= p_{j_1} - (1-\varepsilon)\eta\langle\eta,(p_{j_1}-p_{j_2})\rangle, \\ p^*_{j_2} &= p_{j_2} + (1-\varepsilon)\eta\langle\eta,(p_{j_1}-p_{j_2})\rangle, \end{aligned} \tag{4}$$

where $\varepsilon = \frac{1-e}{2}\in[0,\frac{1}{2})$ and $e\in(0,1]$ is a restitution coefficient [6].

Let $L^1_\alpha = \oplus_{n=0}^{\infty}\alpha^n L^1_n$ be the space of sequences $f=(f_0,f_1,\dots,f_n,\dots)$ of integrable functions $f_n(x_1,\dots,x_n)$ defined on the phase space of n hard spheres that are symmetric with respect to the permutations of the arguments $x_1,\dots,x_n$, equal to zero on the set of forbidden configurations $\mathbb{W}_n$ and equipped with the norm: $\|f\|_{L^1_\alpha} = \sum_{n=0}^{\infty}\alpha^n \int dx_1\dots dx_n|f_n(x_1,\dots,x_n)|$, where $\alpha>1$ is a real number. We denote by $L^1_0 \subset L^1_\alpha$ the everywhere dense set in L^1_α of finite sequences of continuously differentiable functions with compact supports.

On the space of integrable functions, the semigroup of operators $S^*_n(t), t\geq 0$, adjoint to semigroup of operators (1) in the sense of the continuous linear functional is defined (the functional of mean values of observables):

$$(b,f) = \sum_{n=0}^{\infty}\frac{1}{n!}\int_{(\mathbb{R}^3\times\mathbb{R}^3)^n} dx_1\dots dx_n b_n(x_1,\dots,x_n)f_n(x_1,\dots,x_n).$$

The adjoint semigroup of operators is defined by the Duhamel equation:

$$S^*_n(t,1,\dots,n) = \prod_{i=1}^{n} S^*_1(t,i) + \int_0^t d\tau \prod_{i=1}^{n} S^*_1(t-\tau,i)\sum_{j_1<j_2=1}^{n}\mathcal{L}^*_{\mathrm{int}}(j_1,j_2)S^*_n(\tau,1,\dots,n), \tag{5}$$

where for $t\geq 0$ the operator $\mathcal{L}^*_{\mathrm{int}}(j_1,j_2)$ is determined by the formula

$$\mathcal{L}^*_{\mathrm{int}}(j_1,j_2)f_s \doteq \sigma^2\int_{\mathbb{S}^2_+} d\eta\langle\eta,(p_{j_1}-p_{j_2})\rangle\Big(\frac{1}{(1-2\varepsilon)^2}f_n(x_1,\dots,x^\diamond_{j_1},\dots, x^\diamond_{j_2},\dots,x_n)\delta(q_{j_1}-q_{j_2}+\sigma\eta) - f_n(x_1,\dots,x_n)\delta(q_{j_1}-q_{j_2}-\sigma\eta)\Big). \tag{6}$$

In (6) the notations similar to formula (3) are used, $x^\diamond_j \equiv (q_j,p^\diamond_j)$, and the pre-collision momenta (solutions of equations (4)) are determined as follows:

$$p^{\diamond}_{j_1} = p_{j_1} - \frac{1-\varepsilon}{1-2\varepsilon}\eta\langle\eta,(p_{j_1} - p_{j_2})\rangle,$$
$$p^{\diamond}_{j_2} = p_{j_2} + \frac{1-\varepsilon}{1-2\varepsilon}\eta\langle\eta,(p_{j_1} - p_{j_2})\rangle. \tag{7}$$

Hence an infinitesimal generator of the adjoint semigroup of operators $S^*_n(t)$ is defined on $L^1_{0,n}$ as the operator, $\mathcal{L}^*_n = \sum_{j=1}^n \mathcal{L}^*(j) + \sum_{j_1<j_2=1}^n \mathcal{L}^*_{\mathrm{int}}(j_1,j_2)$, where the operator adjoint to free motion operator (2) $\mathcal{L}^*(j) \doteq -\langle p_j, \frac{\partial}{\partial q_j}\rangle$ was introduced.

On the space L^1_n the one-parameter mapping defined by Eq. (5) is a bounded strong continuous semigroup of operators.

3. The dual hierarchy of evolution equations for observables

It is well known [7] that many-particle systems are described by means of states and observables. The functional for mean value of observables determines a duality of states and observables, and, as a consequence, there exist two equivalent approaches to describing the evolution of systems of many particles. Traditionally, the evolution is described in terms of the evolution of states by means of the BBGKY hierarchy for marginal distribution functions. An equivalent approach to describing evolution is based on marginal observables governed by the dual BBGKY hierarchy. In the same framework, the evolution of particles with the dissipative interaction, namely, hard spheres with inelastic collisions, is described [14].

Within the framework of observables, the evolution of a system of hard spheres is described by the sequences $B(t) = (B_0, B_1(t,x_1), \dots, B_s(t,x_1,\dots,x_s), \dots) \in C_\gamma$ of the marginal observables $B_s(t,x_1,\dots,x_s)$ defined on the phase space of $s \geq 1$ hard spheres that are symmetric with respect to the permutations of the arguments $x_1,\dots,x_n$, equal to zero on the set $\mathbb{W}_s$, and for $t \geq 0$ they are governed by the Cauchy problem of the weak formulation of the dual BBGKY hierarchy [14]:

$$\frac{\partial}{\partial t}B_s(t,x_1,\dots,x_s) = \left(\sum_{j=1}^{s}\mathcal{L}(j)B_s(t) + \sum_{j_1<j_2=1}^{s}\mathcal{L}_{\mathrm{int}}(j_1,j_2)B_s(t)\right)(x_1,\dots,x_s)$$
$$+ \sum_{j_1\neq j_2=1}^{s}(\mathcal{L}_{\mathrm{int}}(j_1,j_2)B_{s-1}(t))(x_1,\dots,x_{j_1-1},x_{j_1+1},\dots,x_s), \tag{8}$$

$$B_s(t,x_1,\dots,x_s)|_{t=0} = B^0_s(x_1,\dots,x_s), \quad s \geq 1, \tag{9}$$

where on the set $C_{s,0} \subset C_s$ the free motion operator $\mathcal{L}(j)$ and the operator of inelastic collisions $\mathcal{L}_{\mathrm{int}}(j_1,j_2)$ are defined by formulas (2) and (3), respectively. We refer to recurrence evolution equation (8) as the dual BBGKY hierarchy for hard spheres with inelastic collisions.

The solution $B(t) = (B_0, B_1(t,x_1), \dots, B_s(t,x_1,\dots,x_s), \dots)$ of the Cauchy problem (8),(9) is determined by the expansions [10]:

$$B_s(t,x_1,\dots,x_s) = \sum_{n=0}^{s-1}\frac{1}{n!}\sum_{j_1\neq\dots\neq j_n=1}^{s}\mathfrak{A}_{1+n}(t,\{Y\backslash Z\},Z)B^0_{s-n}$$
$$(x_1,\dots,x_{j_1-1},x_{j_1+1},\dots,x_{j_n-1},x_{j_n+1},\dots,x_s), \tag{10}$$

where the $(1+n)th$-order cumulant of semigroups of operators (1) of hard spheres with inelastic collisions is defined by the formula

$$\mathfrak{A}_{1+n}(t, \{Y\backslash X\}, X) \doteq \sum_{\mathrm{P}:(\{Y\backslash X\}, X)=\cup_i X_i} (-1)^{|\mathrm{P}|-1}(|\mathrm{P}|-1)! \prod_{X_i \subset \mathrm{P}} S_{|\theta(X_i)|}(t, \theta(X_i)), \tag{11}$$

and $Y \equiv (1, \dots, s), Z \equiv (j_1, \dots, j_n) \subset Y$, $\{Y\backslash Z\}$ is the set consisting of one element $Y\backslash Z = (1, \dots, j_1 - 1, j_1 + 1, \dots, j_n - 1, j_n + 1, \dots, s)$, i.e., this set is a connected subset of the partition P such that $|\mathrm{P}| = 1$; the mapping $\theta(\cdot)$ is a declusterization operator defined by the formula: $\theta(\{Y\backslash Z\}) = Y\backslash Z$.

We note that one component sequences of marginal observables correspond to observables of certain structure, namely, the marginal observable $B^{(1)} = (0, b_1(x_1), 0, \dots)$ corresponds to the additive-type observable, and the marginal observable $B^{(k)} = (0, \dots, 0, b_k(x_1, \dots, x_k), 0, \dots)$ corresponds to the k-ary-type observable. If as initial data (9) we consider the marginal observable of additive type, then the decomposition structure of solution (10) is simplified and takes the form

$$B_s^{(1)}(t, x_1, \dots, x_s) = \mathfrak{A}_s(t, 1, \dots, s) \sum_{j=1}^{s} b_1(x_j), \quad s \geq 1.$$

On the space C_γ for abstract initial-value problem (8) and (9), the following statement is true. If $B(0) = (B_0, B_1^0, \dots, B_s^0, \dots) \in C_\gamma^0 \subset C_\gamma$ is finite sequence of infinitely differentiable functions with compact supports, then the sequence of functions (10) is a classical solution, and for arbitrary initial data $B(0) \in C_\gamma$, it is a generalized solution.

We remark that expansion (10) can be also represented in the form of the weak formulation of the perturbation (iteration) series as a result of the applying of analogs of the Duhamel equation to cumulants of semigroups of operators (11).

The mean value of the marginal observable $B(t) = (B_0, B_1(t), \dots, B_s(t), \dots) \in \mathcal{C}_\gamma$ in initial state specified by a sequence of marginal distribution functions $F(0) = (1, F_1^0, \dots, F_s^0, \dots) \in L_\alpha^1 = \oplus_{s=0}^\infty \alpha^s L_s^1$ is determined by the following functional:

$$(B(t), F(0)) = \sum_{s=0}^{\infty} \frac{1}{s!} \int_{(\mathbb{R}^3 \times \mathbb{R}^3)^s} dx_1 \dots dx_s B_s(t, x_1, \dots, x_s) F_s^0(x_1, \dots, x_s). \tag{12}$$

In particular, functional (12) of mean values of the additive-type marginal observables $B^{(1)}(0) = \left(0, B_1^{(1)}(0, x_1), 0, \dots\right)$ takes the form:

$$\left(B^{(1)}(t), F(0)\right) = \left(B^{(1)}(0), F(t)\right) = \int_{\mathbb{R}^3 \times \mathbb{R}^3} dx_1 B_1^{(1)}(0, x_1) F_1(t, x_1),$$

where the one-particle marginal distribution function $F_1(t, x_1)$ is determined by the series expansion [10]

$$F_1(t, x_1) = \sum_{n=0}^{\infty} \frac{1}{n!} \int_{(\mathbb{R}^3 \times \mathbb{R}^3)^n} dx_2 \dots dx_{n+1} \mathfrak{A}_{1+n}^*(t) F_{1+n}^0(x_1, \dots, x_{n+1}),$$

and the generating operator $\mathfrak{A}_{1+n}^*(t)$ of this series is the $(1+n)th$-order cumulant of adjoint semigroups of hard spheres with inelastic collisions. In the general case for mean values of marginal observables, the following equality is true:

$$(B(t), F(0)) = (B(0), F(t)),$$

where the sequence $F(t) = (1, F_1(t), \ldots, F_s(t), \ldots)$ is a solution of the Cauchy problem of the BBGKY hierarchy of hard spheres with inelastic collisions [14]. The last equality signifies the equivalence of two pictures of the description of the evolution of hard spheres by means of the BBGKY hierarchy [7] and the dual BBGKY hierarchy (8).

Hereinafter we consider initial states of hard spheres specified by a one-particle marginal distribution function, namely,

$$F_s^{(c)}(x_1, \ldots, x_s) = \prod_{i=1}^{s} F_1^0(x_i) \mathcal{X}_{\mathbb{R}^{3s} \setminus \mathbb{W}_s}, \quad s \geq 1, \tag{13}$$

where $\mathcal{X}_{\mathbb{R}^{3s} \setminus \mathbb{W}_s} \equiv \mathcal{X}_s(q_1, \ldots, q_s)$ is a characteristic function of allowed configurations $\mathbb{R}^{3s} \setminus \mathbb{W}_s$ of s hard spheres and $F_1^0 \in L^1(\mathbb{R}^3 \times \mathbb{R}^3)$. Initial data (13) is intrinsic for the kinetic description of many-particle systems because in this case all possible states are described by means of a one-particle marginal distribution function.

4. The non-Markovian Enskog kinetic equation

In the case of initial states (13), the dual picture of the evolution to the picture of the evolution by means of observables of a system of hard spheres with inelastic collisions governed by the dual BBGKY hierarchy (8) for marginal observables is the evolution of states described by means of the non-Markovian Enskog kinetic equation and a sequence of explicitly defined functionals of a solution of such kinetic equation.

Indeed, in view of the fact that the initial state is completely specified by a one-particle marginal distribution function on allowed configurations (13), for mean value functional (12), the following representation holds [14, 15]:

$$\left(B(t), F^{(c)}\right) = (B(0), F(t|F_1(t))),$$

where $F^{(c)} = \left(1, F_1^{(c)}, \ldots, F_s^{(c)}, \ldots\right)$ is the sequence of initial marginal distribution functions (13) and the sequence $F(t|F_1(t)) = (1, F_1(t), F_2(t|F_1(t)), \ldots, F_s(t|F_1(t)))$ is a sequence of the marginal functionals of the state $F_s(t, x_1, \ldots, x_s|F_1(t))$ represented by the series expansions over the products with respect to the one-particle marginal distribution function $F_1(t)$:

$$F_s(t, x_1, \ldots, x_s|F_1(t)) \doteq \sum_{n=0}^{\infty} \frac{1}{n!} \int_{(\mathbb{R}^3 \times \mathbb{R}^3)^n} dx_{s+1} \ldots dx_{s+n} \mathfrak{V}_{1+n}(t, \{Y\}, X \setminus Y) \prod_{i=1}^{s+n} F_1(t, x_i), \quad s \geq 2. \tag{14}$$

In series (14) we used the notations $Y \equiv (1, \ldots, s)$, $X \equiv (1, \ldots, s+n)$, and the $(n+1)th$-order generating operator $\mathfrak{V}_{1+n}(t), n \geq 0$ is defined as follows [15]:

$$\mathfrak{V}_{1+n}(t, \{Y\}, X \setminus Y) \doteq \sum_{k=0}^{n} (-1)^k \sum_{m_1=1}^{n} \ldots \sum_{m_k=1}^{n-m_1-\ldots-m_{k-1}} \frac{n!}{(n-m_1-\ldots-m_k)!} \times$$

$$\widehat{\mathfrak{A}}_{1+n-m_1-\ldots-m_k}(t, \{Y\}, s+1, \ldots, s+n-m_1-\ldots-m_k) \prod_{j=1}^{k} \sum_{k_2^j=0}^{m_j} \ldots$$

$$\sum_{k^j_{n-m_1-\dots-m_j+s}=0}^{k^j_{n-m_1-\dots-m_j+s-1}} \prod_{i_j=1}^{s+n-m_1-\dots-m_j} \frac{1}{\left(k^j_{n-m_1-\dots-m_j+s+1-i_j} - k^j_{n-m_1-\dots-m_j+s+2-i_j}\right)!} \times$$
$$\widehat{\mathfrak{A}}_{1+k^j_{n-m_1-\dots-m_j+s+1-i_j}-k^j_{n-m_1-\dots-m_j+s+2-i_j}} (t, i_j, s+n-m_1-\dots-m_j+1+$$
$$k^j_{s+n-m_1-\dots-m_j+2-i_j}, \dots, s+n-m_1-\dots-m_j+k^j_{s+n-m_1-\dots-m_j+1-i_j}),$$

where it means that $k^j_1 \equiv m_j, k^j_{n-m_1-\dots-m_j+s+1} \equiv 0$, and we denote the $(1+n)th$-order scattering cumulant by the operator $\widehat{\mathfrak{A}}_{1+n}(t)$:

$$\widehat{\mathfrak{A}}_{1+n}(t, \{Y\}, X \backslash Y) \doteq \mathfrak{A}^*_{1+n}(t, \{Y\}, X \backslash Y) \mathcal{X}_{\mathbb{R}^{3(s+n)} \backslash \mathbb{W}_{s+n}} \prod_{i=1}^{s+n} \mathfrak{A}^*_1(t, i)^{-1},$$

and the operator $\mathfrak{A}^*_{1+n}(t)$ is the $(1+n)th$-order cumulant of adjoint semigroups of hard spheres with inelastic collisions.

We emphasize that in fact functionals (14) characterize the correlations generated by dynamics of a hard sphere system with inelastic collisions.

The second element of the sequence $F(t|F_1(t))$, i.e., the one-particle marginal distribution function $F_1(t)$, is determined by the following series expansion:

$$F_1(t, x_1) = \sum_{n=0}^{\infty} \frac{1}{n!} \int_{(\mathbb{R}^3 \times \mathbb{R}^3)^n} dx_2 \dots dx_{n+1} \mathfrak{A}^*_{1+n}(t) \mathcal{X}_{\mathbb{R}^{3(1+n)} \backslash \mathbb{W}_{1+n}} \prod_{i=1}^{n+1} F^0_1(x_i), \qquad (15)$$

where the generating operator $\mathfrak{A}^*_{1+n}(t) \equiv \mathfrak{A}^*_{1+n}(t, 1, \dots, n+1)$ is the $(1+n)th$-order cumulant of adjoint semigroups of hard spheres with inelastic collisions.

For $t \geq 0$ the one-particle marginal distribution function (15) is a solution of the following Cauchy problem of the non-Markovian Enskog kinetic equation [14, 15]:

$$\frac{\partial}{\partial t} F_1(t, q_1, p_1) = -\left\langle p_1, \frac{\partial}{\partial q_1} \right\rangle F_1(t, q_1, p_1)$$
$$+\sigma^2 \int_{\mathbb{R}^3 \times \mathbb{S}^2_+} dp_2 d\eta \langle \eta, (p_1 - p_2) \rangle \ \frac{1}{(1-2\varepsilon)^2} F_2(t, q_1, p^\diamond_1, q_1 - \sigma\eta, p^\diamond_2 | F_1(t))$$
$$-F_2(t, q_1, p_1, q_1 + \sigma\eta, p_2 | F_1(t))), \qquad (16)$$

$$F_1(t)|_{t=0} = F^0_1, \qquad (17)$$

where the collision integral is determined by the marginal functional of the state (14) in the case of $s = 2$, and the expressions $p^\diamond_1$ and $p^\diamond_2$ are the pre-collision momenta of hard spheres with inelastic collisions (7), i.e., solutions of Eq. (4).

We note that the structure of collision integral of the non-Markovian Enskog equation for granular gases (16) is such that the first term of its expansion is the collision integral of the Boltzmann–Enskog kinetic equation and the next terms describe all possible correlations which are created by hard sphere dynamics with inelastic collisions and by the propagation of initial correlations connected with the forbidden configurations.

We remark also that based on the non-Markovian Enskog equation (16), we can formulate the Markovian Enskog kinetic equation with inelastic collisions [14].

For the abstract Cauchy problem of the non-Markovian Enskog kinetic equation (16), (17) in the space of integrable functions , the following statement is true [14]. A global in time solution of the Cauchy problem of the non-Markovian Enskog equation (16) is determined by function (15). For small densities and $F_1^0 \in L_0^1(\mathbb{R}^3 \times \mathbb{R}^3)$, function (15) is a strong solution, and for an arbitrary initial data $F_1^0 \in L^1(\mathbb{R}^3 \times \mathbb{R}^3)$, it is a weak solution.

Thus, if initial state is specified by a one-particle marginal distribution function on allowed configurations, then the evolution, describing by marginal observables governed by the dual BBGKY hierarchy (8), can be also described by means of the non-Markovian kinetic equation (16) and a sequence of marginal functionals of the state (14). In other words, for mentioned initial states, the evolution of all possible states of a hard sphere system with inelastic collisions at arbitrary moment of time can be described by means of a one-particle distribution function without any approximations.

5. The Boltzmann kinetic equation for granular gases

It is known [7, 8] the Boltzmann kinetic equation describes the evolution of many hard spheres in the Boltzmann–Grad (or low-density) approximation. In this section the possible approaches to the rigorous derivation of the Boltzmann kinetic equation from dynamics of hard spheres with inelastic collisions are outlined.

One approach to deriving the Boltzmann kinetic equation for hard spheres with inelastic collisions, which was developed in [10] for a system of hard spheres with elastic collisions, is based on constructing the Boltzmann–Grad asymptotic behavior of marginal observables governed by the dual BBGKY hierarchy (8). A such scaling limit is governed by the set of recurrence evolution equations, namely, by the dual Boltzmann hierarchy for hard spheres with inelastic collisions [14]. Then for initial states specified by a one-particle distribution function (13), the evolution of additive-type marginal observables governed by the dual Boltzmann hierarchy is equivalent to a solution of the Boltzmann kinetic equation for granular gases [12], and the evolution of nonadditive-type marginal observables is equivalent to the property of the propagation of initial chaos for states [10].

One more approach to the description of the kinetic evolution of hard spheres with inelastic collisions is based on the non-Markovian generalization of the Enskog equation (16).

Let the dimensionless one-particle distribution function $F_1^{\epsilon,0}$, specifying initial state (13), satisfy the condition, $|F_1^{\epsilon,0}(x_1)| \leq c e^{-\frac{\beta}{2}p_1^2}$, where $\epsilon > 0$ is a scaling parameter (the ratio of the diameter $\sigma > 0$ to the mean free path of hard spheres), $\beta > 0$ is a parameter, and $c < \infty$ is some constant, and there exists the following limit in the sense of a weak convergence: $\mathrm{w}-\lim_{\epsilon\to 0}(\epsilon^2 F_1^{\epsilon,0}(x_1) - f_1^0(x_1)) = 0$. Then for finite time interval the Boltzmann–Grad limit of dimensionless solution (15) of the Cauchy problem of the non-Markovian Enskog kinetic equation (16) and (17) exists in the same sense, namely, $\mathrm{w}-\lim\limits_{\epsilon\to 0}(\epsilon^2 F_1(t,x_1) - f_1(t,x_1)) = 0$, where the limit one-particle distribution function is a weak solution of the Cauchy problem of the Boltzmann kinetic equation for granular gases [6, 12]:

$$\frac{\partial}{\partial t} f_1(t,x_1) = -\Big\langle p_1, \frac{\partial}{\partial q_1}\Big\rangle f_1(t,x_1) + \int_{\mathbb{R}^3\times\mathbb{S}_+^2} dp_2 d\eta \langle \eta, (p_1 - p_2)\rangle \Big(\frac{1}{(1-2\varepsilon)^2} f_1(t,q_1,p_1^\diamond) f_1(t,q_1,p_2^\diamond) - f_1(t,x_1) f_1(t,q_1,p_2)\Big), \tag{18}$$

$$f_1(t,x_1)\big|_{t=0} = f_1^0(x_1), \tag{19}$$

where the momenta $p_1^\diamond$ and $p_2^\diamond$ are pre-collision momenta of hard spheres with inelastic collisions (7).

As noted above, all possible correlations of a system of hard spheres with inelastic collisions are described by marginal functionals of the state (14). Taking into consideration the fact of the existence of the Boltzmann–Grad scaling limit of a solution of the non-Markovian Enskog kinetic equation (16), for marginal functionals of the state (14), the following statement holds:

$$\mathrm{w}-\lim_{\epsilon\to 0}\left(\epsilon^{2s}F_s(t,x_1,\dots,x_s|F_1(t))-\prod_{j=1}^{s}f_1(t,x_j)\right)=0,$$

where the limit one-particle distribution function $f_1(t)$ is governed by the Boltzmann kinetic equation for granular gases (18). This property of marginal functionals of the state (14) means the propagation of the initial chaos [7].

It should be emphasized that the Boltzmann–Grad asymptotics of a solution of the non-Markovian Enskog equation (16) in a multidimensional space are analogous of the Boltzmann–Grad asymptotic behavior of a hard sphere system with the elastic collisions [10]. Such asymptotic behavior is governed by the Boltzmann equation for a granular gas (18), and the asymptotics of marginal functionals of the state (14) are the product of its solution (this property is interpreted as the propagation of the initial chaos).

6. One-dimensional granular gases

As is known, the evolution of a one-dimensional system of hard spheres with elastic collisions is trivial (free motion or Knudsen flow) in the Boltzmann–Grad scaling limit [7], but, as it was taken notice in paper [16], in this approximation the kinetics of inelastically interacting hard spheres (rods) is not trivial, and it is governed by the Boltzmann kinetic equation for one-dimensional granular gases [16–19]. Below the approach to the rigorous derivation of Boltzmann-type equation for one-dimensional granular gases will be outlined. It should be emphasized that a system of many hard rods with inelastic collisions displays the basic properties of granular gases inasmuch as under the inelastic collisions only the normal component of relative velocities dissipates in a multidimensional case.

In case of a one-dimensional granular gas for $t\geq 0$ in dimensionless form, the Cauchy problem (16),(17) takes the form [20]:

$$\begin{aligned}
\frac{\partial}{\partial t}F_1(t,q_1,p_1) &= -p_1\frac{\partial}{\partial q_1}F_1(t,q_1,p_1)+\\
&\int_0^\infty dPP\ \frac{1}{(1-2\varepsilon)^2}F_2(t,q_1,p_1^\diamond(p_1,P),q_1-\epsilon,p_2^\diamond(p_1,P)|F_1(t))-\\
&F_2(t,q_1,p_1,q_1-\epsilon,p_1+P|F_1(t))\Big)+ \qquad (20)\\
&\int_0^\infty dPP\ \frac{1}{(1-2\varepsilon)^2}F_2(t,q_1,\tilde{p}_1^\diamond(p_1,P),q_1+\epsilon,\tilde{p}_2^\diamond(p_1,P)|F_1(t))-\\
&F_2(t,q_1,p_1,q_1+\epsilon,p_1-P|F_1(t))\Big),
\end{aligned}$$

$$F_1(t)|_{t=0}=F_1^{\epsilon,0}, \qquad (21)$$

where $\epsilon > 0$ is a scaling parameter (the ratio of a hard sphere diameter (the length) $\sigma > 0$ to the mean free path), the collision integral is determined by marginal functional (14) of the state $F_1(t)$ in the case of $s = 2$, and the expressions:

$$p_1^\diamond(p_1, P) = p_1 - P + \frac{\varepsilon}{2\varepsilon - 1}P,$$
$$p_2^\diamond(p_1, P) = p_1 - \frac{\varepsilon}{2\varepsilon - 1}P$$

and

$$\tilde{p}_1^\diamond(p_1, P) = p_1 + P - \frac{\varepsilon}{2\varepsilon - 1}P,$$
$$\tilde{p}_2^\diamond(p_1, P) = p_1 + \frac{\varepsilon}{2\varepsilon - 1}P,$$

are transformed pre-collision momenta in a one-dimensional space.

If initial one-particle marginal distribution functions satisfy the following condition: $|F_1^{\epsilon,0}(x_1)| \leq Ce^{-\frac{\beta}{2}p_1^2}$, where $\beta > 0$ is a parameter, $C < \infty$ is some constant, then every term of the series

$$F_1^\epsilon(t, x_1) = \sum_{n=0}^{\infty} \frac{1}{n!} \int_{(\mathbb{R}\times\mathbb{R})^n} dx_2 \ldots dx_{n+1} \mathfrak{A}_{1+n}^*(t) \prod_{i=1}^{n+1} F_1^{\epsilon,0}(x_i) \mathcal{X}_{\mathbb{R}^{(1+n)} \setminus \mathbb{W}_{1+n}}, \tag{22}$$

exists, for finite time interval function (23) is the uniformly convergent series with respect to x_1 from arbitrary compact, and it is determined a weak solution of the Cauchy problem of the non-Markovian Enskog equation (20), (22). Let in the sense of a weak convergence there exist the following limit:

$$\mathrm{w} - \lim_{\epsilon\to 0} \left(F_1^{\epsilon,0}(x_1) - f_1^0(x_1)\right) = 0,$$

then for finite time interval there exists the Boltzmann–Grad limit of solution (23) of the Cauchy problem of the non-Markovian Enskog equation for one-dimensional granular gas (20) in the sense of a weak convergence:

$$\mathrm{w} - \lim_{\epsilon\to 0} \left(F_1^\epsilon(t, x_1) - f_1(t, x_1)\right) = 0, \tag{23}$$

where the limit one-particle marginal distribution function is defined by uniformly convergent arbitrary compact set series:

$$f_1(t, x_1) = \sum_{n=0}^{\infty} \frac{1}{n!} \int_{(\mathbb{R}\times\mathbb{R})^n} dx_2 \ldots dx_{n+1} \mathfrak{A}_{1+n}^0(t) \prod_{i=1}^{n+1} f_1^0(x_i), \tag{24}$$

and the generating operator $\mathfrak{A}_{1+n}^0(t) \equiv \mathfrak{A}_{1+n}^0(t, 1, \ldots, n+1)$ is the $(n+1)$ th-order cumulant of adjoint semigroups $S_n^{*,0}(t)$ of point particles with inelastic collisions. An infinitesimal generator of the semigroup of operators $S_n^{*,0}(t)$ is defined as the operator:

$$\begin{aligned}(\mathcal{L}_n^{*,0} f_n)(x_1, \ldots, x_n) = &- \sum_{j=1}^{n} p_j \frac{\partial}{\partial q_j} f_n(x_1, \ldots, x_n) \\ &+ \sum_{j_1 < j_2 = 1}^{n} |p_{j_2} - p_{j_1}| \left(\frac{1}{(1-2\varepsilon)^2} f_n\left(x_1, \ldots, x_{j_1}^\diamond, \ldots, x_{j_2}^\diamond, \ldots, x_n\right) - \right. \\ &\left. f_n(x_1, \ldots, x_n) \right) \delta\left(q_{j_1} - q_{j_2}\right),\end{aligned}$$

where $x_j^\diamond \equiv \left(q_j, p_j^\diamond\right)$ and the pre-collision momenta $p_{j_1}^\diamond, p_{j_2}^\diamond$ are determined by the following expressions:

$$p_{j_1}^\diamond = p_{j_2} + \frac{\varepsilon}{2\varepsilon - 1}\left(p_{j_1} - p_{j_2}\right),$$
$$p_{j_2}^\diamond = p_{j_1} - \frac{\varepsilon}{2\varepsilon - 1}\left(p_{j_1} - p_{j_2}\right).$$

For $t \geq 0$ the limit one-particle distribution function represented by series (25) is a weak solution of the Cauchy problem of the Boltzmann-type kinetic equation of point particles with inelastic collisions [20]

$$\frac{\partial}{\partial t} f_1(t,q,p) = -p\frac{\partial}{\partial q} f_1(t,q,p) + \int_{-\infty}^{+\infty} dp_1 |p - p_1| \times \left(\frac{1}{(1-2\varepsilon)^2} f_1(t,q,p^\diamond) f_1(t,q,p_1^\diamond) - f_1(t,q,p) f_1(t,q,p_1)\right) + \sum_{n=1}^{\infty} \mathcal{I}_0^{(n)}. \tag{25}$$

In kinetic equation (26) the remainder $\sum_{n=1}^{\infty} \mathcal{I}_0^{(n)}$ of the collision integral is determined by the expressions

$$\mathcal{I}_0^{(n)} \equiv \frac{1}{n!}\int_0^{\infty} dP P \int_{(\mathbb{R}\times\mathbb{R})^n} dq_3 dp_3 \ldots dq_{n+2} dp_{n+2} \mathfrak{V}_{1+n}(t) \frac{1}{(1-2\varepsilon)^2} f_1(t,q,p_1^\diamond(p,P))$$
$$\times f_1(t,q,p_2^\diamond(p,P)) - f_1(t,q,p) f_1(t,q,p+P)) \prod_{i=3}^{n+2} f_1(t,q_i,p_i)$$
$$+ \int_0^{\infty} dP P \int_{(\mathbb{R}\times\mathbb{R})^n} dq_3 dp_3 \ldots dq_{n+2} dp_{n+2} \mathfrak{V}_{1+n}(t) \left(\frac{1}{(1-2\varepsilon)^2} f_1(t,q,\tilde{p}_1^\diamond(p,P))\right.$$
$$\times f_1(t,q,\tilde{p}_2^\diamond(p,P)) - f_1(t,q,p) f_1(t,q,p-P)) \prod_{i=3}^{n+2} F_1(t,q_i,p_i),$$

where the generating operators $\mathfrak{V}_{1+n}(t) \equiv \mathfrak{V}_{1+n}(t, \{1,2\}, 3, \ldots, n+2), n \geq 0$, are represented by expansions (15) with respect to the cumulants of semigroups of scattering operators of point hard rods with inelastic collisions in a one-dimensional space:

$$\widehat{S}_n^0(t,1,\ldots,n) \doteq S_n^{*,0}(t,1,\ldots,s) \prod_{i=1}^{n} S_1^{*,0}(t,i)^{-1}. \tag{26}$$

In fact, the series expansions for the collision integral of the non-Markovian Enskog equation for a granular gas or solution (23) are represented as the power series over the density so that the terms $\mathcal{I}_0^{(n)}, n \geq 1$, of the collision integral in kinetic equation (18) are corrections with respect to the density to the Boltzmann collision integral for one-dimensional granular gases stated in [17, 21].

Since the scattering operator of point hard rods is an identity operator in the approximation of elastic collisions, namely, in the limit $\varepsilon \to 0$, the collision integral of the Boltzmann kinetic equation (26) in a one-dimensional space is identical to zero. In the quasi-elastic limit [21], the limit one-particle distribution function (25)

$$\lim_{\varepsilon \to 0} \varepsilon f_1(t,q,v) = f^0(t,q,v),$$

satisfies the nonlinear friction kinetic equation for granular gases of the following form [16, 21]:

$$\frac{\partial}{\partial t} f^0(t,q,v) = -v\frac{\partial}{\partial q} f^0(t,q,v) + \frac{\partial}{\partial v}\int_{-\infty}^{\infty} dv_1 |v_1 - v|(v_1 - v) f^0(t,q,v_1) f^0(t,q,v).$$

Taking into consideration result (24) on the Boltzmann–Grad asymptotic behavior of the non-Markovian Enskog equation (16), for marginal functionals of the state (14) in a one-dimensional space, the following statement is true [20]:

$$\mathrm{w} - \lim_{\epsilon\to 0} \left(F_s\left(t, x_1, \ldots, x_s | F_1^\epsilon(t)\right) - f_s\left(t, x_1, \ldots, x_s | f_1(t)\right)\right) = 0, \quad s \geq 2,$$

where the limit marginal functionals $f_s(t|f_1(t)), s \geq 2$, with respect to limit one-particle distribution function (25) are determined by the series expansions with the structure similar to series (14) and the generating operators represented by expansions (15) over the cumulants of semigroups of scattering operators (27) of point hard rods with inelastic collisions in a one-dimensional space.

As mentioned above, in the case of a system of hard rods with elastic collisions, the limit marginal functionals of the state are the product of the limit one-particle distribution functions, describing the free motion of point hard rods.

Thus, the Boltzmann–Grad asymptotic behavior of solution (23) of the non-Markovian Enskog equation (20) is governed by the Boltzmann kinetic equation for a one-dimensional granular gas (18). Moreover, the limit marginal functionals of the state are represented by the appropriate series with respect to limit one-particle distribution function (25) that describe the propagation of initial chaos in one-dimensional granular gases.

7. Conclusions

In this chapter the origin of the kinetic description of the evolution of observables of a system of hard spheres with inelastic collisions was considered.

It was established that for initial states (13) specified by a one-particle distribution function, solution (10) of the Cauchy problem of the dual BBGKY hierarchy (8) and (9) and a solution of the Cauchy problem of the non-Markovian Enskog equation (16) and (17) together with marginal functionals of the state (14), give two equivalent approaches to the description of the evolution of states of a hard sphere system with inelastic collisions. In fact, the rigorous justification of the Enskog kinetic equation for granular gases (16) is a consequence of the validity of equality (14).

We note that the developed approach is also related to the problem of a rigorous derivation of the non-Markovian kinetic-type equations from underlying many-particle dynamics which make it possible to describe the memory effects of granular gases.

One more advantage also is that the considered approach gives the possibility to construct the kinetic equations in scaling limits, involving correlations at initial time which can characterize the condensed states of a hard sphere system with inelastic collisions [10].

Finally, it should be emphasized that the developed approach to the derivation of the Boltzmann equation for granular gases from the dynamics governed by the non-Markovian Enskog kinetic equation (16) also allows us to construct higher-order corrections to the collision integral compared to the Boltzmann–Grad approximation.

Author details

Viktor Gerasimenko
Institute of Mathematics of the NAS of Ukraine, Kyiv, Ukraine

*Address all correspondence to: gerasym@imath.kiev.ua

References

[1] Cercignani C. The Boltzmann equation approach to the shear flow of a granular material. Philosophical Transactions of the Royal Society of London. 2002;**360**(1792):407-414. DOI: 10.1098/rsta.2001.0939

[2] Goldhirsch I. Scales and kinetics of granular flows. Chaos. 1999;**9**:659-672. DOI: 10.1063/1.166440

[3] Mehta A. Granular Physics. Cambridge, UK: Cambridge University Press; 2007. 362 p. DOI: 10.1017/CBO9780511535314

[4] Pöschel T, Brilliantov NV, editors. Granular gas dynamics. In: Lecture Notes in Phys. 624. Berlin, Heidelberg: Springer-Verlag; 2003. 369 p. ISBN 978-3-540-20110-6

[5] Brilliantov NV, Pöschel T. Kinetic Theory of Granular Gases. Oxford: Oxford University Press; 2004. 329 p. ISBN-13 978-0199588138

[6] Brey JJ, Dufty JW, Santos A. Dissipative dynamics for hard spheres. Journal of Statistical Physics. 1997;**87**(5-6):1051-1066. DOI: 10.1007/BF02181270

[7] Cercignani C, Gerasimenko VI, Petrina DY. Many-Particle Dynamics and Kinetic Equations. The Netherlands: Springer; 2012. 247 p. ISBN 978-94- 010-6342-5

[8] Gallagher I, Saint-Raymond L, Texier B. From Newton to Boltzmann: Hard Spheres and Short-range Potentials. Zürich Lectures in Advanced Mathematics: EMS Publ House; 2014. 146 p. ISBN-10: 3037191295

[9] Pulvirenti M, Simonella S. The Boltzmann–Grad limit of a hard sphere system: Analysis of the correlation error. Inventiones Mathematicae. 2017;**207**(3): 1135-1237. DOI: 10.1007/s00222-016-0682-4

[10] Gerasimenko VI, Gapyak IV. Low-density asymptotic behavior of observables of hard sphere fluids. Advances in Mathematical Physics. 2018;**2018**:6252919. DOI: 10.1155/2018/6252919

[11] Pareschi L, Russo G, Toscani G, editors. Modelling and Numerics of Kinetic Dissipative Systems. N.Y.: Nova Science Publ. Inc.; 2006. 220 p. ISBN-13 978-1594545030

[12] Villani C. Mathematics of granular materials. Journal of Statistical Physics. 2006;**124**(2-4):781-822. DOI: 10.1007/s10955-006-9038-6

[13] Capriz G, Mariano PM, Giovine P, editors. Mathematical Models of Granular Matter. (Lecture Notes in Math. 1937). Berlin Heidelberg: Springer-Verlag; 2008. 228 p. ISBN 978-3-540-78276-6

[14] Gerasimenko VI, Borovchenkova MS. On the non-Markovian Enskog equation for granular gases. Journal of Physics A: Mathematical and Theoretical. 2014;**47**(3):035001. DOI: 10.1088/1751-8113/47/3/035001

[15] Gerasimenko VI, Gapyak IV. Hard sphere dynamics and the Enskog equation. Kinetic and Related Models. 2012;**5**(3):459-484. DOI: 10.3934/krm.2012.5.459

[16] Mac Namara S, Young WR. Kinetics of a one-dimensional granular medium in the quasielastic limit. Physics of Fluids A. 1993;**5**(1):34-45. DOI: 10.1063/1.858896

[17] Bellomo N, Pulvirenti M, editors. Modeling in applied sciences. In:

Modeling and Simulation in Science, Engineering and Technology. Boston: Birkhäuser; 2000. 433 p. ISBN 978-1-4612-6797-3

[18] Williams DRM, MacKintosh FC. Driven granular media in one dimension: Correlations and equation of states. Physical Review E. 1996;**R9**(R): 54. DOI: 10.1103/PhysRevE.54.R9

[19] Toscani G. One-dimensional kinetic models with dissipative collisions. Mathematical Modelling and Numerical Analysis. 2000;**34**(6): 1277-1291. DOI: 10.1051/m2an:2000127

[20] Gerasimenko VI, Borovchenkova MS. The Boltzmann–Grad limit of the Enskog equation of one-dimensional granular gases. Reports of the NAS of Ukraine. 2013;**10**:11-17

[21] Toscani G. Kinetic and hydrodynamic models of nearly elastic granular flows. Monatschefte für Mathematik. 2004;**142**:179-192. DOI: 10.1007/s00605-004-0241-8

13

Evolutions of Growing Waves in Complex Plasma Medium

Sukhmander Singh

Abstract

The purpose of this chapter to discuss the waves and turbulence (instabilities) supported by dusty plasma. Plasmas support many growing modes and instabilities. Wave phenomena are important in heating plasmas, instabilities, diagnostics, etc. Waves in dusty plasma are governed by the dynamics of electrons, ions and dust particles. Disturbances in solar wind, shocks and magnetospheres are the sources of generation of plasma waves. The strong interest in complex plasma provides us better understanding of physics of dusty universe, solar winds, shocks, magnetospheres, dust control in plasma processing units and surface modifications of materials. The theory of linearization of fluid equation for small oscillation has been introduced. The concept of fine particles in complex plasma and its importance is also explained. The expressions for the growth rate of the instabilities in turbulence plasma have been derived.

Keywords: plasma oscillations, dispersion, turbulence, instabilities, dusty plasma, fine particles, Hall thrusters, resistive plasma, growth rate

1. Introduction to dusty plasma

The presence of fine particles of mass 10^{-10} to 10^{-15} kg and size 1 to 50 micrometer in an electron-ion plasma is called dusty plasma. Dusty plasma also termed complex plasma, plasma crystals, colloidal crystals, fine particle plasma, coulomb crystal or aerosol plasma and has been found in naturally in solar system, planetary rings, interplanetary space, interstellar medium, molecular clouds, circum-stellar clouds, comets, Earth's environments, etc. Manmade plasmas are ordinary flames, dust in fusion devices, rocket exhaust, thermonuclear fusion, Hall thruster, atmospheric aerosols etc. [1–7]. The detail of existence of dusty plasma is given in **Table 1**. Earlier works shows that dust in plasmas has been considered as unwanted constituents and researchers had tried various methods to eliminate dust particles from plasma-processing units. For the moment, Positive aspects of dusty plasmas emerged, dust particles are playing various positive roles in plasma processing devices. The dust particles experience different forces in plasma. The Gravitational force, drag force, electromagnetic forces, polarization force and radiation pressure [1–8].

2. Physical processes in dusty plasma

There are many circumstances when astrophysical plasma and dust particles are found to coexist together. The study of dusty plasmas systems has an exciting

Cosmic dusty plasmas	Dusty plasmas in the solar system	Dusty plasmas on the earth	Man-made dusty plasmas
Solar nebulae	Cometary tails and comae	Ordinary flames	Rocket exhaust
Planetary nebulae	Planetary ring Saturn's rings	Atmospheric aerosols	Dust on surfaces of space vehicle
Supernova shells	Dust streams ejected from Jupiter	charged snow	Microelectronic fabrication
Interplanetary medium	Zodiacal light	lightning on volcanoes	Dust in fusion devices
Molecular clouds	Cometary tails and comae		Thermonuclear fireballs
Circumsolar rings			Dust precipitators used to remove pollution from
Asteroids			

Table 1.
Classification of dusty plasmas.

properties which has attracted researchers over the world. These fine particles acquire some charges from the electrons to get charged. Moreover, in ordinary plasma, the charge considered to be constant on each particle, whereas, the charge on the dust particle varies with time and position [9, 10]. The charge on the dust particle generally depends on the type of dust grain, the surface properties of dust grain, the dust dynamics, the temperature, density of plasma and the wave motion in the medium. The plasma environments around these particles determine the nature of the charge (positive or negative) of these dusty plasmas. Although, most of the cases, these charged dust particles are negatively charged through different charging process. Their electric charge is determined by the size and composition of the grains [9, 10]. The fact that the frequencies associated with dust particles are smaller than those with electrons -ions and presences of fine particles modifies the dynamics of plasma motions and give rise to new types of propagating modes. For Dust acoustic waves, where ions and electrons are supposed to be inertia less pressure as gradient is balanced by the electric force, leading to Boltzmann electron and ion number density perturbations, whereas the mass of the dust play an important role in dust dynamics. In the dust acoustic wave the inertia is provided by the massive dust particles and the electrons and ions provide the restoring force. The effect of dust is to increase the phase velocity of the ion acoustic waves. This can be interpreted formally as an increase in the effective electron temperature which has important consequences for wave excitation. The dusty plasma also has a trend to oscillate at its plasma frequency [9, 10].

3. Parameters of dusty plasma

Dusty plasma and ordinary (electron-ion) plasma are different from each other due to the charge to mass ratio difference.

3.1 Dust plasma frequency

The electron and ion plasma frequency is much greater than the dust plasma frequency and it is defined as

$$\omega_{pd} = \sqrt{\frac{Z^2 e^2 n_{d0}}{\varepsilon_o m_d}} < < \omega_{pi}, \omega_{pe} \quad (1)$$

3.2 Gyro frequency in dusty plasma

When charged dust grain/particle executes a spiral motion about the magnetic lines of force, then dust particle moves perpendicular to the magnetic field with Gyro frequency of the dust particle. The centrifugal force is balanced by the Lorentz force. In mathematically,

$$\frac{m_d v^2}{r_d} = eZvB \quad (2)$$

The radius of gyration is

$$r_d = \frac{m_d v}{eZB} = \frac{v}{\Omega_{cd}} \quad (3)$$

where, $\Omega_{cd} = \frac{eZB}{m_d}$, is called the dust cyclotron frequency.

3.3 Macroscopic neutrality

The quasi-neutrality condition is obtained for the negatively charged dusty plasma by $n_{e0} = n_{i0} + Zn_{d0}$, here n_{d0} is equilibrium dust particle density and Z is the electric charge number on the dust particles.

3.4 Strongly vs. weakly coupled dusty plasma (Coulomb correlation parameter)

The property of dust particles in the plasma is expressed by coupling parameter Γ. It is the ratio of the interparticle Coulomb potential energy to the thermal energy of the particles. When the value of Γ exceeds unity, the species are termed to be strongly coupled otherwise weakly coupled dusty plasma. if r_d is the average interparticle separation between particles, then coupling parameter

$$\Gamma = \frac{e^2 Z^2}{4\pi\varepsilon_0 r_d k_B T_d} \quad (4)$$

The interparticle separation can be found out by the relation $r_d = \left(\frac{4\pi n_d}{3}\right)^{-\frac{1}{3}}$. For the typical values of $Ze = 5000e$, $k_B T_d = 0.05$ eV, $n_d = 10^{10}$ m^{-3}, the coupling parameter comes out to be 1500. It is also experienced that, when $\Gamma \sim 170$, the dust particles are found in arranged fashion and said to be Coulomb crystals.

4. Applications of ordinary plasma

Plasma technology is safe, less costly and playing important roles in every fields of daily life. Some of these applications are discussed in **Table 2**.

Fields	Applications
Telecommunication	The Global Positioning System (GPS) use ionosphere's plasma layer to reflect the signal transmitted by GPS satellite for further communication usage.
Sterilization	To sterilize the surgical equipments, which are directly connected with patient's immune system, where cleanliness is difficult
Medical treatment	Plasma treatment is contact-free, painless hardly damage tissue.
In dentistry	Plasmas treatment are used inside the root canal to kill the bacteria
Pollution controlling	Plasma technology is used to control gaseous and solid pollutions.
Water Purification	for destroying viruses and bacteria in a water, Ozone (O3) generated by plasma technology is more effective and less costly at large scale than existing chlorination method
Etching and cleaning of materials	To removes contaminants and thin layers of the substratum by bombarding with the plasma species which break the covalent bonds. It is also used to control the weight of the exposed substrate.
fusion research	Plasma is used to achieve high temperature to run the controlled thermonuclear fusion reactors
nanotechnology	Plasma discharges are helpful in growing the nanoparticles for nano world.

Table 2.
Applications of plasma in different fields.

4.1 Applications of dusty plasma

The presence of dust particles in a system also has positive impacts and has many applications in nanotechnology to synthesize the desired shape and size of the particles by controlling the dynamics of charged dust grains. Surface properties of the exposed materials could be improved by coating with plasma enhanced chemical vapor deposition method. Methane plasma is used to synthesize productive Carbon nanostructures which have like high hardness and chemical inertia. Dusty plasmas are also used for the fabrication of semiconductor chips, solar cells and flat panel displays.

5. Current status of the research

As we know that plasma support electrostatic as well as electromagnetic waves because of the motions of the charged particle. Studies of these waves provide the useful information about the state of the system. The resonance frequencies of plasmas waves can be used as diagnostics tool to characterize the plasma parameters. Plasma waves are generated for acceleration of energetic particles and heating plasmas. The exponential growing waves and modes in plasma removes the free energy from the system and permit the system to become unstable. The study of dusty plasma has gained interest in the last few decades due to its observations [1–10]and applications in the space and laboratory [1−10]. Many authors studied the linear and nonlinear electrostatic wave in the presence and absence of the external magnetic field [11–13]. Sharma and Sugawa studied the effect of ion beam in dusty plasma on ion cyclotron wave instability [13]. The presence of the charged dust grains in the plasma modifies the collective behavior of a plasma and excites the new modes [12–14].

The present charged dust particles introduces dust acoustic and dust ion acoustic waves in the plasma after altering the dynamics of electrostatic and electromagnetic waves of ordinary plasma [11–14]. The charged dust grain also introduces growing and damped modes. Tribeche and Zerguini studied the dust ion-acoustic waves in collisional dusty plasma [15]. Rao et al. [16] predicted the existence of dust-acoustic wave in an unmagnetized plasma that has inertial dust and Maxwellian distributed electrons and ions. Shukla and Silin [17] showed the existence of dust-ion acoustic wave in a plasma. Barkan experimentally investigated that negatively charged dust grains enhances the growth rate of the electrostatic ion cyclotron instability [18]. Akhtar et al. [19] studied the dust-acoustic solitary waves in the presence of hot and cold dust grains. The existence of dust-acoustic wave and dust ion acoustic wave has been confirmed by many investigators in a laboratory experiments [18–20]. Ali [21] reported the electrostatic potential due to a test-charge particle in a positive dusty plasma. Bhukhari et al. derived generalized dielectric response function for twisted electrostatic waves in unmagnetized dusty plasmas [22]. Mendonça et al. showed that a modified Jeans instability lead to the formation of photonbubble in a dusty plasma which in turn form two different kinds of dust density perturbations [23]. Pandey and Vranjes predicted that growth rate of the instability is proportional to the whistler frequency in a magnetized dusty plasma [24].

6. Plasma model and basic equations

Phase and group velocity can be calculated by finding the relation between ω and k. This relation $\omega = \omega(k)$, is called the dispersion relation and contains all the physical parameters of the given medium in which wave propagates. If the frequency has an imaginary part, that indicates an instability. Plasma instability involves some growing modes, whose amplitude increases exponentially. In other words instability represents the ability of the plasma to escape from a configuration of fields [8, 9].

6.1 Electrostatic and electromagnetic waves in ordinary (electron-ion) plasma

Charged particles in a plasmas couples to electromagnetic field. Because of this effect various kinds of waves are formed in plasmas. Plasma waves are electrostatic or electromagnetic based on perturbed (oscillated) magnetic field. If there is a perturbed magnetic field ($\vec{B}_1 \neq 0$), plasma support electromagnetic waves. If the oscillating magnetic field associated with the wave is absent ($\vec{B}_1 = 0$), then only electrostatic waves are supported by plasma. In addition, Electrostatic waves may have longitudinal and transverse component depending on the direction of propagation with the perturbed electric field [8, 9].

A thermal unmagnetized plasma support many modes as discussed by Tonks and Langmuir in 1929. One is transverse waves in a plasma have dielectric constant $\varepsilon_r(\omega) = \frac{\varepsilon(\omega)}{\varepsilon_0} = \left(1 - \frac{{\omega_{pe}}^2}{\omega^2}\right)$. In case of a lower frequency wave ($\omega < \omega_{pe}$), the dielectric constant would be negative. It turn out that if refractive index become imaginary, then waves cannot propagate but are damped (absorbed). Therefore plasma behaves like a waveguide with propagation and cut-off regions depending on the range of frequencies [8, 9].

The electron plasma wave and ion acoustic wave are the examples of electrostatic longitudinal modes, that is particle oscillate parallel to the direction of wave propagation. Ion acoustic waves are electrostatic waves, when both ions and electrons are allowed to oscillate in the wave-field. IA waves are low frequency longitudinal wave and we can use the plasma approximation, $n_{e1} \approx n_{i1} \approx n_0$.

The electron plasma wave (Langmuir mode) satisfy the dispersion relation $\omega(k) = \sqrt{\omega_{pe}{}^2 + 3k^2 \frac{k_B T_e}{m_e}}$, whereas the ion acoustic mode satisfy the dispersion relation $\omega(k) = \frac{k}{\sqrt{1+\lambda_{De}{}^2 k^2}} \frac{k_B T_e}{m_i}$, here λ_{De} is the Debye length of electron [8, 9]. The propagation of electromagnetic waves in the unmagnetized plasma yield the dispersion relation $\omega(k) = \sqrt{\omega_{pe}{}^2 + k^2 c^2}$.

7. Theoretical formulation for the studies of waves in dusty plasma

We consider a unmagnetized collisionless plasma consisting of electrons, ions and dust particles. Here we use the fluid equations and Maxwell's equations to derive the dispersion relations in dusty plasma corresponding to ordinary electron – ion plasma. We denote $\vec{v}_\alpha$ and n_α are the plasma velocity and density of the different species $(\alpha = e, i, d)$ having mass m_α, temperature T_α in electron-volt. We write the equations of continuity and equation of motion of particles to derive the dispersion relation. Then the equations of motion governing the plasma can be written as

$$\frac{\partial n_\alpha}{\partial t} + \vec{\nabla} \cdot \left(\vec{v}_\alpha n_\alpha \right) = 0 \tag{5}$$

$$\frac{d\vec{v}_\alpha}{dt} = \left\{ \frac{\partial}{\partial t} + \left(\vec{v}_\alpha \cdot \vec{\nabla} \right) \right\} \vec{v}_\alpha = \frac{Q}{m_\alpha} \left(\vec{E} + \vec{v}_\alpha \times \vec{B} \right) - \frac{T_\alpha \vec{\nabla} n_\alpha}{m_\alpha n_\alpha} \tag{6}$$

If we define thermal velocities $V_{T\alpha} = \sqrt{\frac{T_\alpha}{m_\alpha}}$.

In the above equation, the derivative $\frac{d\vec{v}_\alpha}{dt}$ is called the convective derivative. $\frac{d\vec{v}_\alpha}{dt}$ can be viewed as the time derivative of $\vec{v}_\alpha$ taken in a "fluid" frame of reference moving with a velocity of $\vec{v}_\alpha$ relative to a rest frame. $\frac{\partial \vec{v}_\alpha}{\partial t}$ represents the rate of change of $\vec{v}_\alpha$ at a fixed point in space and $\left(\vec{v}_\alpha \cdot \vec{\nabla} \right) \vec{v}_\alpha$ represents the change of $\vec{v}_\alpha$ measured by an observer moving in the fluid frame into a region where $\vec{v}_\alpha$ is inhomogeneous.

7.1 Linearization of fluid equations

We consider the perturbed density $n_{\alpha 1}$ and velocity $\vec{v}_{\alpha 1}$ indicated by subscript 1 along with their unperturbed density $n_{\alpha 0}$ and velocity $u_{\alpha 0}$. The unperturbed electric field (magnetic field) as $\vec{E}_0$ ($\vec{B}_0$) and the perturbed value of the electric field (magnetic field) is taken as $\vec{E}_1$ ($\vec{B}_1$). To linearize all the equations, let us write $n_\alpha = n_{\alpha 0} + n_{\alpha 1}$, $\vec{v}_\alpha = \vec{v}_{\alpha 1} + \vec{v}_{\alpha 0}$ and $\vec{E} = \vec{E}_1 + \vec{E}_0$. If the amplitude is chosen to be much smaller than the wavelength of the instability, the equations of motion can be linearized. The perturbed quantities $f_{\alpha 1}$ are much smaller than their unperturbed values $f_{\alpha 0}$, that is $f_{\alpha 1} << f_{\alpha 0}$. If $\vec{v}_{\alpha 0}$ and $n_{\alpha 0}$ are constant, the terms $\left(\vec{v}_{\alpha 0} \cdot \vec{\nabla} \right) n_{\alpha 0}$, $n_{\alpha 0} \left(\vec{\nabla} \cdot \vec{v}_{\alpha 0} \right)$ and $n_{\alpha 1} \left(\vec{\nabla} \cdot \vec{v}_{\alpha 0} \right)$ are equal to be zero. Further the terms $\left(\vec{v}_{\alpha 1} \cdot \vec{\nabla} \right) n_{\alpha 1}$, and $n_{\alpha 1} \left(\vec{\nabla} \cdot \vec{v}_{\alpha 1} \right)$ are neglected as they are quadratic in perturbation. The linearized form of the fluid equations can be written as

$$\frac{\partial n_{\alpha 1}}{\partial t} + n_{\alpha 0}\vec{\nabla}\cdot(v_{\alpha 1}) + v_{\alpha 0}.\vec{\nabla}n_{\alpha 1} = 0 \quad (7)$$

$$\frac{\partial v_{\alpha_1}}{\partial t} + v_{\alpha 0}\left(\vec{\nabla}\cdot v_{\alpha 1}\right) + \frac{V^2{}_{th\alpha}}{n_{\alpha_0}}\vec{\nabla}n_{\alpha 1} = \frac{Q}{m_\alpha}\left(-\vec{\nabla}\varphi_1 + \vec{v}_{\alpha 1}\times\vec{B}_0\right) \quad (8)$$

$$\varepsilon_0\nabla^2\varphi_1 = \rho = e(n_{e1} + Zn_{d1} - n_{i1}) \quad (9)$$

Let us define $\delta = \frac{n_{d0}}{n_{i0}}$ is the relative dust density, then quasi-neutrality condition follow

$$\frac{n_{e0}}{n_{i0}} = Z\delta + 1 \quad (10)$$

Thus, the assumptions of small oscillation give a set of linear equations.

7.2 Dust-acoustic waves (DAW)

The DAW is an electrostatic wave generated in dusty plasma, where inertia is provided by the dust grains. It is same to the ion-acoustic wave in general plasma, where inertia is provided by the ions. The frequency ω_{pd} of dust acoustic wave is very low due to the high dust mass than the ion (electron) plasma frequency $(\omega_{pi}, \omega_{pe})$. That why dusty plasma supports low frequencies waves. In mathematically, $\omega_{pd} = \sqrt{\frac{e^2Z^2n_{d0}}{\varepsilon_o m_d}} < < \omega_{pi}, \omega_{pe}$, where n_{d0} is the equilibrium dust density.

Let us consider a situation, when dust density is get disturbed. This change will alter the charge on the dust particles and results to an enhancing negative space charge due to the process of negative dust charging. This total space charge density of dust $\rho_d = eZn_{d0}$ is shielded by the surrounding plasma ions and electrons. Therefore an electric field is generated due to the space charge by the fluctuations of dust charge density. This oscillating electric field imparts the force on the dust particle, which further pushes the fluctuations in the direction of the electric field and thus the wave propagates.

We consider that fluctuations are plane wave, which propagating inside the dusty plasma having the form $f = f_o \exp\left\{i\left(\vec{k}\cdot\vec{r} - \omega t\right)\right\}$. Them the time derivative $(\partial/\partial t)$ can be replaced by $-i\omega$ and the gradient $\vec{\nabla}$ by ik. Here $f_1 \equiv n_{\alpha 1}$, $\vec{v}_{\alpha 1}$, $\vec{E}_1$, $\vec{B}_1$. The electrons and ions are assumed to inertia less as compared with mass of the dust grains and should have a Boltzmann distribution, namely

$$n_e = n_{e0}\exp\left(\frac{e\phi_1}{T_e}\right) \cong n_{e0}\left(1 + \frac{e\phi_1}{T_e}\right) \quad (11)$$

$$n_i = n_{i0}\exp\left(-\frac{e\phi_1}{T_i}\right) \cong n_{i0}\left(1 - \frac{e\phi_1}{T_i}\right) \quad (12)$$

Here, n_{e0} and n_{i0} denote the unperturbed values of the electron and ion density respectively. Let us limit that, all the unperturbed velocities are zero, then equation of continuity and equations of motion follows.

The above three equations can be written as

$$-i\omega\, n_{d1} + ikn_{d0}v_{d1} = 0 \quad (13)$$

$$-i\omega\, v_{d1} + ik\frac{V^2{}_{thd}}{n_{d0}}n_{d1} = \frac{Ze}{m_d}ik\phi_1 \quad (14)$$

$$-k^2\phi_1 = \frac{e}{\varepsilon_0}(n_{e0} + n_{e1} + Zn_{d0} + Zn_{d1} - n_{i0} - n_{i1}) \tag{15}$$

Eqs. (13) and (14) gives

$$n_{d1} = \frac{k^2 n_{d0}\phi_1 Ze}{m_d\left(k^2 V^2{}_{thd} - \omega^2\right)} \tag{16}$$

After substituting into Poisson's equation, we obtain

$$-k^2\phi_1 = \frac{e}{\varepsilon_0}(n_{e0} + Zn_{d0} - n_{i0}) + \frac{e}{\varepsilon_0}\left\{n_{e0}\left(1 + \frac{e\phi_1}{T_e}\right) - n_{i0}\left(1 - \frac{e\phi_1}{T_i}\right)\right\} + Z\frac{e}{\varepsilon_0}n_{d1} \tag{17}$$

The first term reduces to zero under the quasi-neutrality condition ($n_{i0} = n_{e0} + Zn_{d0}$). Let, the relative dust density is defined by $\delta = n_{d0}/n_{i0}$, then we get

$$-k^2\phi_1 = \frac{e^2 n_{i0}}{\varepsilon_0 T_{i0}}\left(1 + \frac{T_i}{T_e}(1 - \delta Z_d)\right)\phi_1 + Z\frac{e}{\varepsilon_0} \times \frac{k^2 n_{d0}\phi_1 Ze}{m_d\left(k^2 V^2{}_{thd} - \omega^2\right)} \tag{18}$$

After simplification for the nontrivial solution, we readily obtain

$$\omega^2 = k^2 V^2{}_{thd} + \frac{\omega_{pd}{}^2}{1 + \frac{1}{\lambda_{Di}{}^2 k^2}\left(1 + \frac{T_i}{T_e}(1 - \delta Z_d)\right)} \tag{19}$$

the dispersion relation of the DAW shows depends on dust density, temperature of electron, ion and dust. It also shows depends on the inertia of electron and ion.

7.2.1 Limiting cases

In the limit of cold dust ($T_d = 0$) and cold ions ($T_i << T_e$). Then, the dispersion relation simplifies into the dispersion relation of ion-acoustic wave in an ordinary plasma

$$\omega = \frac{k\lambda_{Di}\omega_{pd}}{\sqrt{1 + \lambda_{Di}{}^2 k^2}}, \tag{20}$$

which is the same as for the ion-acoustic wave in classical plasma.

7.2.2 Behavior at low wave length

For small wave numbers $k^2\lambda_{D,i}^2 << 1$ the wave is acoustic $\omega = kC_{DAW}$ with the dust-acoustic wave speed

$$C_{DAW} = \sqrt{\varepsilon Z_d^2 \frac{kT_i}{m_d}} \tag{21}$$

7.2.3 Behavior at high wave length

For large wave numbers $k^2\lambda_{D,i}^2 >> 1$, the wave is not propagating and just oscillates at the dust plasma frequency.

7.3 Dust ion acoustic wave (DIAW)

In the previous expression of ion-acoustic wave, the wave speed depends on ion temperature and on the mass of the dust. In the DIAW, the dust particles are supposed to immobile. We write the equation of motion, continuity and Poisson's equation

$$\frac{\partial n_i}{\partial t} + \vec{\nabla} \cdot (v_i n_i) = 0 \tag{22}$$

$$\frac{\partial v_i}{\partial t} + v_i \frac{\partial v_i}{\partial x} = \frac{e}{m_i}\frac{\partial \phi}{\partial x} \tag{23}$$

$$\frac{\partial^2 \phi}{\partial x^2} = -\frac{e}{\varepsilon_0}(n_i - n_e) \tag{24}$$

The Poisson's equation contains the perturbed electron and ion densities. The electrons are treated as Boltzmann distributed as follow

$$n_e = n_{e0} \exp\left(\frac{e\varphi}{kT_e}\right) \tag{25}$$

The only place, where the dust properties enter is the quasi-neutrality condition

$$n_{i0} = n_{e0} + Z_d n_{d0} \tag{26}$$

Eqs. (24), (25) and (26) gives the dispersion relation for the DIAW as

$$\omega^2 = \frac{\omega_{pi}^2 k^2 \lambda_{D,e}^2}{1 + k^2 \lambda_{D,e}^2} = \left(\frac{n_{i0}}{n_{e0}}\right) \frac{kT_e}{m_i} \frac{k^2}{1 + k^2 \lambda_{D,i}^2} \tag{27}$$

Using Eq. (25)

$$\omega^2 = \frac{\omega_{pi}^2 k^2 \lambda_{D,e}^2}{1 + k^2 \lambda_{D,e}^2} = \left(1 + \frac{Z n_{d0}}{n_{e0}}\right) \frac{kT_e}{m_i} \frac{k^2}{1 + k^2 \lambda_{D,i}^2} \tag{28}$$

It is clear from Eq. (27), that phase speed of the DIAW is increase as dust charge density increases. But the electron density has opposite effect on the speed of the DIAW.

8. Dissipative turbulence/instabilities in Hall thruster plasma

The section is devoted to the existing instabilities in a Hall thruster plasma. The principle of thrusters is the ionization of a Noble gas (propellant) in a crossed filed discharge channel. The accelerated heavy ions of inert gas are used to generate a thrust by the use of electrostatic forces. Xenon is used as an ion thruster propellant because of its low reactivity with the chamber and high molecular weight [25–33]. These types of devices support many waves and instabilities because of the turbulence nature of the plasma. These instabilities affect the performance and the efficiency of the device. In order to control these instabilities and further consequences, it has become necessary to study the growth rate of these instabilities. In a Hall thruster, the electrons experiences force along the azimuthal direction because of $\vec{E} \times \vec{B}$ drift. The collision momentum transfer frequency (v) between the electrons and neutral atoms are also taken into account to see the resistive effects in the plasma. Since ions do not feel magnetic field because of their larger larmor radius compared to length of the device. There equation of motion for ions can be written as

$$M\left\{\frac{\partial}{\partial t}+\left(\vec{v}_i\cdot\vec{\nabla}\right)\right\}\vec{v}_i = e\vec{E} \tag{29}$$

Motion of electrons under the electric and magnetic fields

$$mn_e\left\{\frac{\partial}{\partial t}+\left(\vec{v}_e\cdot\vec{\nabla}\right)+v\right\}\vec{v}_e = -en_e\left(\vec{E}+\vec{v}_e\times\vec{B}\right)-\vec{\nabla}p_e \tag{30}$$

8.1 Linearization of fluid equations

Let us denote the perturbed densities for ions and electrons as n_{i1} and n_{e1} velocities as $\vec{v}_{i1}$ and $\vec{v}_{e1}$ respectively. The unperturbed velocities v_0 and u_0 are taken in the x- and y-direction respectively. The amplitude of oscillations of the perturbed densities are taken small enough. The linearized form of Eq. (29) and Eq. (30) are written as

$$M\left(\frac{\partial}{\partial t}+v_0\frac{\partial}{\partial x}\right)\vec{v}_{i1} = -e\vec{\nabla}\phi_1 \tag{31}$$

$$m\left(\frac{\partial}{\partial t}+u_0\frac{\partial}{\partial y}+v\right)\vec{v}_{e1} = e\left(\vec{\nabla}\phi_1-\vec{v}_{e1}\times\vec{B}_0\right)-\frac{T_e\vec{\nabla}n_{e1}}{n_0} \tag{32}$$

The continuity equations of electrons and ions can be linearized as below

$$\left(\frac{\partial}{\partial t}+v_0\frac{\partial}{\partial x}\right)n_{i1}+n_0\left(\vec{\nabla}\cdot\vec{v}_{i1}\right)=0 \tag{33}$$

$$\left(\frac{\partial}{\partial t}+u_0\frac{\partial}{\partial y}\right)n_{e1}+n_0\left(\vec{\nabla}\cdot\vec{v}_{e1}\right)=0 \tag{34}$$

Fourier analysis: We seek the sinusoidal solution of the above equations, therefore the perturbed quantities are taken as $f_1 \sim f_0 \exp\left(i\omega t - i\vec{k}\cdot\vec{r}\right)$. Them the time derivative $(\partial/\partial t)$ can be replaced by $i\omega$ and the gradient $\vec{\nabla}$ by ik, here $f_1 \equiv n_{i1}$, n_{e1}, ϕ_1, $\vec{v}_{i1}$, $\vec{v}_{e1}$, $\vec{E}_1$ together with ω as the frequency of oscillations and $\vec{k}$ as the propagation vector.

By using Fourier analysis from Eq. (31)–(34), the perturbed ion and electron densities are given as follows,

$$n_{i1} = \frac{ek^2 n_0\phi_1}{M(\omega-k_x v_0)^2} \tag{35}$$

$$n_{e1} = \frac{en_0(\omega-k_y u_0-iv)k^2\phi_1}{m\Omega^2(\omega-k_y u_0)+m(\omega-k_y u_0-iv)k^2{V_{th}}^2} \tag{36}$$

The expression for the electron density n_{e1} is derived under the assumptions that $\Omega >> \omega$, $k_y u_0$ and v in view of the oscillations observed in Hall thrusters.

8.2 Dispersion equation and growth rate of electrostatic oscillations

Finally, we use the expressions for the perturbed ion density n_{i1} and electron density n_{e1} in the Poisson's equation $\varepsilon_0 \nabla^2 \phi = e(n_{e1} - n_{i1})$ in order to obtain

$$-k^2\phi_1 = \frac{\omega_e{}^2\hat{\omega}k^2\phi_1}{\Omega^2(\omega - k_y u_0) + \hat{\omega}k^2 V_{th}{}^2} - \frac{\omega_i{}^2 k^2 \phi_1}{(\omega - k_x v_0)^2} \tag{37}$$

For the nontrivial solution of the above equation, the perturbed potential $\phi_1 \neq 0$, we have from Eq. (37)

$$\frac{\omega_e{}^2\hat{\omega}}{\Omega^2(\omega - k_y u_0) + \hat{\omega}k^2 V_{thE}{}^2} + \frac{(\omega - k_x v_0)^2 - \omega_i{}^2}{(\omega - k_x v_0)^2} = 0 \tag{38}$$

This is the dispersion relation that governs the electrostatic waves in the Hall thruster's channel. In the above equations, we introduced parameter $\omega_{e(i)} = \sqrt{\frac{e^2 n_0}{m(M)\varepsilon_0}}$, $\Omega = \frac{eB_0}{m}$ and $V_{th} = \sqrt{\frac{Y_e T_e}{m}}$.

After simplification of Eq. (38) we obtain

$$\begin{aligned}
&\omega^3(\omega_e{}^2 + \Omega^2 + k^2 V_{th}{}^2) - \omega^2\left[(\omega_e{}^2 + \Omega^2 + k^2 V_{th}{}^2)(k_y u_0 + 2k_x v_0) + iv(\omega_e{}^2 + k^2 V_{th}{}^2)\right] \\
&+\omega\left[k_x{}^2 v_0{}^2 \omega_e{}^2 + 2k_x v_0(k_y u_0 \omega_e{}^2 + iv\omega_e{}^2 + ivk^2 V_{th}{}^2) + (k^2 V_{th}{}^2 + \Omega^2)(2k_x v_0 k_y u_0 + k_x{}^2 v_0{}^2 - \omega_i{}^2)\right] \\
&-(k_x{}^2 v_0{}^2 - \omega_i{}^2)\left[k_y u_0(\Omega^2 + k^2 V_{th}{}^2) + ivk^2 V_{th}{}^2\right] - \omega_e{}^2 k_x{}^2 v_0{}^2 (k_y u_0 + iv) = 0
\end{aligned} \tag{39}$$

This is the dispersion equation that governs the electrostatic waves in the Hall thruster's channel. It is clear from the above equation that Hall thruster support different waves and instabilities which satisfies the dispersion relation (39).

8.3 Results and discussion

To estimate the growth rates of the instability, we numerically solve Eq. (39) by giving typical values of all parameters used for the thruster [25–32]. Therefore, for investigating the growths of the waves, we plot the negative imaginary parts of the complex roots (correspond to the instabilities) in **Figures 1–3**.

It is found that the growth of the wave is enhanced for the larger values of the wave number of the oscillations as shown in **Figure 1**. In others words, the oscillations of larger wavelengths are stable. The findings are consistent to the results predicted by Kapulkin et al. [34]. In **Figure 2**, the variation of growth of the wave with the momentum transfer collision frequency is shown and it has been depicted that instability grow at faster rates in the presence of more electron collisions. This is mainly due to the resistive coupling with electrons' drift's in the presence of more collisions. The similar results are also predicted by Fernandez et al. [35] in the simulation studies of dissipative instability which shows that the growth of the instability is directly proportional to the square root of the collision frequency.

In **Figure 3** it is noted that the wave grows faster if the electrons carry higher temperature. The present finding matched with the same results observed by other investigators [36, 37] of dissipative instability in an plasma.

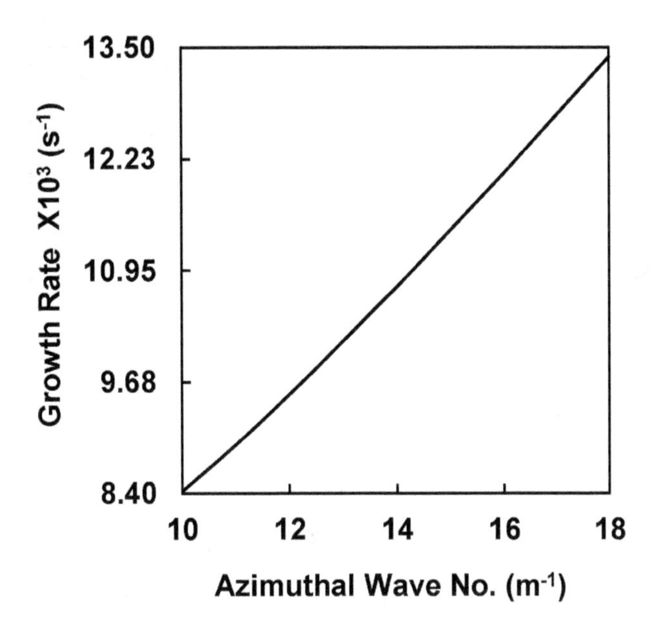

Figure 1.
Growth rate versus azimuthal wave number, with parameters of Hall plasma thrusters as $B_0 \sim 100 - 200\ G$, $n_0 \sim 5 \times 10^{17}\ 10^{18}\ m^3$, $T_e = 10 - 15\ eV$, $u_0 \sim 10^6\ m/s$, $v \sim 10^6\ /s$ *and* $v_0 \sim 2 \times 10^4 - 5 \times 10^4\ m/s$.

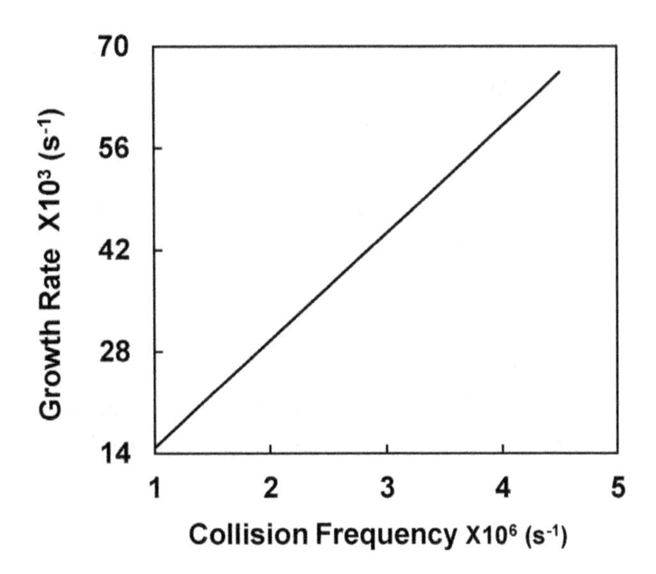

Figure 2.
Growth rate versus collision frequency.

8.4 Conclusion

In summary, we can say that dusty plasma physics has vital role in novel material processing and diagnostics tools. The nanoparticles of desired shape can be synthesized by controlling the dynamics of charged particles in the semiconductor industry. For example, the rotation of the dust particles can extract the electron flux in the magnetron sputtering unit. Thus, dusty plasma is a remarkable field in all areas of natural sciences. Though, the elimination of dust particles in the semiconductor industry is still a main alarm. The advanced development tools to learning

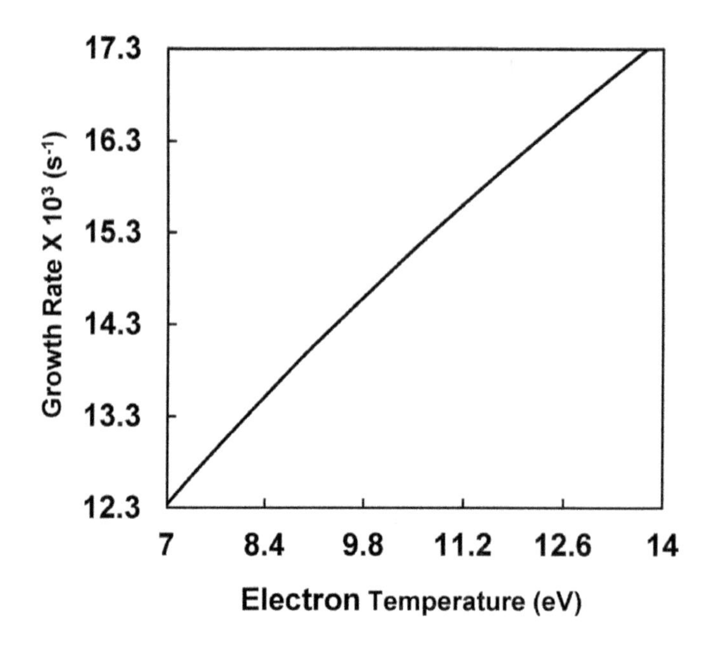

Figure 3.
Variation of growth rate with electron temperature.

dusty plasma will lead to additional discoveries about the astrophysical events and new scientific findings. The different dispersion relations are derived to see the behavior of the wave with wave number in complex plasma. The theory of linearization has been used for the smaller amplitude of the oscillations to derive the perturbed quantities. In the last section, the growth rate of dissipative instability has been depicted in a Hall thruster turbulence plasma. The instability grow faster with the collision frequency, azimuthal wave number and the electron temperature.

Acknowledgements

The University Grants Commission (UGC), New Delhi, India is thankfully acknowledged for providing the startup Grant (No. F. 30-356/2017/BSR).

Author details

Sukhmander Singh
Plasma Waves and Electric Propulsion Laboratory, Department of Physics, Central University of Rajasthan, Ajmer, Kishangarh, India

*Address all correspondence to: sukhmandersingh@curaj.ac.in

References

[1] Shukla PK, Mamun AA. Series in Plasma Physics: Introduction to Dusty Plasma Physics. Bristol: Taylor & Francis Group, CRC Press; 2001

[2] Kersten H, Thieme G, Fröhlich M, Bojic D, Tung D H, Quaas M, Wulff H, Hippler R. Complex (dusty) plasmas: Examples for applications and observation of magnetron induced phenomena 2005 Pure and Applied Chemistry. 77 415. DOI: 10.1351/pac200577020415

[3] Merlino RL. Dusty plasmas and applications in space and industry. Plasma Physics Applied. 2006. ISBN: 81-7895-230-0

[4] Bonitz M, Henning C, Block D. Complex plasmas: A laboratory for strong correlations. Reports on Progress in Physics 2010;**73**(6):066501. DOI: 10.1088/0034-4885/73/6/066501

[5] Dusty Plasmas in the Laboratory, Industry, and space physics. Today. 2004;**57**(7):32. DOI: 10.1063/1.1784300

[6] Kersten H, Wolter M. Complex (dusty) plasmas: Application in material processing and tools for plasma diagnostics. Introduction to Complex Plasmas. 2010 (pp. 395–442). Springer, Berlin, Heidelberg.

[7] Shukla PK. Dusty Plasmas: Physics, Chemistry and Technological Impacts in Plasma Processing. In: Bouchoule A. Wiley, New York. 2000

[8] Bellan PM. Fundamentals of Plasma Physics. 1st ed. UK: Cambridge University Press; 2008

[9] Chen FF. Introduction to Plasma Physics and Controlled Fusion. 2nd ed. New York: Springer-Verlag. 2006; p. 200

[10] Vladimir EF, Gregor EM. Complex and Dusty Plasmas: From Laboratory to Space. Bosa Roca, United States: Taylor & Francis Inc, CRC Press Inc.; 2010

[11] Varma RK, Shukla PK, Krishan V. Electrostatic oscillations in the presence of grain-charge perturbations in dusty plasmas. Physical Review E. 1993;**47**(5): 3612. DOI: 10.1103/PhysRevE.47.3612

[12] Cui C, Goree J. Fluctuations of the charge on a dust grain in a plasma. IEEE Transactions on Plasma Science. 1994; **22**(2):151–158. DOI: 10.1109/27.279018

[13] Sharma SC, Sugawa M. The effect of dust charge fluctuations on ion cyclotron wave instability in the presence of an ion beam in a plasma cylinder. Physics of Plasmas. 1999;**6**(2):444–448 (1999).

[14] Merlino RL. Current-driven dust ion-acoustic instability in a collisional dusty plasma. IEEE Transactions on Plasma Science. 1997;**25**(1):60–65. DOI: 10.1109/27.557486

[15] Tribeche M, Zerguini TH. Current-driven dust ion-acoustic instability in a collisional dusty plasma with charge fluctuations. Physics of Plasmas. 2001;**8** (2):394–398. DOI: 10.1063/1.1335586

[16] Rao NN, Shukla PK, Yu MY. Dust-acoustic waves in dusty plasmas. Planetary and Space Science. 1990;**38** (4):543–546. DOI: 10.1016/0032-0633(90)90147-I

[17] Shukla PK, Silin VP. Dust ion-acoustic wave. Physica Scripta. 1992;**45**: 508. DOI: 10.1088/0031-8949/45/5/015

[18] Barkan A, Merlino RL, D'angelo N. Laboratory observation of the dust-acoustic wave mode. Physics of Plasmas. 1995;**2**(10):3563–3565. DOI: 10.1063/1.871121

[19] Morfill GE, Thomas H. Plasma crystal. Journal of Vacuum Science & Technology, A: Vacuum, Surfaces, and

Films. 1996;**14**(2):490–495. DOI: 10.1116/1.580113

[20] Barkan A, D'angelo N, Merlino RL. Experiments on ion-acoustic waves in dusty plasmas. Planetary and Space Science. 1996;**44**(3):239–242. DOI: 10.1016/0032-0633(95)00109-3

[21] Ali S. Potential distribution around a test charge in a positive dust-electron plasma. Frontiers of Physics. 2016;**11**(3): 115201. DOI: 10.1007/ s11467-015- 0545-2

[22] Bukhari SS. Ali, Rafique M, Mendonca JT. Twisted electrostatic waves in a self-gravitating dusty plasma. Contributions to Plasma Physics. 2017; **57**:404–413. DOI: 10.1002/ctpp.201700063

[23] Mendonça JT, Guerreiro A, Ali S. Photon bubbles in a self-gravitating dust gas: Collective dust interactions. The Astrophysical Journal. 2019;**142**(6pp): 872. DOI: 10.3847/1538-4357/aafe7e

[24] Pandey BP, Vranjes J. Physics of the dusty Hall plasmas. Physics of Plasmas. 2006;**13**(12):122106. DOI: 10.1063/ 1.2402148

[25] Kaufman HR. Technology of closed drift thrusters. AIAA Journal. 2012;**23** (1):78–86. DOI: 10.2514/3.8874

[26] Goebel DM, Katz I. Fundamentals of Electric Propulsion: Ion and Hall Thrusters. New York: Wiley; 2008

[27] Jahn RG. Physics of Electric Propulsion. New York: McGraw-Hill; 1968.

[28] Singh S, Malik HK, Nishida Y. High frequency electromagnetic resistive instability in a Hall thruster under the effect of ionization. Physics of Plasmas 2013;**20**:102–109 (1–7).

[29] Singh S, Malik HK. Growth of low frequency electrostatic and electromagnetic instabilities in a Hall thruster. IEEE Transactions on Plasma Science. 2011;**39**:1910–1918

[30] Singh S, Malik HK. Resistive instabilities in a Hall thruster under the presence of collisions and thermal motion of electrons. The Open Plasma Physics Journal. 2011;**4**:16–23

[31] Malik HK, Singh S. Resistive instability in a Hall plasma discharge under ionization effect. Physics of Plasmas. 2013;**20**:052115(1–8)

[32] Singh S, Malik HK. Role of ionization and electron drift velocity profile to Rayleigh instability in a Hall thruster plasma: Cutoff frequency of oscillations. Journal of Applied Physics. 2012;**112**:013307 (1–7)

[33] Malik HK, Singh S. Conditions and growth rate of Rayleigh instability in a Hall thruster under the effect of ion temperature. Physical Review E. 2011; **83**:036406 (1–8)

[34] Kapulkin A, Kogan A and Guelman M. Noncontact emergency diagnostics of SPT in flight. Acta Astronautica. 2004;**55**:109–119

[35] Fernandez E, Scharfe MK, Thomas CA, Gascon N, and Cappelli MA. Growth of resistive instabilities in $\vec{E} \times \vec{B}$ plasma discharge simulations. Physics of Plasmas. 2008;**15**:012102(1-10)

[36] Alcock M W and Keen B E. Experimental observation of the drift dissipative instability in an afterglow plasma. Physical Review A. 1971;**3**:1087–1096

[37] A. Kapulkin and M. M. Guelman. Low-frequency instability in near-anode region of Hall thruster. IEEE Transactions on Plasma Science. 2008; **36**:2082–2087

14

Studies of Self-Diffusion Coefficient in Electrorheological Complex Plasmas through Molecular Dynamics Simulations

Muhammad Asif Shakoori, Maogang He, Aamir Shahzad and Misbah Khan

Abstract

A molecular dynamics (MD) simulation method has been proposed for three-dimensional (3D) electrorheological complex (dusty) plasmas (ER-CDPs). The velocity autocorrelation function (VACF) and self-diffusion coefficient (D) have been investigated through Green-Kubo expressions by using equilibrium MD simulations. The effect of uniaxial electric field (M_T) on the VACF and D of dust particles has been computed along with different combinations of plasma Coulomb coupling (Γ) and Debye screening (κ) parameters. The new simulation results reflect diffusion motion for lower-intermediate to higher plasma coupling (Γ) for the sufficient strength of $0.0 < M \geq 1.5$. The simulation outcomes show that the M_T significantly affects VACF and D. It is observed that the strength of M_T increases with increasing the Γ and up to $\kappa = 2$. Furthermore, it is found that the increasing trend in D for the external applied M_T significantly depends on the combination of plasma parameters (Γ, κ). For the lower values of Γ, the proposed method works only for the low strength of M_T; at higher Γ, the simulation scheme works for lower to intermediate $M_{T,}$ and D increased almost 160%. The present results are in fair agreement with parts of other MD data in the literature, with our values generally overpredicting the diffusion motion in ER-CDPs. The investigations show that the present algorithm more effective for the liquids-like and solid-like state of ER-CDPs. Thus, current equilibrium MD techniques can be employed to compute the thermophysical properties and also helps to understand the microscopic mechanism in ER-CDPs.

Keywords: Diffusion coefficient, electrorheological complex plasma, Molecular dynamics simulations, uniaxial electric field

1. Introduction

1.1 Electrorheological fluids

The dielectric fluids consisting of micro-sized (0.1-100 μm) solid particles, which display particular characteristics under the influence of the external applied electric field, are known as electrorheological fluids (ERFs). The dielectric fluids, such as olive oil, silicon oil hydrocarbon, etc., have low permittivity, conductivity and viscosity. The solid particles immersed in dielectric fluids are mostly polymers, metal oxide silica, and alumina silicates. These particles maintained the low viscosity of carrier fluids in the absence of external electric field strength. Without an external electric field, particles behave as a liquid. When the external electric field is turned on, these particles behave like solids due to changes in viscosity. The ERFs change their physical properties for the application of the external electric field. When the electric field is turned on the suspension of dielectric fluids, the solid particles are polarized and make a thin chain (string) along the direction of the applied electric field. The thickness of the particles depends on the intensity of the electric field. The viscosity of ERFs increased with increases an external applied electric field. If the electric field was turned off, the fluids reverse from solid to liquids within milliseconds. These fluids are also known as intelligent fluids. There are various types of rheological fluids, such as electrorheological fluids, magnetorheological fluids, positive electrorheological, negative electrorheological fluids etc. These fluids rapidly respond to the electric field and change their physical properties such as shear stress, elastic modulus and viscosity. These fluids are used in vehicle engineering, such as valves, clutches and breaks etc. The conventional ERFs have various industrial applications, such as vibration control in smart materials. Changes in the microstructure and physical properties are used in medical to control ultrasonic transmissions and sound transmission with low losses [1, 2].

1.2 Plasma

Plasma is a partially ionized gas that contains electrons, ions and neutral radicals. In our universe, 99% of physical matter is in the plasma state. In space, the most visible matter is in the plasma state; the sun and stars are the main examples of plasma in our universe. Plasma species show collective behaviors when any external perturbation is applied. The whole plasma is perturbed when an external force has applied this behavior of particles called collective behavior of plasma. Classification of plasma depends on the species temperatures such as hot plasma, cold plasma, ultracold plasma, ideal and non-ideal plasma etc. [3].

1.2.1 Complex (dusty) plasmas

Complex (dusty) plasmas (CDPs) contained micro to submicron-sized conductive, and dielectric particles called grain in addition to plasma species (neutral atom, electron, positive or negative ions). The conductive grain has a $3e^{-11}$ kg mass and about e^4 eV charge. Mostly having a negative charge but in the rear case also have positive charge depend on charging phenomena. Dust particles are naturally present in the plasma and can be manually inserted in a plasma medium through sputtering, etching and chemical reaction—the dust particles made by a single element or composition of different elements. The dust particles like a swimmer in the sea of electrons and ions and respond to electromagnetic forces. There are different mechanisms of charging dust particles. The charge amount depends on the charging phenomena. Dusty plasmas illustrate an astrophysical matter in white dwarfs, neutron stars, giant planetary interiors and supernova core. In laboratory ultra-cold plasma, charged stabilized collides and electrolytes, laser-cooled ions in cryogenic traps, and dusty plasma. The warm dense matter and strongly coupled complex (dusty) plasmas (SCCDPs) are relevant models for nuclear fusion devices [3–6].
The CDPs are classified according to the energies of interacting charged dust particles. When interacting particle's potential energy exceeds their kinetic energy,

then CDPs are called as SCCDPs. This system appears in a wide variety of physical systems. The Weakly coupled complex (dusty) plasmas (WCCDPs) are inverts of SCCDPs. The SCCDPs also in high order structural form or exist in the crystalline state at low temperature with high density. The WCCDPs are mostly remained in the gaseous state of the plasma and having high temperature with low density. Various states of dusty plasma are easily observed through a video camera under laboratory conditions. SCCDPs are found in nature and also in laboratory experiments [7, 8]. The phase transition (condensation) can be observed by reducing the temperatures of CDPs. The structural order of dust particles is formed under some external conditions. The dusty plasmas are encountered in astrophysics and are extensively believed to play a significant role in cosmology to reveal the structure and origin of the universe and its galaxies. The transport and thermodynamic properties are well studied for CDPs through experiment, theoretical and computational methods. Thermophysical properties are well investigated in the recent decade, such as thermal conductivity [9–11], diffusion phenomena [12, 13], shear viscosity [13], thermodynamic properties [14], dynamical structure factor [4, 15] and propagation of different waves [3, 5].

1.3 Applications of complex dusty plasmas

Initially, CDPs originate in astrophysics and space physics; nowadays, it becomes a fascinating, applicable field in space physics as an analogue to unravel issues like the role of dust accumulation super high speed crashed in space and the formation of planets and many industries on different scales. The CDPs play an essential role as an analogue system for investigating multifaceted cross-disciplinary phenomena, such as an experimental study of non-linear dynamics and long-range interaction in strongly correlated systems. These systems often have challenging investigations because they generally require extreme temperature and pressure conditions, such as very low temperature and high density. These systems belong to different research institutes. The CDPs allow the study and formation of the strongly coupled systems under laboratory conditions at room temperature and for easily attainable pressure. The CDPs analogue was recently used to model crystallization dynamics in 2D and excitation of quantum dots, viscoelastic material, shock and non-linear waves and recently electrorheological fluids [16].

In industrial applications, CDPs are directly applicable in the processing of microelectronics devices. It is used for deposition integrated circuits, masking, and stripping. The capacitors and transistors make on the silicon wafer chip millions of transistors put on the Pentium chip with the help of plasma to safe from contaminations. Dust particles have both advantages and disadvantages in the technologies. It plays a vital role in the scientific research of various technologies and industries. It plays a significant role during the thin-film depositions, processing of ceramic, insulation, filtration processes, petroleum industry, biomechanics, paper industry, packed bed reactors etc. Dust particles are helped to enhance the efficiency and stability of solar cells, LED (light emitted diode), improving lighting source, display, and laser technology. Through plasma processing, the coating of particles has been produced. It has grown or modified existing materials in the semiconductor industry [17]. The scientific communities currently focus on controlling nuclear fusion reactions and developing devices such as tokamak to produce efficient and carbon-free energy.

In the field of medicine and healthcare, the CDPs have become an emerging field, which combines plasma physics, life science, and clinical medicine. In the area of life science, it is directly used in biological medicine such as double-helix molecular interaction, sterilizing surgical instruments and implants, wound healing, cancer therapy, break DNA damage for human prostate cancer [18]. The CDPs were found to help kill cancer cells without affecting the healthy tissues. It is also used to inactive the bacterial in the field of health care. It is also used for food storage technologies. Furthermore, the CDPs are helpful as a diagnostic technique for a precise calculation of plasma parameters such as non-linear laser spectroscopy, spectrally resolved nanosecond imaging, phase-resolved optical emission spectroscopy, and laser-induced fluorescence VUV, UV, VIS and IR, etc. [19].

1.4 Electrorheological complex dusty plasma

CDPs have electrorheological characteristics like conventional electrorheological fluids when an external ac electric field is applied. The dust particles respond quickly to the external applied electric field and make a chain (string) sheet-like as conventional ERFs so-called electrorheological CDPs (ER-CDPs). The first time ER-CDPs were observed by Ivlev A. V. et al., in 2008 during the microgravity experiments (PK-3 plus laboratory) and MD simulation. They show the phase transition of CDPs from isotropic to anisotropic with increasing the external applied electric field [20]. Under the influence of electric field dc mode, attractive attractions between charged dust particles are introduced by ions streaming. In this way, wake potential behind the dust grains produced, which make particles strings (sheet), in the dc discharge plasma sheath, where charged dust particles levitated against the electric field's gravity force. Such a type of system is non-Hamiltonian, and wake potential is asymmetric. The ac electric fields have a much smaller frequency than ion plasma frequency, but larger than dust plasma frequency was applied to the complex plasma. In this way, the wake potential is symmetric, and a system known as Hamiltonian due to the electric field, dipole–dipole attractions increased in ER-CDPs, particles arrange themselves in sheet, string, or crystalline structure same as conventional ERFs. The phase transition in ER-CDPs has studied experiments and simulations method [20–23].

After discovering ER-CDPs, it opens up new dimensions of research for plasma science and technologies communities. There is little literature available to understand the macroscopic phenomena of ER-CDPs. Ivlev A. V. et al., done a PK-3 plus experiments under microgravity conditions and molecular dynamics simulations, observed phase transition from an isotropic to string with increasing external ac electric field [24]. Later they have done *PK*-4 dc discharge experiment and observed an anisotropic structure under the influence of an external ac electric field [22]. Yaroshenko V. V. theoretically studied the propagations of dust lattice waves along the electric field in a one-dimensional string. He found the instability leads to spontaneous excitations of compressional waves [25]. Rosenberg *M.* theoretically studied the formation and excitation of waves in one-dimensional ER-CDPs under the ac electric field [26]. Kana *et al.* explored the phase transition in ER-CDPs using MD simulation and observed the anisotropic structure under the influence of ac electric field and did not find the anisotropic structure for the dc electric field mode [21]. Sukhinin *et al.* used a Monte Carlo (MC) simulation of plasma polarization around dust particles in an external applied electric field. They mentioned that due to induced dipole potential, the formation of dust particles alignment, chain (string) multi-layered structure, and coagulation of charged dust particles [27]. A self-consistent model was developed for plasma anisotropy (string) of charged dust particles under the external electric field's action by Sukhinin *et al.* [28]. No evidence has been found of thermophysical properties in ER-CDPs. In the future, for precise tailoring, new materials may be modeled with the help of ER-CDPs. The ER-CDPs can play a significant role in modeled new smart materials. It is also be used to generate negative dipolar interparticle interactions [24].

1.5 Diffusion coefficient

The diffusion coefficient is one of the transport properties of gas, liquids. During the diffusion process, the particles migrate from high concentrations to low concentrations. The diffusion rate can be controlled by variations of external parameters such as temperature, concentration, external electromagnetic forces (electric and magnetic). Investigations of diffusion have different purposes in liquids, such as exploring the dynamical properties and understanding the microscopic phenomena. For the CDPs research direction, the investigations of diffusion motion for applying external fields such as electric and magnetic become a hot topic in current research. Different types of diffusion motion exist in dusty plasma regimes depends on temperature and forces. Diffusion also plays an essential role in exploring the dynamical properties (structure, waves, and instabilities) of many biological, physical, and chemical systems. The diffusion motion of dust particles in CDPs continues as one of the active research topics, an essential consequence of understanding the transport and dynamical properties [12, 29–33]. It is one of the primary sources to lose energy (stopping power) in CDPs. Therefore, we can easily understand the microscopic phenomena of particles in the different states [34]. An extensive amount of previous studies have been made to understand the diffusion motion in CDPs. Molecular dynamics simulations were performed and investigate the velocity autocorrelation function for 2D WCCDPs [12], Langevin dynamics simulation for 2D SCCDPs [35] and diffusion coefficient [36]. For the investigations of self-diffusion coefficient MD simulations were performed for 3D CDPs [30, 33, 37], one-component plasma [31] and ionic mixtures [32].

The main objective of the present book chapter is to give an overview of electrorheological (ER) fluids and ER-CDPs. We have computed the effect of the external applied uniaxial electric field on the velocity autocorrelation function and self-diffusion coefficient of 3D ER-CDPs using MD simulations over a wide range of input CDPs parameters.

2. MD simulation algorithm and parameters

The Computer simulation provides a linkage between theoretical and experimental research work. In thermal fluids, science and engineering, the installations of experimental setups are very complex and high cost. In the age of modern technology, the cost-effective and low time consumption high computational power is most prominent. Nowadays, it is a trend that before starting the experimentations, first test with computational tools than verified with experiments. There are various computer algorithms and techniques are used for the calculation of various properties in various materials. We perform computer simulations to test a theoretical model, verify experimental data, and also for comparison purposes. [38]. Different computational techniques can be designed for extreme conditions, such as for very low temperature and high density. It also acts as a bridge between microscopic length, time scale, and macroscopic worlds. MD and Monte Carlo (MC), and Langevin dynamics simulations are the main methods used to compute CDPs' physical properties. Different software's are also available to compute various physical properties and build a new model by following one simulation scheme.

MD simulations have become a prominent computational tool to investigate various properties such as thermophysical and dynamical properties in different types of fluids and materials. The MD simulations consist of the numerical solution of the Newton equation of motion for the spherical particles system [9]. This section explained the MD simulation algorithm to calculate the self-diffusion coefficient (D). Equilibrium MD simulations are used to investigate the D in ER-CDPs. Yukawa potential (screened Coulomb) is the most commonly used potential for CDPs systems, while other physical systems such as physics of chemical and polymer, medicine and biology systems, astrophysics, and environmental research are adopted. Most scholars used Yukawa potential to screen charged dust particles in the dusty plasma results in isotropic interactions. In the present research used applied a uniaxial ac electric field for additional dipole–dipole-like interactions. This idea was taken from Pk-3 plus experiment under the microgravity conditions [20], theoretical approach [39], MD simulation study [24], and *Pk*-4 experiments [23]. For ER-CDPs, the charged dust particles interact with each other through Yukawa potential and Quadrupole interactions due to external applied ac electric field: the equation for ER-CDPs becomes as

$$W(r,\theta) = Q_d^2\left[\frac{1}{4\pi\varepsilon_o}\frac{\exp^{-r/\lambda_D}}{r} - 0.43\frac{M_T^2\lambda_D^2}{r^3}\left(3\cos^2\theta - 1\right)\right] \tag{1}$$

First-term in the above equation presents pair-wise Yukawa potential in the absence of an electric field, and the second term shows the Quadrupole interaction between particles due to uniaxial ac external electric field [22]. Where λ_D is Debye length, Q_d is the charge on dust particles, r distance between interacting particles, and ϵ_o permittivity of space. The θ is the angle between the electric field ($\boldsymbol{E}$) and interacting dust particles. In the present study, we take $\theta = 0°$ for uniaxial ac electric field for anisotropic interactions. The $\boldsymbol{r}$ is the distance between the interacting particles, $\boldsymbol{r} = \boldsymbol{r}_j - \boldsymbol{r}_i$. The M_T is thermal Mach number normalized with the thermal velocity of charged dust particles $M_T^2 = \langle u_d^2\rangle/v_T^2$ where $v_T^2 = T_d/m_d$ the drift velocity is proportional to an electric field ($u_d \propto E$), an electric field can be measured in the unit of M_T. For small values of M_T, in ER-CDPs, the interactions of particles are the same as dipolar interaction in conventional ERFs. In the prior study, strings (chain) of dust particles were observed in typical conditions when an external ac electric field was applied.

There are two central (dimensionless) parameters, which are fully-characterized complex plasma systems. The first parameter is the plasma coupling parameter (define same as Coulomb systems) $\Gamma = Q_d^2/4\pi\epsilon_o a_{ws} k_B T$, a_{ws} Wigner-Seitz radius defined as $(3/4\pi n)^{\frac{1}{3}}\pi n^{\frac{-1}{2}}$ Where n is the equilibrium dust number density, k_B the Boltzmann constant, and T is the system's absolute temperature. The plasma coupling (Γ) parameter is the ratio of average potential energy to the average kinetic energy of interacting dust particles. The SCCDPs associated with $\Gamma > 1$ inversely account as weakly coupled dusty plasma ($\Gamma < 1$). Another essential parameter is the Debye screening parameter, $\kappa = a_{ws}/\lambda_D$. The plasma dust frequency $\omega_{pd} = \left(Q_d^2/3\pi\epsilon_0 m a^3\right)^2$ describes the time scale of a complex plasma system; here, m is the dust particle's mass.

2.1 Green-Kubo relation of diffusion coefficient

Green-Kubo integral formula were used to calculate the self-diffusion coefficient (D) for 3D CDPs [30, 33], one component plasma [31] and ionic mixtures [32]. Here we have used the same numerical models for the investigations of D in 3D ER-CDPs; the Green-Kubo relation is given as following.

$$D = \frac{1}{3N}\int_0^\infty Z(t)dt \tag{2}$$

In Eq. (2), $Z(t)$ is known as the velocity autocorrelation function (VACF), 3 shows the 3D system, and N represents the number of simulated dust particles. The D is the integral of VACF calculated over all the particle velocity product's ensemble average segments at a time (t) and initial time (t_0). If $Z(t)$ decays too slowly, Yukawa particles' motion is described as anomalous diffusion for the integral equation one converges. The transport coefficient's existence, autocorrelation function must rapidly decay enough for integral to convergence. Different types of diffusion motion were analyzed through VACF [12], defined as

$$Z(t) = \langle v_j(t).v_j(0) \rangle \tag{3}$$

Where $v_j(t)$ denote the jth particle's velocity at simulation time (t) and (0), the brackets $\langle . \ldots \rangle$ represent the canonical ensemble average all over particles. $Z(t)$ demonstrates the decay that is characterized by plasma frequency. The existence of the transport coefficient, autocorrelation function must rapidly decay enough for integral convergence [29, 33, 34].

In this method, classical molecular dynamics simulations are used to map the trajectories of $N = 500$ and calculated the diffusion coefficient. Periodic boundary conditions (PBCs) were used to be constrained enforced during the N-particles simulation under statistical microcanonical ensemble (NVE). The edge length of the 3D cubic simulation box is L with $12.79a$. Particles are placed randomly in the simulation cubic cell for the beginning of the simulations [15]. An appropriate time step ($dt = 0.001\omega_{pd}$) was selected to calculate particle position and velocity at each step. The leapfrog integration method is used to integrate the equation of motion and calculate each particle's force, acceleration, and position for every time step. The whole scheme in MD dynamically simulates the equation of motion of N-dust particles with interaction through Yukawa potential, and Quadrupole interaction was given in Eq. (1). The Ewald summation method was used for calculations of Yukawa and dipolar potential energies and forces. Here we select the Ewald convergence parameter ($\gamma = 5.6/L$) for the Yukawa and anisotropic interactions. Suitable input parameters are selected to get the accuracy and consistency of the model. The appropriate system temperature $((T = 1/\Gamma)$ Yukawa Ewald summation method, screening length, Mach number, time step, number of particles, simulation run time, etc., are selected to provide 3D ER-CDPs diffusion motion. Negative derivatives of Eq. (1) calculate forces on each interacting charged dusty particle in the Yukawa system. MD simulations are performed $2 \times 10^5/\omega_{pd}$ time unit to record the SDC (Γ, κ, M_T). The 12-core-processor takes about three hours to perform each simulation. Total 156 simulations are performed and calculated the VACF and D for a wide range of plasma parameters and external applied uniaxial electric field. For the present chapter, the EMD method is used of ER-CDPs and calculated the D over a wide range of $\Gamma = 2$ –500, $1 < \kappa \geq 3$, and $N = 500$. In addition, the conservation of total energies and momentum is checked and verify that system has self-consistency.

3. Simulation results and discussions

This book chapter used the equilibrium MD simulation method using Yukawa (screened Coulomb) potential and dipolar interactions. Without external forces (electric and magnetic), Yukawa potential was used and calculated thermophysical properties. In this work, we add dipolar interactions with the same numerical schemes and analyses the diffusion motion through VACF and D. First of all; we compute the VACF for Screening parameters $(1 < \kappa \geq 3)$, plasma coupling parameters ($1 < \Gamma \geq 500$), uniaxial electric field strength ($0.0 \leq M_T \geq 1.50$) for the same number of particles (N = 500). After that, D is calculated for nearly the same parameters used for the VACF.

The single-particle motions are computed for 3D ER-CDPs through the equilibrium MD simulation using VACF for diffusion by employing Eq. (3). VACF are analyzed and discussed for plasma screening κ = 1.4, 2 and 3, plasma coupling (Γ =2–500), number of spherical charged dust particles (N = 500) with the variation of the external uniaxial electric field. The VACF has been widely used for 3D CDPs over a wide combination range of plasma parameters (κ, Γ) [12, 30, 36]. The VACF was computed at M_T = 0.0 for comparison purposes with earlier MD results. It has been shown that MD outcomes have comparable fair agreements with previous results. Here we defined the critical strength of the applied external electric field (M_{CT}). The M_{CT} is directly proportional to plasma coupling strengths or inversely proportional to system temperature. We have been found that above the M_{CT} system goes out of thermal equilibration. We have also got noisy results above the M_{CT} strength of the external applied uniaxial electric field. The critical strength of the uniaxial electric field is different for the different combinations of CDPs parameters.

The simulation outcomes of VACF in 3D ER-CDPs as a function of the simulation time scale for four different plasma coupling values ($\Gamma \equiv$ 20, 50, 100, and 250) at constant κ = 1.40 and N = 500 are displayed in **Figure 1**. The effects of variation uniaxial electric field ($M_T \equiv$ 0.01–1.30) on VACF are computed to analyze the diffusion motion of charged dust particles. It has been shown that M_T significantly depends on plasma temperature and the screening of charged particles. At Higher plasma temperature (Γ = 20), one can easily observe the rapid decay of VACF with very weak particle oscillations. This regime M_T does not have significant effects on the dynamics of dust particles. **Figure 1(b)** shows the oscillatory damped motion and slightly decreasing amplitude of VACF up to uniaxial electric field strength (M_T = 0.90). For higher plasma coupling strength ($\Gamma \equiv$ 100,250), results of VACF show the higher oscillations with slightly damping phenomena. **Figure 2** demonstrated the VACF at κ = 2.0 for four different regimes of strongly coupled CDPs (Γ = 20, 50, 200, 400) N = 500 with a variation of uniaxial electric field ($0.0 < M_T \geq 1.50$) It is observed that the effect of M_T on VACF is nearly the same as was observed in **Figure 1**. It has also been shown that the M_{CT} increases with increases in the screening strength. Furthermore, we increased the screening strength κ = 3.0 and Γ = 30, 75, 250, and 500 are shown in **Figure 6**. The trend of damping and oscillations of dust particles was observed the same as in **Figures 1** and **2**. Therefore, the VACF below the M_{CT} simulation outcomes is founded in good agreement with previous data [13, 30, 35, 36].

It is observed that at higher temperatures (low coupling, Γ), the M_T does not affect the dynamics and oscillation of the particles. So what we can say that at higher system temperature, the radius between interacting charged dust particles is large. Due to the large distance, Eq. (1) remains invalid for higher uniaxial electric field strength. The motion of a particle in this regime indicates only thermal motion. Complete damped oscillations of dust particles were observed for a long time at intermediate values of Γ. The magnitude of oscillations slightly decreased with increasing M_T, increasing the diffusion at a low rate of thermal motion. In ER-CDPs, the effect of M_T on VACF can be easily observed from intermediate to higher plasma coupling (Γ) strength. The simulation time length scales, maximum at intermediate values of Γ and further increasing Γ, the length scale slightly decreased. It was observed that the oscillation was not fully damped at higher Γ. It is concluded that a uniaxial electric field can affect diffusion motion in ER-CDPs for the sufficient strength of M_T., so what can say the presented numerical method gives us precise and reliable data for comparative study.

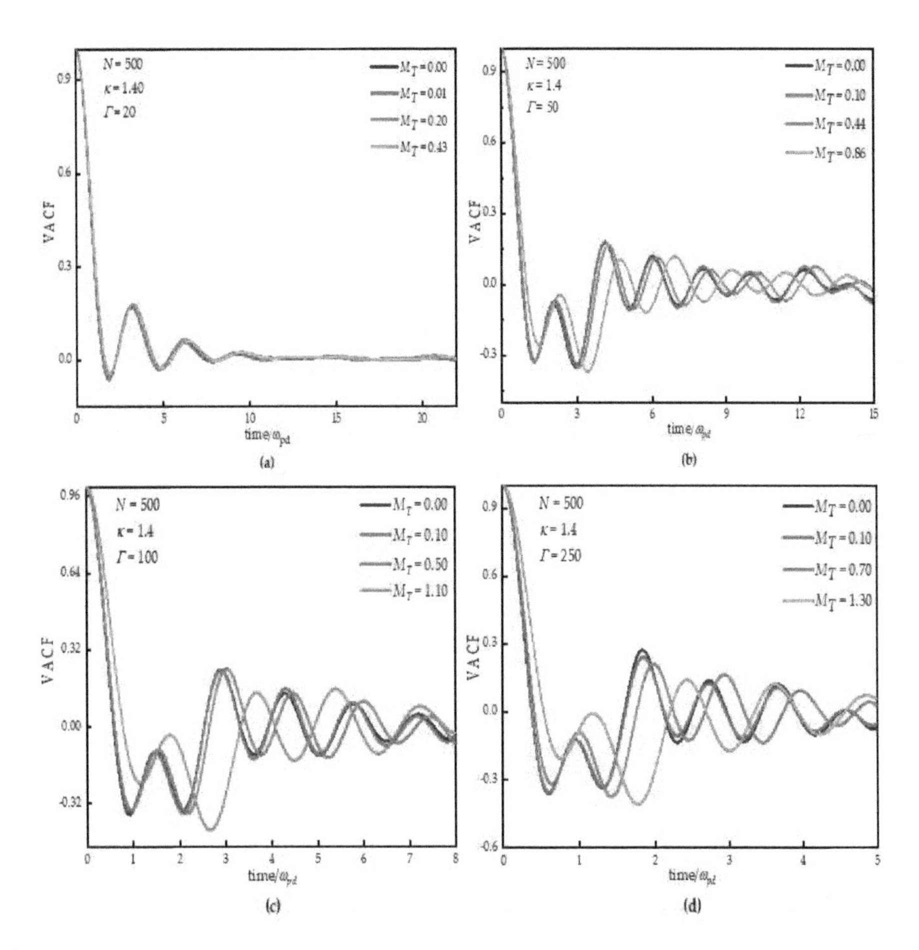

Figure 1.
VACF as a function of molecular dynamics simulation time scale for constant plasma screening parameter (κ = 1.40). MD simulations results obtained for N = 500 *simulated dust particles with the variation of uniaxial external electric field (0.0 <* $M_T \leq$ *1.30) and four different plasma regimes (a)* Γ = 20, *(b)* Γ = 50, *(c)* Γ = 100 *and (d)* Γ = 250.

Now we focused on the primary outcomes through the equilibrium MD method. The simulation data of this work are that the self-diffusion coefficient (D) of ER-CDPs can be obtained with minimum statistical error at M_T = 0.0 by equilibrium MD. Moreover, it is shown that the MD with GKR has a fair good agreement with previous simulation outcomes of Ohta and Hamaguchi [30], Daligault [31, 32], and Begum and Das [33, 37] at M_T = 0.0. **Figures 3–5** display the primary outcomes of D measured from the equilibrium MD method for ER-CDPs at κ = 1.4, 2, and 3, respectively. D of ER-CDPs was calculated for different uniaxial electric field strengths ($0 < M_T < 1.5$). Here we have explained the D for a wide range of combinations of plasma parameters (Γ, κ) over acceptable M_T ranges below the M_{CT}. This section focused on the variation of plasma coupling ($1/T$), electric field, Debye screening length, and constant N = 500 simulated particles.

The effect of M_T on D in 3D ER-CDPs is calculated for plasma coupling ($\Gamma \equiv 20$, 50, 100, and 200), Debye screening length ($\kappa \equiv 1.4$), uniaxial electric field ($0.0 \leq M_T \leq 1.30$), and N = 500 number of particles shown in **Figure 3**. We have performed several MD simulations at constant κ = 1.4 and N = 500 to analyze the effect of M_T on D. The reported data of D shows good agreements for small M_T values with previous results [30–32]. With respective of Γ different regimes are under consideration. It is found that the M_{CT} values are the same as in **Figures 1, 2, 6**

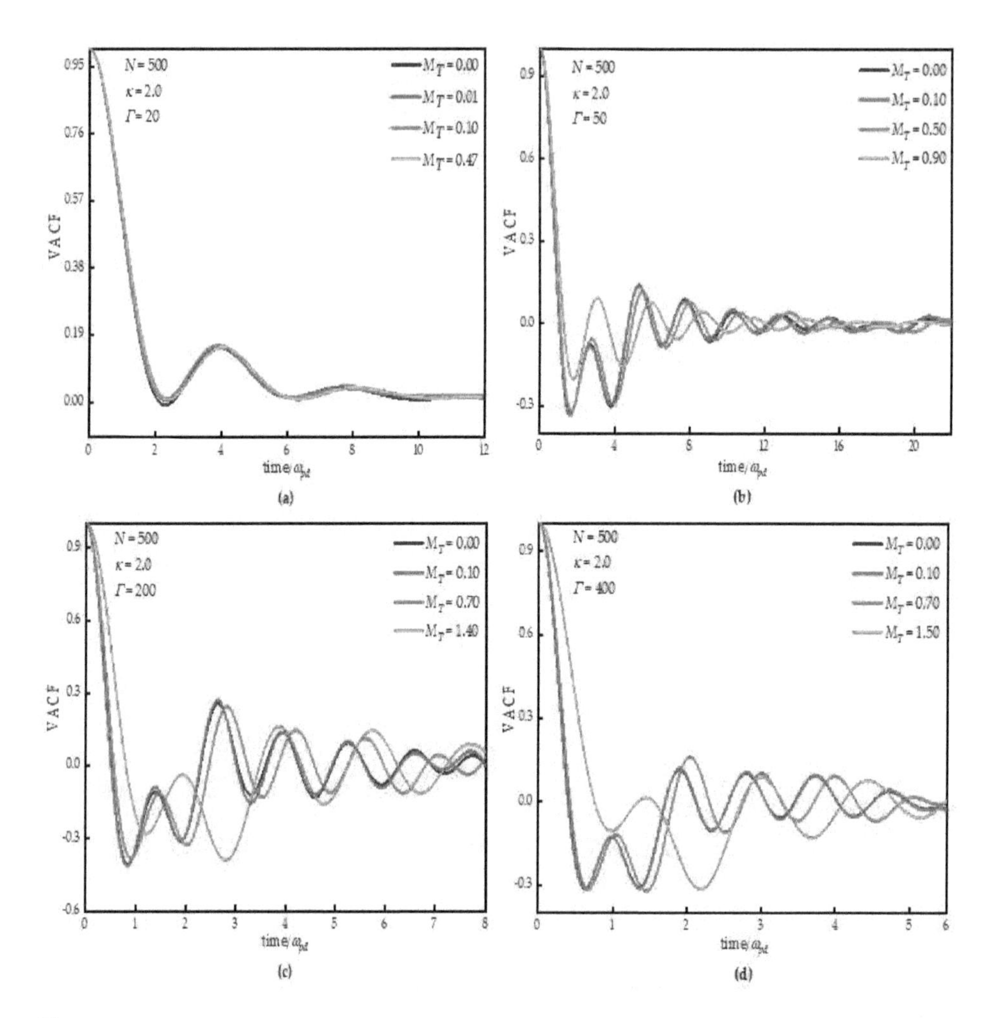

Figure 2.
VACF as a function of molecular dynamics simulation time scale for constant plasma screening parameter (κ = 2.0). MD simulations results obtained for N = 500 simulated dust particles with the variation of uniaxial external electric field ($0.0 < M_T \leq 1.30$) and four different plasma regimes (a) Γ = 20, (b) Γ = 50, (c) Γ = 200 and (d) Γ = 400.

for the same combinations of plasma parameters. It is noted that for plasma coupling (Γ = 20), the critical values ($M_{CT} > 0.42$) for κ = 1.4 display in **Figure 3**. Wehave found the increasing trend of M_T with increases the plasma coupling values or decreasing the system temperature such as Γ = 50, 100and 200 critical $M_{CT} \geq 0.88$, 1.12, 1.25 respectively. The possible reason for the low strength of M_T in high system temperature is the large interparticle distance show with the term ($1/\mathbf{r}^3$) in Eq. (1). The increasing values of D with respective of Γ at κ = 1.4 can be easily observed from four panels of **Figure 3**. Upon further increasing M_T, the MD simulation gives an error in the form of out of thermal equilibrium. At lower plasma coupling strength (Γ =20),the$D$ does not significantly increase under the applied external electric field's influence; only 0.03% observed the integral values of VACF. The increasing of diffusion was observed for Γ =50 are 20%, Γ = 100 are 75% and Γ = 200 are 160%. The electric field effect on D in ER-CDPs can be found in the same as conventional electrorheological fluids and ionic liquids [40].

The simulation outcomes of D in 3D ER-CDPs at constant N = 500, different values of Γ as a function of the uniaxial electric field (M_T) for κ = 2 and 3 are shown in **Figures 4** and **5**, respectively. The effect of M_T was computed in four different

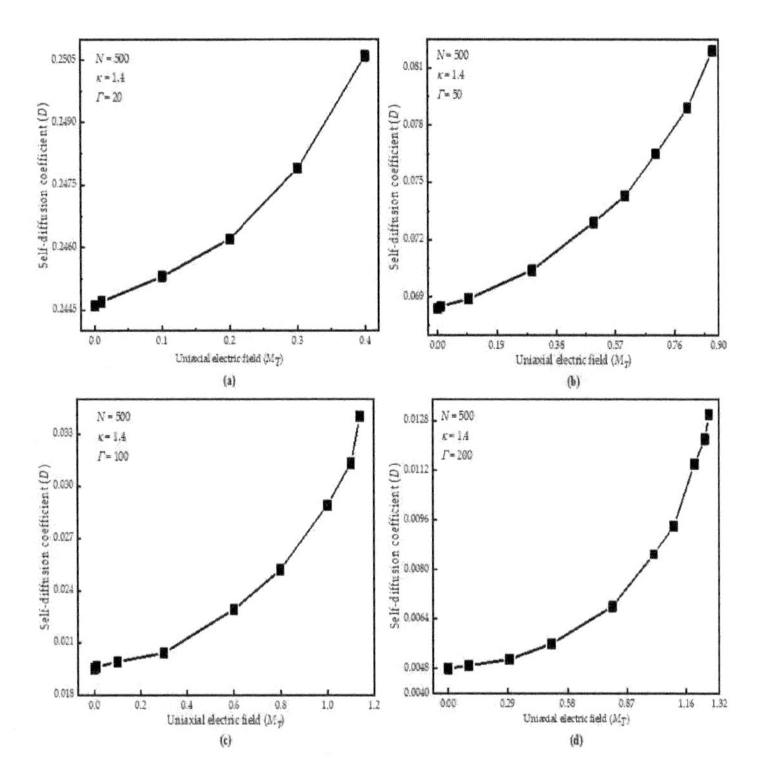

Figure 3.
Self-diffusion coefficient as a function of external applied uniaxial electric field (0.0 < M_T < 1.30), for constant plasma screening (κ = 1.4) and a number of particles (N = 500), considered four different ER-CDPs states (a) Γ = 20, (b) Γ = 50, (c) Γ = 100 and (d) Γ = 200.

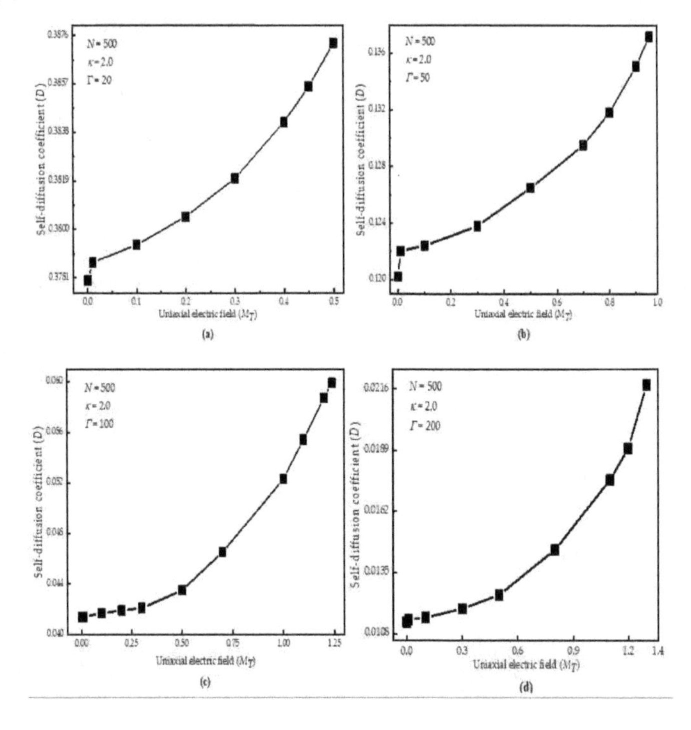

Figure 4.
Self-diffusion coefficient as a function of external applied uniaxial electric field (0.0 < M_T < 1.40), for constant plasma screening (κ = 2) and a number of particles (N = 500), considered four different ER-CDPs states (a)Γ = 20, (b)Γ = 50, (c)Γ = 100 and (d)Γ = 200.

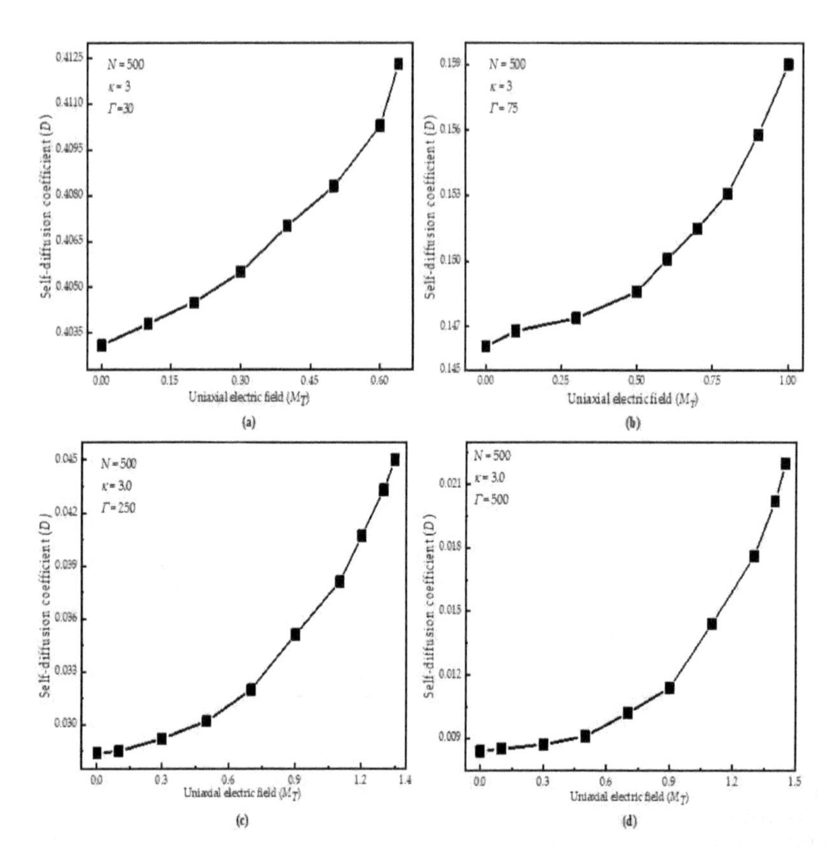

Figure 5.
Self-diffusion coefficient as a function of external applied uniaxial electric field (0.0 < M_T < 1.50), for constant plasma screening (κ = 3) and a number of particles (N = 500), considered four different ER-CDPs states (a) Γ = 30, (b) Γ = 75, (c) Γ = 250 and (d) Γ = 500.

states of ER-CDPs by performed almost thirty-three simulations. The M_{CT} values for $\kappa = 2.0$ at $\Gamma = 20$, $M_T > 0.45$, $\Gamma = 50$, $M_{CT} > 0.90$, $\Gamma = 100$, $M_{CT} > 1.23$, and $\Gamma = 200$, $M_{CT} > 1.35$. The increase in D under the action of external applied uniaxial electric field for $\Gamma = 20$, 50, 100, and 200 are 1.7, 11, 42, and 115%. The increases in the D for $\kappa = 3.0$ same N were observed nearly the same as said above for $\kappa = 2.0$. The effect of M_T on the D is prominent at higher values of Γ parameters. It was also noted that M_T does not significantly affect D lower to intermediate values of Γ. From **Figures 3–5**, we can conclude that the proposed MD simulation method worked for the limited strength of M_T. It was noted that the system's critical values above the system go out of thermal equilibrations, and kinetic energy increased very largely up to 23 orders of magnitude, but the potential energy does not change higher order of magnitude. Below the critical strength, the potential energy of interacting Yukawa dust particles increases with M_T.

4. Conclusion

An equilibrium MD simulation has been performed to report the velocity autocorrelation function (VACF) and self-diffusion coefficient (D) of three-dimensional (3D) electrorheological complex (dusty) plasmas (ER-CDPs) for the analysis of diffusion motion of dust particles. The interactions and forces between dust particles were modeled by Yukawa potential and Quadrupole interactions of charged dust particles. This paper highlights the outcomes of VACF and D for 3D ER-

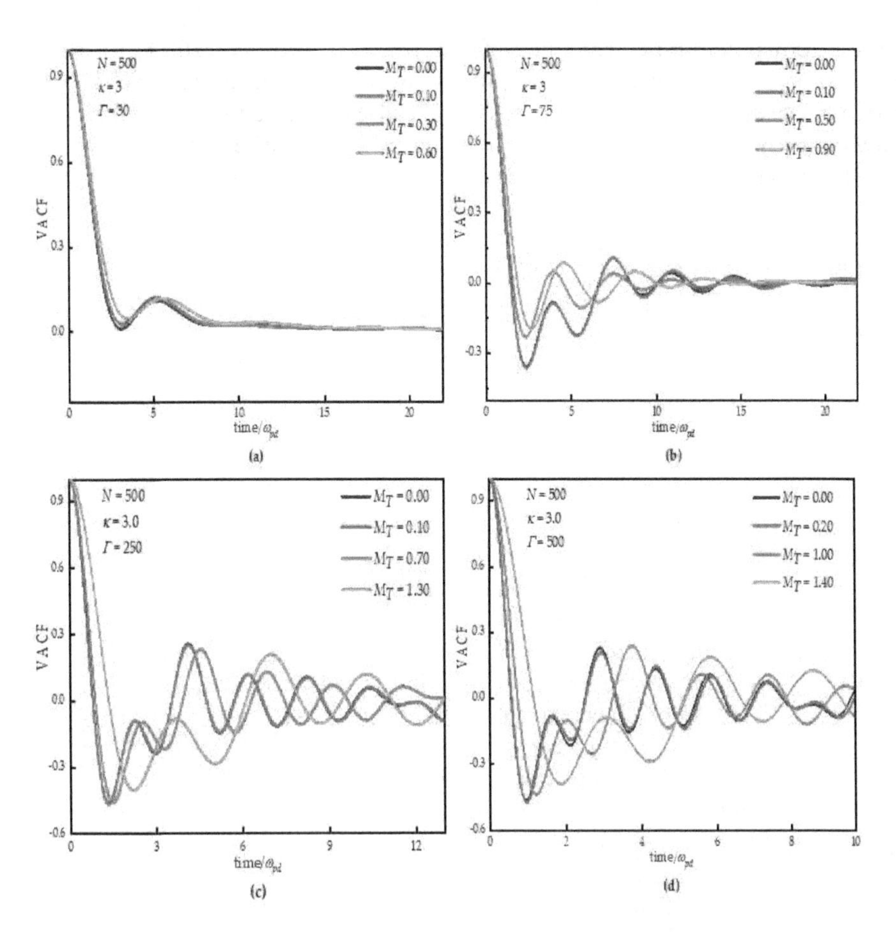

Figure 6.
VACF as a function of molecular dynamics simulation time scale for constant plasma screening parameter ($\kappa = 3.0$)*. MD simulations results obtained for* N = 500 *simulated dust particles with the variation of uniaxial external electric field* ($0.0 < M_T \leq 1.30$) *and four different plasma regimes (a)* $\Gamma = 30$, *(b)* $\Gamma = 75$, *(c)* $\Gamma = 250$ *and (d)* $\Gamma = 500$.

CDPs in the presence of a uniaxial electric field (M_T). We used Green-Kubo expression to calculate VACF and D over a wide range of CDPs Coulomb coupling (Γ)and Debye screening (κ) parameters. The VACF and D are significantly dependent on the plasma parameters (κ, Γ) and M_T. The calculated results are highly consistent with other MD data in the absence of M_T. It was observed that D decreased with increasing the Γ. The M_T significantly affects the diffusion motion in ER-CDPs from intermediate to higher values of Γ and does not affect low Γ. The D increases with increasing the M_T and κ. A new investigation gives more comprehensive and reliable data for VACF and D over given plasma parameters. It has been demonstrated that the presented numerical results for a given range of plasma parameters (Γ, κ) and the number of particles are good performances. The proposed numerical model is suit-able for liquid-like and solid-like states for 3D CDPs. We can be concluded that the developed MD simulations approach will be employed to investigate the thermophysical properties of ER-CDPs.

Acknowledgements

The support provided by the National Science Fund for Distinguished Young Scholars of China (No. 51525604), the Foundation for Innovative Research Groups of the National Natural Science Foundation of China (No. 51721004) for the completion of the present work is gratefully acknowledged. The authors would also like to thank Dr. X. D. Zhang at the Network Information Center of Xi'an Jiaotong University for supporting the HPC platform and the National Centre for Physics (NCP) Islamabad allocation computational power to check and run the MD code.

Abbreviation and symbol

EMD	Equilibrium molecular dynamic
ERFs	Electrorheological fluids
κ	Screening strength
Γ	Coulomb coupling
M_T	Thermal mach number
N	Number of particles
CDP	Complex (dusty) plasma
SCCDPs	Strongly coupled complex (dusty) plasmas
WCCDPs	Weakly coupled complex (dusty) plasmas
3D	Three dimensional
PK-3 plus	Plasma Kristall-3 plus
VACF	Velocity autocorrelation function
D	Self-diffusion coefficient
GKR	Green-Kubo relation
PBCs	Periodic boundary conditions
k_B	Boltzmann constant
ω_{pd}	Plasma frequency
T	Plasma temperature
a_{ws}	Wigner Seitz radius
θ	The angle between the electric field and particles vector

Author details

Muhammad Asif Shakoori[1], Maogang He[1*], Aamir Shahzad[2] and Misbah Khan[3]

1 Key laboratory of Thermal Fluid Science and Engineering, Ministry of Education (MOE), School of Energy and Power Engineering, Xi'an Jiaotong University, Xi'an, China

2 Molecular Modeling and Simulation Laboratory, Department of Physics, Government College University Faisalabad (GCUF), Faisalabad, Pakistan

3 Department of Refrigeration and Cryogenics Engineering, School of Energy and Power Engineering, Xi'an Jiaotong University, Xi'an, China

*Address all correspondence to: mghe@xjtu.edu.cn

References

[1] Szary ML, Noras M. Experimental study of sound transmission loss in electrorheological liquids. ASME Int. Mech. Eng. Congr. Expo. Proc. 2002; **240**:137–142. DOI:10.1115/IMECE2002-33349

[2] Stanway R, Sproston JL, El-Wahed AK. Applications of electrorheological fluids in vibration control: A survey. Smart Mater. Struct. 1996;5:464–482. DOI:10.1088/0964-1726/5/4/011

[3] Shahzad A, Shakoori MA, He M-G, Bashir S. Sound Waves in Complex (Dusty) Plasmas. *Comput. Exp. Stud. Acoust. Waves, Intech open*.2018. DOI: 10.5772/intechopen.71203

[4] Shahzad A, Shakoori, MA, He M-G, Feng Y. Numerical Approach to Dynamical Structure Factor of Dusty Plasmas. Book entitled "Plasma Science and Technology Basic Fundamentals and Modern Applications". *Intech Open.* 2020; DOI:org/10.5772/intechopen. 78334

[5] Shahzad A, Shakoori MA, He M-G. Wave Spectra in Dusty Plasmas of Nuclear Fusion Devices. Fusion energy. *Intech Open*. 2020.DOI:org/10.5772/intechopen.91371

[6] Shahzad A, He M-G, shakoori MA. Thermal Transport and Non-Newtonian behaviors of 3D Complex Liquids Using Molecular Simulations. Proceeding of the Proceedings of 14th International Bhurban Conference on Applied Sciences & Technology (IBCAST); 10th - 14th January 2017; Islamabad, Pakistan, 2017.p.472–474.

[7] Shahzad A, Khan MQ, Shakoori MA, He M-G Feng Y. Thermal Conductivity of Dusty Plasmas through Molecular Dynamics Simulations. *Intech Open. 2020;* DOI:http://dx.doi.org/10.5772/ 57353

[8] Shahzad A, He M-G. Calculations of thermal conductivity of complex (dusty) plasmas using homogenous nonequilibrium molecular simulations. Radiat. Eff. Defects Solids. *2015;***170**:758–770.DOI:10.1080/10420150.2015.1108316

[9] Shahzad A, Kashif M, Munir T, He M-G, Tu X. Thermal conductivity analysis of two-dimensional complex plasma liquids and crystals. Phys. Plasmas. *2020*;**27**:103702.DOI:10.1063/5.0018537

[10] Shahzad A, He M-G. Thermal conductivity calculation of complex (dusty) plasmas. Phys. Plasmas.2012;**19**: 083707.DOI:10.1063/1.4748526

[11] Shahzad A, He M-G, Haider SI, Feng Y. Studies of force field effects on thermal conductivity of complex plasmas. Phys. Plasmas. 2017;**24**:093701. DOI: 10.1063/1.4993992

[12] Shahzad A, He M-G, He K. Diffusion motion of two-dimensional weakly coupled complex (dusty) plasmas. Phys. Scr. 2013;**87**:035501.DOI: 10.1088/0031-8949/87/03/035501

[13] Shahzad A, He M-G. Shear viscosity and diffusion motion of two-dimensional dusty plasma liquids. Phys. Scr. 2012;**86**:015502.DOI:10.1088/0031-8949/86/01/015502

[14] Shahzad A, He M-G. Thermodynamic characteristics of dusty plasma studied by using molecular dynamics simulation. Plasma Sci. Technol. 2012;**14**:771–777.DOI: 10.1088/1009-0630/14/9/01

[15] Shahzad A, Shakoori MA, He M-G, Feng Y. Dynamical structure factor of complex plasmas for varying wave vectors. Phys. Plasmas. 2019; **26**: 023704.DOI: 10.1063/1.5056261

[16] Donkó I, Hartmann P, Donkó Z. Molecular dynamics simulation of a two-dimensional dusty plasma. Am. J.

Phys. 2019;**87**:986–993.DOI: 10.1088/0031-8949/86/01/015502

[17] Adamovich I. *et al.* The 2017 Plasma Roadmap: Low temperature plasma science and technology. J. Phys. D. Appl. Phys. 2017;**50**:323001 20.DOI: 10.1088/ 1361-6463/aa76f5

[18] Hirst AM, *et al.* Low-temperature plasma treatment induces DNA damage leading to necrotic cell death in primary prostate epithelial cells. Br. J. Cancer. 2015**112**:1536–1545.DOI:10.1038/bjc.2015.113

[19] Laroussi M. Cold Plasma in Medicine and Healthcare: The New Frontier in Low Temperature Plasma Applications. Front. Phys. 2020;**8**:1–7. DOI: 10.3389/fphy.2020.00074

[20] Ivlev AV, *et al.* First observation of electrorheological plasmas. Phys. Rev. Lett. 2008;**100**:1–4.DOI:10.1103/PhysRevLett.100.095003

[21] Kana D, Dietz C, Thoma MH. Simulation of electrorheological plasmas with superthermal ion drift. Phys. Plasmas. *2020;***27**:103703. DOI:10.1063/5.0010021

[22] Ivlev AV, Thoma MH, Räth C, Joyce G, Morfill GE. Complex plasmas in external fields: The role of non-hamiltonian interactions. Phys. Rev. Lett. 2011;**106**:155001.DOI:10.1103/PhysRevLett.106.155001

[23] Schwabe M *et al.* Slowing of acoustic waves in electrorheological and string-fluid complex plasmas. *New* J. Phys. 2020;**22**:083079.DOI: 10.1080/10420150.2015.1108316

[24] Ivlev AV *et al.* Electrorheological complex plasmas. IEEE Trans. Plasma Sci. 2010;**38**:733–740.DOI: 10.1109/TPS.2009.2037716

[25] Yaroshenko VV. Charge-gradient instability of compressional dust lattice waves in electrorheological plasmas. Phys. Plasmas. 2019;**26**:083701.DOI: 10.1063/1.5115346

[26] Rosenberg M. Waves in a 1D electrorheological dusty plasma lattice. J. Plasma Phys. 2015;**81**:905810407.DOI: 10.1017/S0022377815000422

[27] Sukhinin GI, Fedoseev AV, Khokhlov RO, Suslov SYU. Plasma Polarization Around Dust Particle in an External Electric Field. Contrib. to Plasma Phys. 2012;**52**:62–65.

[28] Sukhinin GI *et al.* Plasma anisotropy around a dust particle placed in an external electric field. Phys. Rev. E. *2017;***95**:063207.DOI:10.1103/PhysRevE.95.063207

[29] Hartmann P, *et al.* Self-diffusion in two-dimensional quasimagnetized rotating dusty plasmas. Phys. Rev. E. *2019;***99**:013203.DOI:10.1103/PhysRevE.99.013203

[30] Ohta H, Hamaguchi S. Molecular dynamics evaluation of self-diffusion in Yukawa systems. Phys. Plasmas. *2000;***7**: 4506–4514.DOI:10.1063/1.1316084

[31] Daligault J. Practical model for the self-diffusion coefficient in Yukawa one-component plasmas. Phys. Rev. E - Stat. Nonlinear, Soft Matter Phys. 2012; **86**:1–2. DOI: 10.1103/PhysRevE.86.047401

[32] Daligault J. Diffusion in ionic mixtures across coupling regimes. Phys. Rev. Lett. 2012; **108**:1–5.DOI:10.1103/PhysRevLett.108.225004

[33] Begum M, Das N. Self-Diffusion of Dust Grains in Strongly Coupled Dusty Plasma Using Molecular Dynamics Simulation. 2016;**8**:49–54.DOI:10.9790/4861-0806024954

[34] Ott T, Bonitz M, Donkó Z, Hartmann P. Superdiffusion in quasi-two-dimensional Yukawa liquids. Phys.

Rev. E - Stat. Nonlinear, Soft Matter Phys. 2008;**78**:1–6.DOI:10.1103/PhysRevE.78.026409

[35] Dzhumagulova KN, Masheeva RU, Ramazanov TS, Donkó Z. Effect of magnetic field on the velocity autocorrelation and the caging of particles in two-dimensional Yukawa liquids. *Phys. Rev. E - Stat. Nonlinear, Soft Matter Phys.* 2014;**89**:1–7.DOI: 10.1103/PhysRevE.89.033104

[36] Dzhumagulova KN, Ramazanov T S, Masheeva RU. Velocity Autocorrelation Functions and Diffusion Coefficient of Dusty Component in Complex Plasmas. Contrib. to Plasma Phys. 2012;**52**:182–185.DOI:10.1002/ctpp.201100070

[37] Begum M, Das N. Self-diffusion as a criterion for melting of dust crystal in the presence of magnetic field. Eur. Phys. J. Plus. 2016;**131**:1–12.DOI: 10.1140/epjp/i2016-16046-2

[38] Hollingsworth SA, Dror RO. Molecular Dynamics Simulation for All. Neuron. 2018; **99**:1129–1143.DOI: 10.1016/j.neuron.2018.08.011

[39] Kompaneets R, Morfill GE, Ivlev AV. Design of new binary interaction classes in complex plasmas. Phys. Plasmas. 2009;**16**:043705.DOI: 10.1063/1.3112703

[40] Clark R. *et al.* Effect of an external electric field on the dynamics and intramolecular structures of ions in an ionic liquid. J. Chem. Phys. 2019;**151**: 164503.DOI: 10.1063/1.5129367

15

Polarized Thermal Conductivity of Two-Dimensional Dusty Plasmas

Aamir Shahzad, Madiha Naheed, Aadil Mahboob, Muhammad Kashif, Alina Manzoor and H.E. Maogang

Abstract

The computation of thermalt properties of dusty plasmas is substantial task in the area of science and technology. The thermal conductivity (λ) has been computed by applying polarization effect through molecular dynamics (MD) simulations of two dimensional (2D) strongly coupled complex dusty plasmas (SCCDPs). The effects of polarization on thermal conductivity have been measured for a wide range of Coulomb coupling (Γ) and Debye screening (κ) parameters using homogeneous non-equilibrium molecular dynamics (HNEMD) method for suitable system sizes. The HNEMD simulation method is employed at constant external force field strength (F^*) and varying polarization effects. The algorithm provides precise results with rapid convergence and minute dimension effects. The outcomes have been compared with earlier available simulation results of molecular dynamics, theoretical predictions and experimental results of complex dusty plasma liquids. The calculations show that the kinetic energy of $SCCDP_S$ depends upon the system temperature ($\equiv 1/\Gamma$) and it is independent of higher screening parameter. Furthermore, it has shown that the presented HNEMD method has more reliable results than those obtained through earlier known numerical methods.

Keywords: Plasma thermal conductivity, complex dusty plasma, Homogenous non-equilibrium molecular dynamics, force field strength, system size, plasma parameters etc.

1. Introduction

Recently, thermophysical properties of complex materials are a major concern in the field of science and engineering. The term thermophysical properties used to pass on both thermodynamic and transport properties. Experimental or theoretical methods to study properties of fluids depend on microscopic and macroscopic categories [1–4]. The conventional macroscopic measurements depend on the state of stress, temperature, and density. Thermodynamic properties are defined by the equilibrium conditions of the system which consist of temperature, heat capacity, entropy, pressure, internal energy, enthalpy, and density, whereas the transport properties comprise thermal conductivity, diffusion viscosity, and waves with their instabilities. For further explanation of the process in detail for these systems, data that is applicable to thermodynamics, transport, optics, transmission, light, and other features are required for non-ideal plasma [5–9]. In this regard, various opinions regarding computer research methods including theoretical and numerical performance have greatly improved for non-ideal Plasma [10]. Determination for some reason, thermal conductivity is also a big problem for thermophysical researchers. Developmental aspects of heat transport in micron and nanoscale materials have shifted to the domain of technical issues as there are other areas, such as phonon heat transfer in semiconductor superlattices, which have received widespread attention from researchers. To study the internal energy of particles, their momentum, and heat transfer thus remains a crucial task. Therefore, thermal management, strategies sustainable high performance, reliability, and service life are main purposes. One such strategy is to develop new therapeutic materials based on dusty plasma that are more effective. Regulation with approval became a significant issue in modern technology [11]. Yet similar interests are present in plasma fusion, and it can be productive radiotoxic dust in plasma-wall reactions. In many ways, this chapter provides an update literature survey on thermal transport as well as heat flow strategies to determine thermal behaviors in two-dimensional (2D) complex liquids. The coefficients were computed through the Green Kubo (GK) equilibrium molecular dynamics (EMD) simulations by Salin and Caillol [12] and variance procedure (VP) estimation used by the Faussurier and Murillo [13]. Donkó and and Hartmann employed the inhomogenous non equilibrium MD (InHNEMD) method to investigate the transport and thermal conductivity [14]. Very recently, a homogeneous NEMD (HNEMD) and homogenous perturbed MD (HPMD) schemes are introduced by Shahzad and He (current authors) for strongly coupled complex dusty plasmas (SCCDPs) to compute the thermal transport and behaviors of SCCDPs [15–17]. For the computation of transport properties, in particular, numerical models are proposed in interest to investigate thermal behavior over a suitable range of system temperature and density values (Γ, κ). Complex fluids (dusty plasma fluids) have been used for many purposes, like power generation, semiconductors industry, cosmetics, paper industry, etc. [18].

1.1 Plasma

As we all know that 99% of matter exist in space is plasma and it is called forth state of matter. Basically plasma occurs in electrified gas form, where atoms dissociated into electrons and positive ions. It is form of matter in different areas of physics such as technical plasma, terrestrial plasma and in astrophysics. Plasma is produced artificially in laboratory used in many technical purposes likely in fluorescent lights, display, fusion energy research and other more. Term "Plasma" first time used by Irving Langmuir [19], who is an American physicist and defined plasma as "plasma is quasi-neutral gas of charged particles which exhibit collective behavior". Quasi-neutral means that gas becomes electrically neutral when number of ions equal to number of electrons ($ni \approx ne \approx n$).Where, ni is ion density, ne is electron density and n is number density. Collective behavior means that charged particles collide with each other due to coulomb potential and electric field. Plasma is extensively used in the field of science and technology. It plays a very significant role in over daily life. Plasma is used in over daily life fields such as laser, sterilizing of medical instruments, lightening, intense power beams, water purification planet and many more.

In 1922, American scientist Irving Langmuir was the only one person who defined plasma for the first time. In 1930, the study of plasma physics was started by some scholars; they are inspired by some particle problems. In 1940, hydromagnetic waves were advanced by Hanes Alfven [19] and these waves are called Alfven waves. Furthermore, he described that these waves would be used for the study of astrophysical plasma. At the start of 1950, the research on magnetic fusion energy was started at the same time in Soviet, Britain and USA. In 1958, the research on magnetic fusion energy was considered the branch of thermonuclear power. Primarily, this research was carried out as confidential but after the realization that controlled fusion research was not liked by military and therefore this research was publicized by above said three countries. Due to the reason, other countries may participate in fusion research based on plasma physics. At the end of 1960, plasma is created with different plasma parameters by Russian Tokomak configuration. In 1970 and 1980, various advanced tokomaks were built and approved the performance of tokomak. Moreover, fusion break almost achieved in tokomak and in 1990, the research on dusty plasma physics had begun. The dusty plasma is defined as "when charged particles absorbed in plasma, becomes four components plasma containing electrons, ions, neutral and dust particles" and dust particles alter the properties of plasma which is called "Dusty Plasma" [19].

1.2 Types of plasma

Plasma has complex characteristics and properties, characterized through temperature of electron and ion, density and degree of ionization. (i) **Hot plasma**: If plasma fulfills $T_e \cong T_i$ this condition then plasma is considered as hot plasma because hot plasma has very high temperature and also thermal equilibrium obtains due to frequent interactions between particles. Hot plasma is also called thermal plasma. It approaches to local thermodynamics equilibrium (LTE) and is created with high gas pressure in discharge tube in the laboratory. Hot plasma is produced by sparks, flames and atmospheric arcs. (ii) **Cold plasma**: When plasma satisfies $T_e > T_i > T_g$ this condition, plasma is called cold plasma. Where T_e,T_i, and T_g represent the temperature of electrons, ions and gas molecules. Cold plasma is created in laboratory with the positive column glow discharge tube. Motion of gas molecules is considered ignore because electron energy is very high as compared with gas molecules. Moreover, nonthermal equilibrium does not exist because collision between gas molecules and electrons is considers as low due to low gas pressure. On this regime, magnetic field is very weak and considered as ignore, only electric field is acted on charged particle. Application of cold plasma is self-decontaminating filter, food processing and sterilizing of tooth. (iii) **Ultracold plasma**: When the temperature of electrons and ions become low as 100mk and $10\mu k$ with density 2×10^9 cm^{-3}, then, plasma is called ultracold plasma. The behavior of ultracold plasma is obtained when Debye screening length becomes smaller than the sample size due to positive ions clouds trapped electrons. Ultacold plasma is considered as strongly coupled plasma because the coulomb interaction energy between the neighbor particles is more than thermal energy of charged particles. Such type of plasma is created in laboratory through pulsed laser and photoionizing laser cooled atoms [20].

1.3 Classification of dusty plasmas

Dusty plasma is characterized by an important parameter, coulomb coupling parameter Γ. The Coulomb coupling parameter is explained as, consider there are two dust particles, having same charge and separated by distance 'a' from each other. The coulomb potential energy of dust particle is $\varepsilon_c = \frac{q_d^2}{a} \exp. (-\frac{a}{\lambda_d})$, where, q_d is the charge on dust particle, a is the distance between dust particles and λ_d is Debye screening length of dust particle. The thermal energy of dust particle is $K_B T_d$. Coulomb coupling parameter is defined as "ratio of coulomb potential energy to thermal energy". On the basis of coulomb coupling parameter, the dusty plasma is classified in ideal plasma (weakly coupled dusty plasma) and non-ideal plasma (strongly coupled dusty plasma) and is represented as Γ_c. (i) **Ideal plasma:** Ideal plasma is defined by plasma parameter called coulomb coupling and denoted as $\Gamma = \frac{P.E}{K.E}$ and is defined as "when kinetic energy of plasma is much larger than

potential energy at low temperature and low density". Ideal plasma is also called weakly coupled dusty plasma and is known by $\Gamma > 1$ this condition. Ideal plasma does not have definite structure due to less collision between particles and low density. Moreover, weakly coupled plasma is defined by plasma parameter called coulomb coupling parameter Γ. When the value of coupling parameter becomes negligible then plasma is called weakly coupled plasma. Weakly coupled plasma is also called hot plasma. When the temperature of electron becomes equal to temperature of ion ($T_e \cong T_i$) then plasma is called hot plasma or ideal plasma. Hot plasma is generated in laboratory in the discharge tube with high gas pressure. Examples of hot plasma are flame, sparks and atmospheric arcs. Weakly coupled dusty plasma has not specific shape because at low density and high temperature and the interaction between interacting particles becomes very low. (ii) **Non-ideal plasma:** Dusty plasma will be strongly coupled when it satisfies this condition $\Gamma \geq 1$. Strongly coupled dusty plasma is also called nonideal plasma. Dust particle in several laboratory plasma systems is strongly coupled due to their small interparticle distance, low temperature and huge electric charge. Moreover, dusty plasma will be nonideal or strongly coupled, if average thermal energy of charged dust particle is much lesser than average potential energy. Examples of non-ideal plasma are laser generated plasma, brown dwarfs, exploding wires, high power electrical fuses, etc. Furthermore, the Yukawa potential or coulomb coupling potential Γ is used to define strongly coupled plasma. The ratio of potential energy to kinetic energy is called coulomb coupling potential. When kinetic energy becomes lower than potential energy i.e., $\Gamma > 1$. Its mean strongly coupled dusty plasma is also called cold plasma because of inter-particle kinetic energy decreases from potential energy and particles in plasma turn into crystalline shape. Crystalline shapes of particles in plasma have examined in many laboratory experiments [1–10]. Food processing and sterilization of tooth are the application of cold plasma. In strongly coupled plasma charge particles are affected by electric field but magnetic field affect is neglected for such type of cold plasma.

1.4 Complex (dusty) plasma and applications

Dusty plasma is generally electron ion plasma containing additional charged particulates. This charged component is sometimes termed as dust particle with size of micron. The properties of dusty plasma become more complex when charged particle immersed in plasma. Due to this reason such plasma is called dusty plasma and dusty plasma is also called complex plasma. Dust particles may be made of ice particles or it may be metallic. Dust particles are heavier than ions and their size ranging from few millimeters to nanometer. When dust particle coexists with plasma (electron, ions, neutral and dust particle) it becomes dusty plasma. Dust particle exists in different shapes and size and it presents in entire universe and also in atmosphere. Usually it is solid form but also exists in liquid and gaseous form. Dust particle can be charge by the flow of electrons and ions. Charged dust particle is affected by electric and magnetic field and their electric potential varies from 1 to 10 V. Dust particle can be grown in laboratory. Dusty plasma has attracted attention of many researchers Transport properties of dusty plasma has played a very important role in the field of science and technology. Mostly the plasma exists in universe is dusty plasma. Dusty plasmas exist in atmosphere of stars, solar wind, sun, galaxies, planetary rings, cosmic radiation, magneto and ionosphere of earth.

Human life is influenced by plasma science. It plays a very significant role in laser developments of fusion energy, sterilizing of medical instruments, plasma processing, intense particle beam, high power energy sources, lightening, high power radiation sources and development of fusion energy controlling. Plasma governs diverse important devices and technological applications. Plasma processing technologies are one of the most important technologies. Plasma processing technologies are playing important role in advance modern technologies of superconductor film growth and diamond film. In addition, the practical application of plasma physics involves the treatment of materials by means plasma technologies. The ionization of system are used to produced particular physical characteristics of plasma, which involve three types of processes, Creation of new materials, Destruction of toxic materials and Superficial modification of existing materials. For industrial process, plasma technology uses two different types of plasma, the cold plasma and thermal plasma. The first type of plasma is cold plasma. Properties of cold plasma are described by electron temperature because electron temperature is greater than ion temperature. The surfaces modification is produced due to plasma particles interact with material, as a result different functional properties of materials are achieved. Cold plasma is produced in vacuum with microwave, dc source or low power rf. The second type of plasma is thermal plasma which is produced at high pressure by radio frequency, microwave source or direct or alternating current. Mostly, thermal plasma is used to devastate toxic materials. Furthermore, plasma has become one of the fast growing research fields which have attracted many researchers. Plasma has advanced applications in the field of industry, textile, plasma chemistry, fusion devices, environmental safety and printing technology and as well as in medical field. In the past decade, plasma physics has become fast developing research in medical field due to increase the atmospheric pressure of plasma sources. Plasma used in medicine has considered the latest developing novel research field with the connection of life science and plasma physics. Moreover, in past ten to fifteen years, World wild research group has set their attention on biological materials with cold atmospheric plasma interactions. Plasma used the field of life sciences in decontamination, in therapeutic medicine and in medical implant technology. The atmospheric pressure of plasma has used to reduce the efficiency of contaminations of food containers and food products. Feed gas humidity is used to adjust the level of contamination. Furthermore, plasma created in polymer tube, used in endoscopes [15–19]. Operating tools such as bone saw blade, neurosurgical and endoscope are sterilized before starting the surgery or dental treatments. Plasma plays a very dominant role in diagnostic system, treatments and in medical instruments. For decontamination of germs and sterilization operation tools, the non-equilibrium discharge plasma is used, is not dangerous for environment and patient as well [21].

2. Molecular dynamics simulations

MD simulation is a powerful technique that can be used to solve many physical problems in atomic material research. MD simulation is handled normally all microscopic information and molecular methods have proven to be the product of applied research. Plasmas and complex liquids have various uses, ranging from semiconductor chips, colloids, thin films, and electrochemistry to biochemical films and other important areas where structures play an important role. MD simulation plays an important role in all the advanced sectors, such as textile science, engineering, physics, plasma physics, astronomy, life sciences or organic sciences, and the chemical industry. Computer simulation has become increasingly important in detecting complex motion systems. Using faster and more sophisticated computer systems, it can be studied the habitat, composition, and behavior of large complex systems. In the 1950s, Alder [19], Wainwright and Rahman [19] used the first MD simulations for liquid argon [1–10], their references herein]. MD simulation has two basic kinds which rely on the properties so far which we can be going to calculate: one is EMD simulations (EMDS) and the other one is NEMD simulations (NEMDS). In the present work, NEMDS is applied to investigate the thermal conductivity of complex plasma at different dusty plasma parameters.

2.1 Numerical model and algorithm

NEMD simulation is used to detect the dust trajectory of an interacting system using Yukawa forces between dust particles. In present case, the HNEMD simulation (HNEMDS) method is used to calculate the thermal flow of complex (dust) plasma formed using Yukawa interaction taking in to account the charged particles with polarization effects and it is given in the form [22]:

$$\phi_{ij}(|\mathbf{r}|) = \frac{Q_{\mathrm{d}}^2}{4\pi\varepsilon_0}\frac{e^{-|\mathbf{r}|/\lambda_D}}{|\mathbf{r}|} + \frac{d^2}{r^3}\left(1+\frac{r}{\lambda_D}\right)e^{-|\mathbf{r}|/\lambda_D}, \tag{1}$$

The first term in Eq. (1) provides the screened charge–charge interaction (form of Yukawa interaction) and the second term gives the screened dipole–dipole interaction. Yukawa potential model of dust particles interaction can be established to take into consideration the polarization effect, the temperature and the screening effects. Here 'r' is the magnitude of interparticle distance, Q is the charge of dust particles, and λ^d is the Debye screening length. We have three normalized (dimensionless) parameters to characterize the Yukawa interaction model $\phi_Y(|\mathbf{r}|)$, the Coulomb coupling parameter $\Gamma = (Q^2/4\pi\varepsilon_0).(1/a_{ws}k_BT)$, where a_{ws} is Wigner Seitz radius and it is equal to $(\mathrm{n}\pi)^{-1/2}$, here n is the number of particles per unit area (N/A). The k_B and T are Boltzmann constant and absolute temperature of the system, and A is the system area. The second is the screening strength (dimensionless inverse) $\kappa = {aws}/{\lambda_D}$, and additional normalized external force field strength, $F^* = (F_Z).(a_{ws}/\mathbf{J}_Q)$. GK relations ($\mathrm{GKR_S}$) for the hydrodynamic transport coefficients of uncharged particles of pure liquids have applied to calculate the thermal conductivity of 2D complex plasma. Here, $\mathbf{J}_\mathbf{Q}$ is the current heat vector at time t of 2D case [1–4].

$$\mathrm{J}_Q(t)A = \sum_{i=1}^{N} E_i\frac{\mathbf{p}i}{m} - \frac{1}{2}\sum_{i\neq j}(\mathbf{r}i - \mathbf{r}j).\left(\frac{\mathbf{p}i}{m}.\mathbf{F}ij\right) \tag{2}$$

where $\mathbf{F}_{ij}$ is the total interparticle force at time t, on particle i due to j, $\mathbf{r}_{ij} = \mathbf{r}_i - \mathbf{r}_j$ are the position vectors (interparticle separation), and $\mathbf{P}_i$ is the momentum vector of the ith particle. Where, E_i is the total energy of particle i.

$$E_i = \frac{\mathbf{p}^2{}_i}{2m} + \frac{1}{2}\sum_{i\neq j}\phi_{ij} \tag{3}$$

$$\dot{\mathbf{r}}_i = \frac{\mathbf{p}_i}{m}, \dot{\mathbf{p}}_i = \sum_{j=1}^{N}\mathbf{F}_i + \mathbf{D}_i(\mathbf{r}_i, \mathbf{p}_i).\mathbf{F}_e(t) - \alpha\mathbf{p}_i. \tag{4}$$

In Eq. (4), $\mathbf{F}_i = -\frac{d\phi_Y(|\mathbf{r}|)}{dr_i}$ is the Yukawa interaction force acting on particle i, where $\phi_Y|\mathbf{r}|$ is given from Eq. (1), and $\mathbf{D}_i = \mathbf{D}_i\{(\mathbf{r}_i, \mathbf{p}_i), i = 1,2, \ldots, N\}$ is the phase space distribution function with $\mathbf{r}_i$ and $\mathbf{p}_i$ being the coordinate and momentum vectors of the ith particle in an N-particle system. A Gaussian thermostat multiplier (α) has been used to maintain the system temperature at equilibrium position and it is given as

$$\alpha = \frac{\sum_{i=1}^{N}\left[\mathbf{F}_i + \mathbf{D}_i(\mathbf{r}_i, \mathbf{p}_i).\mathbf{F}_i(t)\right].\mathbf{p}_i}{\sum_{i=1}^{N} p_i^2/m_i} \tag{5}$$

When an external force field is selected parallel to the z-axis $\mathbf{F}_e(t) = (0, F_z)$, in the limit $t \to \infty$. Then the thermal conductivity is given as [8].

$$\lambda = \frac{1}{2k_B A T^2}\int_0^{\infty}\left\langle \mathbf{J}_{Q_z}(t)\mathbf{J}_{Q_z}(0)\right\rangle dt = \lim_{F_z\to 0}\lim_{t\to\infty}\frac{-\left\langle \mathbf{J}_{Q_z}(t)\right\rangle}{TF_z} \tag{6}$$

where $\mathbf{J}_{Qz}$ is the z-component of the heat flux vector (energy current). All time series data are recorded during HNEMD and used in Eq. (6) to calculate λ. The detail of present scheme with all parameters (Gaussian thermostat multiplier, external force, $\mathbf{D}_i$, F_i, etc) for Yukawa interaction has been reported in our earlier works [8]. The most time consuming part of the algorithm is the calculation of particle interactions (energy and interaction forces). It has been shown in our previous work that the proposed method has the advantage of calculating Yukawa interaction and its associated energy with the appropriate computational power at the right time of the computer simulation. The actual HNEMD simulations are performed between $1.5 \times 10^5/\omega_p$ and $3.0 \times 10^5/\omega_p$ time units in the series of data recording of thermal conductivity (λ). Here, ω_p is the plasma frequency and it is defined as $\omega_{pd} = (Q_d^2/2\pi\varepsilon_0 m_d a^3)^{1/2}$, where m_d is the mass of a dust particle.

3. Simulation results and discussion

In this section, we have discussed the preliminary results obtained through HNMED simulation for Coulomb coupling parameters of Γ (= 10, 100), polarization values Γ_d = (0, 1, 10, 20, 50 and 100) and Debye screening (κ = 1.4, 2 and 3) at constant external force field (F^* = 0.02) of 2D strongly coupled dusty plasmas. We have chosen suitable number of particles (N = 400) in the simulation box with edge length (Lx, Ly). Periodic boundary conditions (PBCs) are applied along with minimum image convection in a simulation box of length L.

There are different conditions to improve the efficient results of thermal conductivity under polarization effects. These conditions include the system size (N), Coulomb coupling (Γ), Debye screening length (κ), system total length (t), simulation time step (dt), and external force field (F^*) strength and polarization values (Γ_d). We have chosen suitable parameters for precise results of thermal conductivity with increasing Γ_d and κ. The simulation data are for a suitable system size (N = 400) with different Γ, which covers the values of strongly coupled plasma states (nonideal gases-liquid to crystalline) at constant force field strength (F^* = 0.02) with varying polarization values Γ_d (0 to 100).

The polarized thermal conductivity of SCCDPs stated here may be scaled as $\lambda_0 = \lambda/nm\omega_p a^2$ (by the plasma frequency). It is demonstrated that the ω_E decreases with increasing κ, $\omega_E \to \omega_p/\sqrt{3}$ as $\kappa \to 0$, for the 3D case [1–10]. Furthermore, for the assessment of appropriate equilibrium range (nearly-equilibrium of external field strength) of 2D HNEMD scheme, various sequences of the polarized thermal conductivity corresponding to an increasing order of an external force field $[\mathbf{F}_e(t) = (0, F_z) \equiv F^* = (F_z)\,(a/J_Q)]$ are planned to measure the nearly equilibrium values of the λ_0. This possible appropriate F^* value gives the steady state λ_0 investigations, which are appropriate for the whole range of plasma states of $\Gamma \equiv (10, 100)$ and $\kappa \equiv (1.4, 3)$.

Figures 1–3 illustrate the normalized polarized thermal conductivity (plasma frequency, ω_p), as a function of Coulomb coupling (system temperature $\equiv 1/\Gamma$) for the cases of $\kappa = 1.4$, 2 and 3, respectively, at constant force F^* with varying six values of polarizations $\Gamma_d = (0, 1, 10, 20, 50$ and $100)$. For three cases, the simulations are performed with setting $N = 400$ for $\kappa = 1.4, 2$ and 3, respectively, at constant

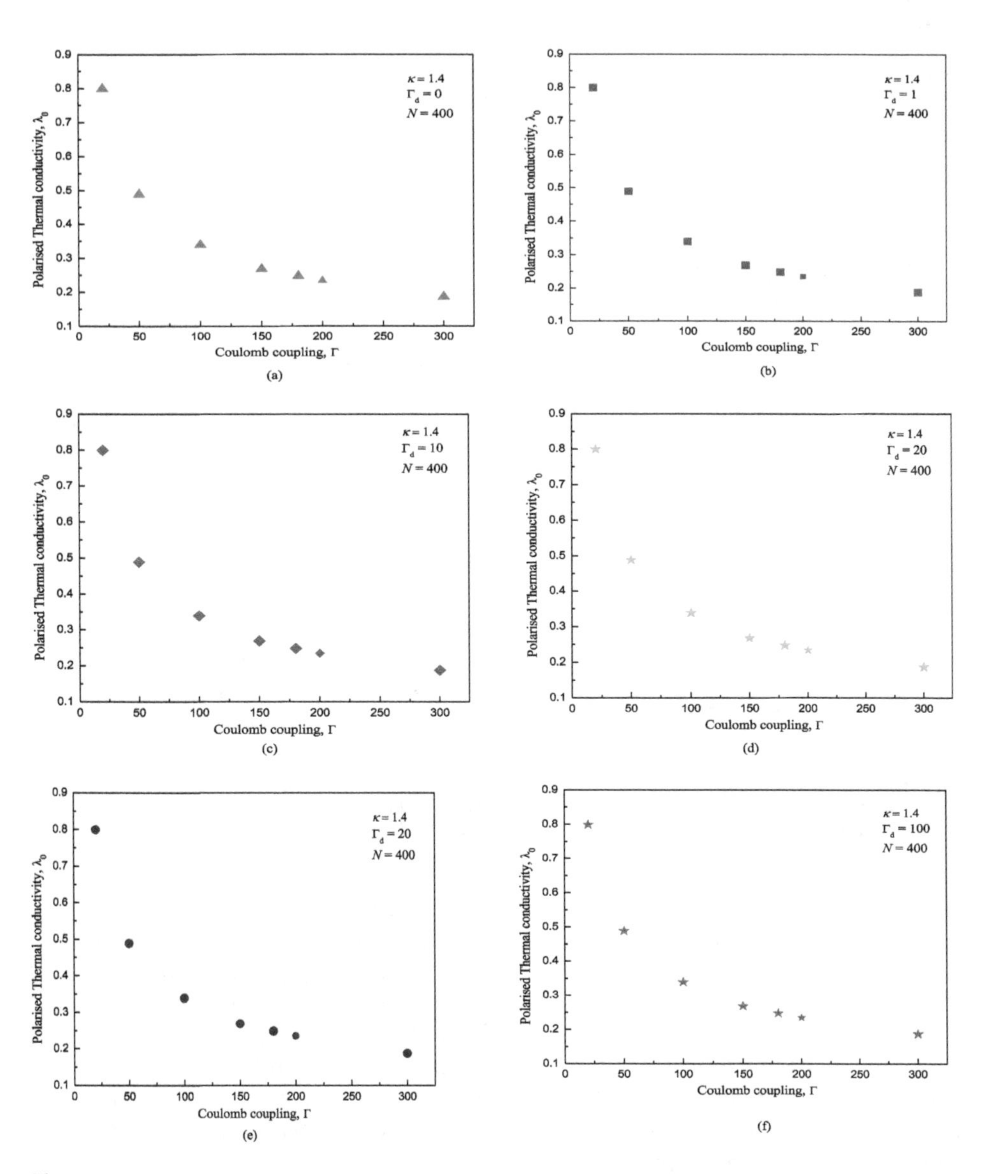

Figure 1.
Variations of thermal conductivity as a function of Coulomb coupling of strongly coupled complex plasma at $\kappa = 1.4$ *with* N = 400 *and (a)* $\Gamma_d = 0$, *(b)* $\Gamma_d = 10$, *(c)* $\Gamma_d = 20$, *(d)* $\Gamma_d = 50$, *(e)* $\Gamma_d = 50$ *and (f)* $\Gamma_d = 100$.

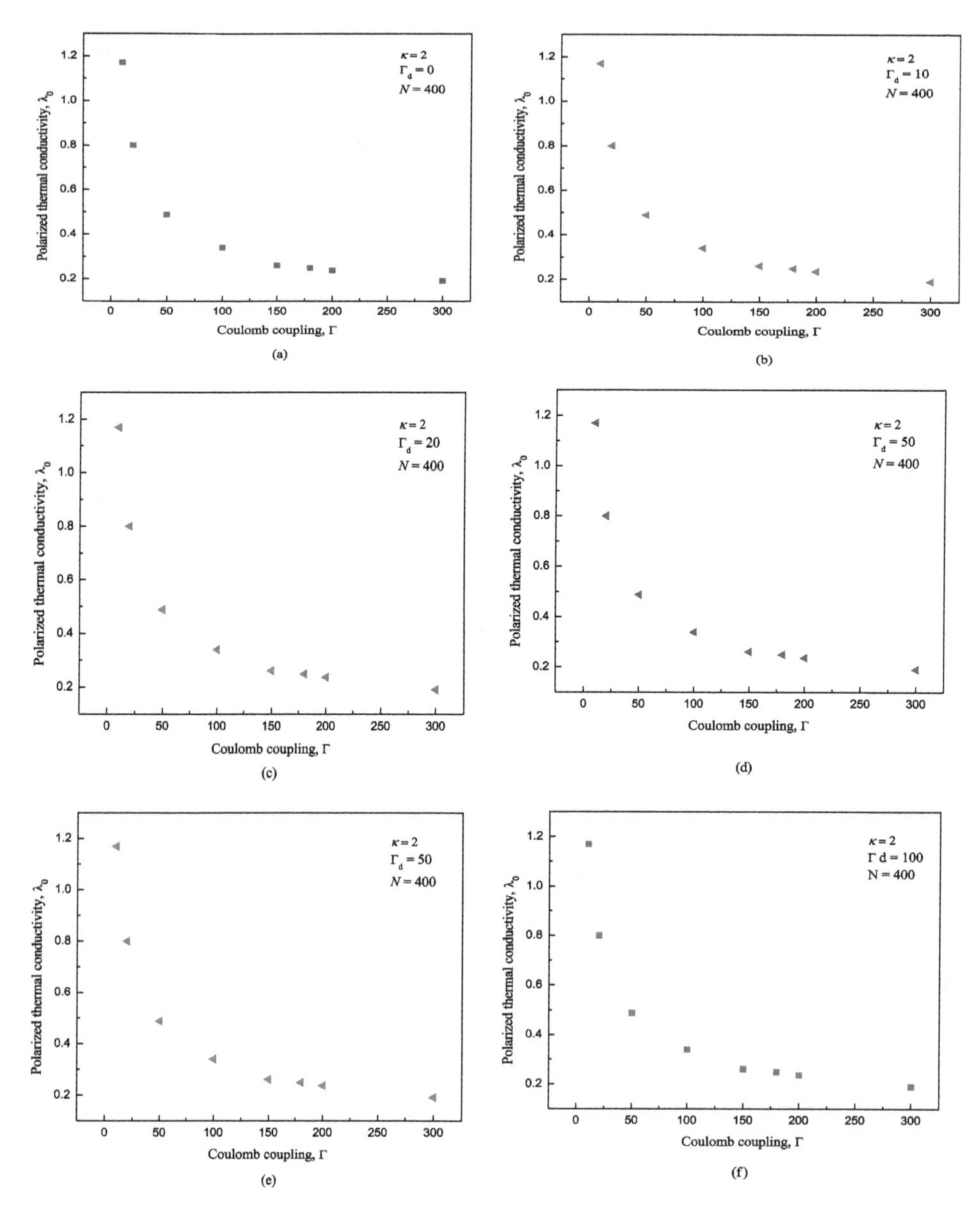

Figure 2.
Variations of thermal conductivity as a function of Coulomb coupling of strongly coupled complex plasma at $\kappa = 2.0$ *with* N = 400 *and (a)* $\Gamma_d = 0$, *(b)* $\Gamma_d = 10$, *(c)* $\Gamma_d = 20$, *(d)* $\Gamma_d = 50$, *(e)* $\Gamma_d = 50$ *and (f)* $\Gamma_d = 100$.

$F^* = 0.02$. Performing HENMD simulations with varying polarizations at constant F^* we examined the efficiency and reliability of the polarized λ_0 measurements. For three cases, we evaluate the six various simulation data sets covering from nonideal state ($\Gamma = 10$) to a strongly coupled liquid regime ($\Gamma = 100$). Figures show that the effects of polarization on the thermal conductivity have no significant changes and it is seen that the thermal conductivity remains constant under varying polarizations. However, the present results of thermal conductivity under varying polarizations are satisfactory agreement with earlier know available numerical data for a complete range of plasma parameters.

Figure 4 shows comparisons with earlier available 2D and 3D numerical data of thermal conductivity with setting $N = 1024$ and $\Gamma_d = 1$. The current results are

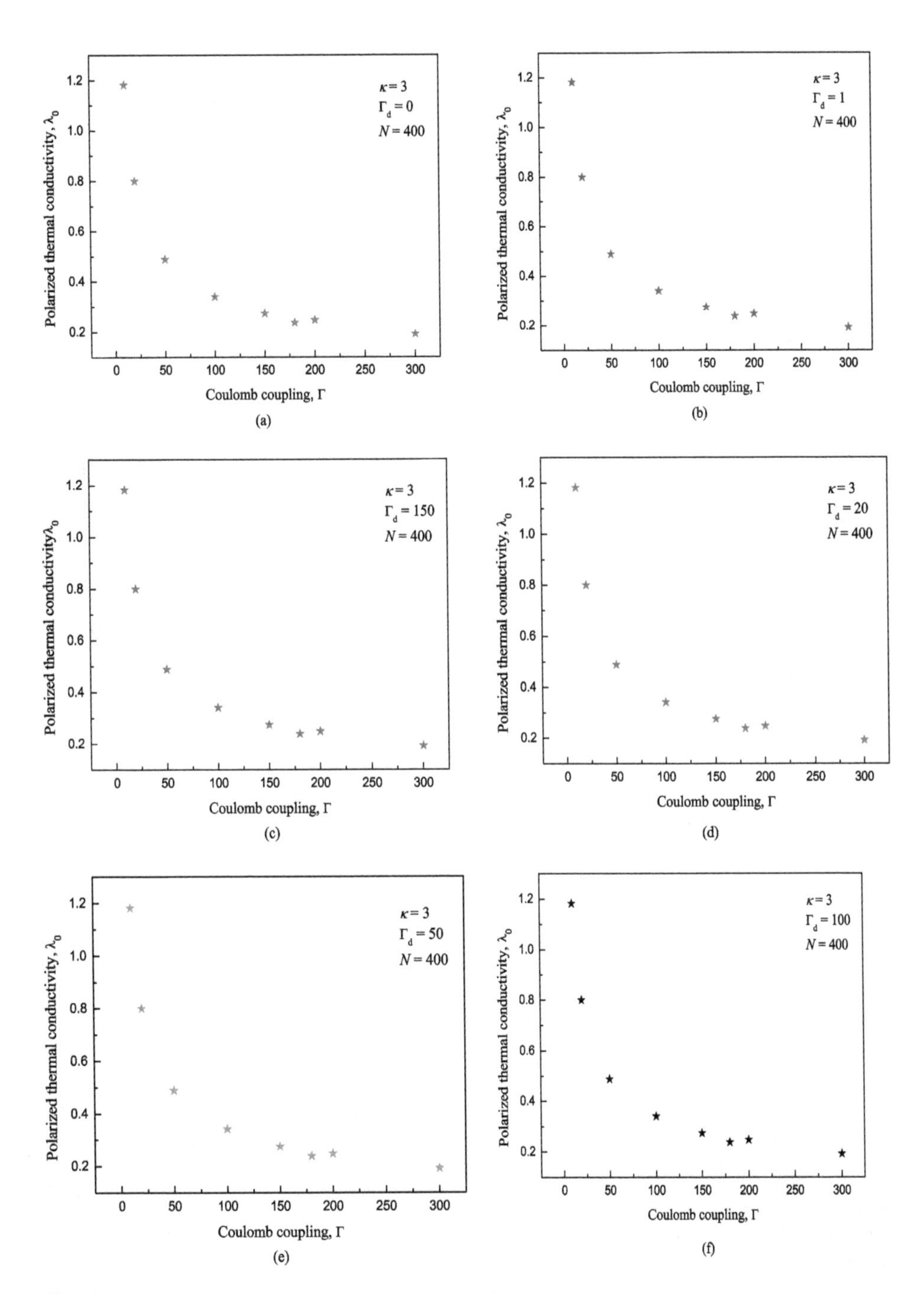

Figure 3.
Variations of thermal conductivity as a function of Coulomb coupling of strongly coupled complex plasma at $\kappa = 3.0$ *with* N = 400 *and (a)* $\Gamma_d = 0$, *(b)* $\Gamma_d = 10$, *(c)* $\Gamma_d = 20$, *(d)* $\Gamma_d = 50$, *(e)* $\Gamma_d = 50$ *and (f)* $\Gamma_d = 100$.

generally excellent agreement for the whole Coulomb coupling range and plot show overall the same behaviors as in the earlier simulation method s of 2D plasma systems. Figure involve the earlier work of 3D HNEMD and HPMD by Shahzad and He [8, 16, 23], EMD investigations of Salin and Caillol [12] and 3D theoretical prediction of Faussurier and Murillo [13] as well as 2D GKR-EMD of Khrustalyov and Vaulina [24].

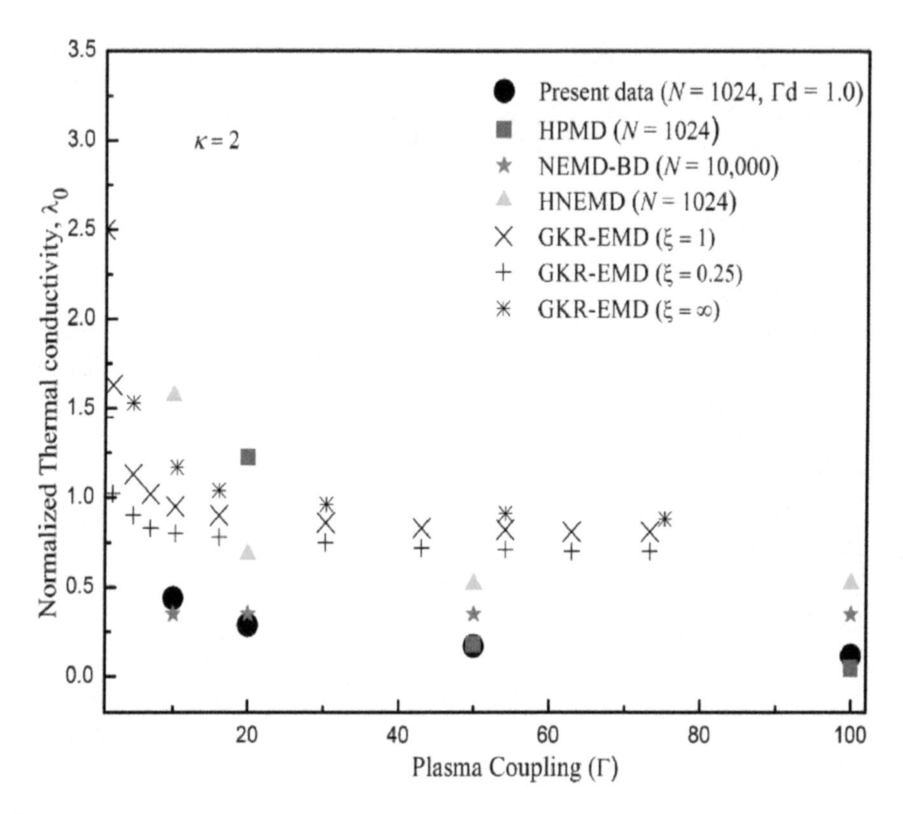

Figure 4.
Comparison of thermal conductivity as a function of Coulomb coupling of strongly coupled complex plasma at $\kappa = 2.0$ *with* $N = 1024$ *and* $\Gamma_d = 1$.

4. Conclusions

The HNEMD scheme is used for the investigation of thermal conductivity under the influence of varying polarization values for various screening lengths κ (= 1.4, 3) and Coulomb couplings Γ (= 10, 100) but constant force field strength. It has been shown that the current HNMED scheme with polarization and earlier HPMD and GKR-EMD techniques have comparable efficiency over the suitable range of plasma parameters, both generating reasonable data of polarized λ_0. New simulation results provide more reliable and ample results for the polarized thermal conductivity for a whole range of (Γ, κ) than previous known simulation data. Thermal conductivity results estimated from this newly developed HNEMD scheme are in well matched with available numerical results generally underpredicting within ±20% - 35%. The current HNEMD data with less statistical noise and best efficiency have been computed by taking different parameters (N, $1/\ \Gamma$, κ and Γd) for the current HNEMD scheme to measure the thermal conductivity of 2D DP systems, at constant $F^* = 0.02$. For future work, the HNEMD method introduced here may easily extend to other physical systems with varying force field with modifications.

Acknowledgements

We are grateful to the National High Performance Computing Center of Xian Jiaotong University and National Advanced Computing Center, National Center of Physics (NCP), Pakistan for allocating computer time to test and run our MD code.

Abbreviations

Strongly coupled complex dusty plasmas	(SCCDP)
Homogeneous non-equilibrium molecular dynamics	(HNEMD)
Coulomb coupling	(Γ)
Debye screening length	(κ)
External force field strength	(F^*)
Homogenous non-equilibrium molecular dynamics	(HNEMD)
Non-equilibrium molecular dynamics	NEMD
Molecular dynamics	(MD)
Inhomogenous non-equilibrium molecular dynamics	(In HNEMD)
Strongly coupled plasma	(SCP)
Equilibrium molecular dynamics	(EMD)
Thermal conductivity	(λ)
Normalized thermal conductivity	(λ_0)
Number of Particles	(N)

Author details

Aamir Shahzad[1*], Madiha Naheed[1], Aadil Mahboob[1], Muhammad Kashif[1], Alina Manzoor[1] and H.E. Maogang[2]

1 Department of Physics, Molecular Modeling and Simulation Laboratory, Government College University Faisalabad (GCUF), Faisalabad, Pakistan

2 Key Laboratory of Thermal Fluid Science and Engineering, Ministry of education (MOE), School of Energy and Power Engineering, Xi'an Jiaotong University, Xi'an, China

*Address all correspondence to: aamirshahzad_8@hotmail.com

References

[1] Shahzad A. Impact of Thermal Conductivity on Energy Technologies. London: IntechOpen; 2018. DOI: 10.5772/intechopen.72471

[2] Shahzad A, Haider SI, He MG, Yang F. Introductory Chapter: A Novel Approach to Compute Thermal Conductivity of Complex System, a Chapter from the Book of Impact of Thermal Conductivity on Energy Technologies. London: IntechOpen; 2018. 3p. DOI: 10.5772/intechopen.75367

[3] Shahzad A, He MG. Interaction contributions in thermal conductivity of three dimensional complex liquids. AIP Conference Proceedings. 2013; 1547: 172-180

[4] Shahzad A, Shakoori MA, He M-G, Bashir S. Sound waves in complex (Dusty) plasmas. In: Computational and Experimental Studies of Acoustic Waves. Rijeka, Croatia: InTech; 2018. DOI: doi.org/10.5772/intechopen.71203

[5] Killian T, Patard T, Pohl T, Rost J. Ultracold neutral plasmas. Physics Reports. 2007; 449: 77. DOI: 10.1016/j.physrep.2007.04.007

[6] Shukla PK, Mamun AA. Introduction to Dusty Plasma Physics. New York: CRC Press; 2015. ISBN 0 7503 0653

[7] Shahzad A, He MG. Thermal conductivity of three-dimensional Yukawa liquids (dusty plasmas). Contributions to Plasma Physics. 2012; 52(8):667. DOI: 10.1002/ctpp.201200002

[8] Shahzad A, He M-G. Numerical experiment of thermal conductivity in two-dimensional Yukawa liquids. Physics of Plasmas. 2015; 22(12):123707 htps://doi.org/10.1063/1.4938275

[9] Shahzad A, Aslam A, He M-G. Equilibrium molecular dynamics simulation of shear viscosity of two-dimensional complex (dusty) plasmas. Radiation Effects and Defects in Solids, 2014; 169(11):931-941 htps://doi.org/10.1080/10420150.2014.968852

[10] Shahzad A, Manzoor A, Wang W, Mahboob A., Kashif M, He M.-G, Dynamic Characteristics of Strongly Coupled Nonideal Plasmas. Arab J Sci Eng. Published: 05 July (2021). https://doi.org/10.1007/s13369-021-05954-4

[11] Shahzad A, Kashif M, Munir T, He M.-G and Tu X, Thermal conductivity analysis of two-dimensional complex plasma liquids and crystals. Phys. Plasmas, 2020; 27:103702. https://doi.org/10.1063/5.0018537

[12] Salin G, Caillol JM. Equilibrium molecular dynamics simulations of the transport coefficients of the Yukawa one component plasma. Physics of Plasmas. 2003; 10 (5):1220-1230

[13] Faussurier G, Murillo MS. Gibbs-Bogolyubov inequality and transport properties for strongly coupled Yukawa fluids. Physical Review E. 2003; 67 (4): 046404

[14] Donkó Z, Hartmann P. The thermal conductivity of strongly coupled Yukawa liquids. Physical Review E. 2004; 69 (1):016405

[15] Shahzad A, He MG. Thermal conductivity calculation of complex (dusty) plasmas. Physics of Plasmas. 2012;19(8):083707

[16] Shahzad A, He MG. Homogeneous nonequilibrium molecular dynamics evaluation of thermal conductivity in 2D Yukawa liquids. International Journal of Thermophysics. 2015, 36 (10 – 11):2565-2576

[17] Shahzad A, He MG, Irfan Haider S, Feng Y. Studies of force field effects on

thermal conductivity of complex plasmas. Physics of Plasmas, 2017; 24(9): 093701

[18] Shahzad A. Impact of Thermal Conductivity on Energy Technologies (IntechOpen, London, 2018), http://dx.doi.org/10.5772/intechopen.72471.

[19] Chen FF. Introduction to plasma physics and controlled fusion. New York: Sppring verlag; 2006, (2nd edition)

[20] Shahzad A, Shakoori MA, He MG, Yang F. Dynamical structure factor of complex plasmas for varying wave vectors. Physics of Plasmas. 2019; 26 023704

[21] Shahzad A, He M-G. Computer Simulation of Complex Plasmas: Molecular Modeling and Elementary Processes in Complex Plasmas. 1st ed. Saarbrücken, Germany: Scholar's Press; 2014. p.170

[22] Ramazanov T. S., Gabdulin A.Zh., Moldabekov Z.A., MD Simulation of Charged Dust Particles With Dipole Moments. IEEE Trans. Plasma Sci. 2015; 43: 4187–4189. 10.1109/TPS.2015.2490282

[23] Shahzad A, Haider S.I., Kashif M., Shifa M.S., Munir T. and He M.-G., Thermal Conductivity of Complex Plasmas Using Novel Evan-Gillan Approach. Commun. Theor. Phys. 2018; 69:704. https://doi.org/10.1088/0253-6102/69/6/704

[24] Khrustalyov YV, Vaulina OS. Numerical simulations of thermal conductivity in dissipative two-dimensional Yukawa systems. Physical Review E. 2012; 85 (4):046405

Permissions

All chapters in this book were first published by InTech Open; hereby published with permission under the Creative Commons Attribution License or equivalent. Every chapter published in this book has been scrutinized by our experts. Their significance has been extensively debated. The topics covered herein carry significant findings which will fuel the growth of the discipline. They may even be implemented as practical applications or may be referred to as a beginning point for another development.

The contributors of this book come from diverse backgrounds, making this book a truly international effort. This book will bring forth new frontiers with its revolutionizing research information and detailed analysis of the nascent developments around the world.

We would like to thank all the contributing authors for lending their expertise to make the book truly unique. They have played a crucial role in the development of this book. Without their invaluable contributions this book wouldn't have been possible. They have made vital efforts to compile up to date information on the varied aspects of this subject to make this book a valuable addition to the collection of many professionals and students.

This book was conceptualized with the vision of imparting up-to-date information and advanced data in this field. To ensure the same, a matchless editorial board was set up. Every individual on the board went through rigorous rounds of assessment to prove their worth. After which they invested a large part of their time researching and compiling the most relevant data for our readers.

The editorial board has been involved in producing this book since its inception. They have spent rigorous hours researching and exploring the diverse topics which have resulted in the successful publishing of this book. They have passed on their knowledge of decades through this book. To expedite this challenging task, the publisher supported the team at every step. A small team of assistant editors was also appointed to further simplify the editing procedure and attain best results for the readers.

Apart from the editorial board, the designing team has also invested a significant amount of their time in understanding the subject and creating the most relevant covers. They scrutinized every image to scout for the most suitable representation of the subject and create an appropriate cover for the book.

The publishing team has been an ardent support to the editorial, designing and production team. Their endless efforts to recruit the best for this project, has resulted in the accomplishment of this book. They are a veteran in the field of academics and their pool of knowledge is as vast as their experience in printing. Their expertise and guidance has proved useful at every step. Their uncompromising quality standards have made this book an exceptional effort. Their encouragement from time to time has been an inspiration for everyone.

The publisher and the editorial board hope that this book will prove to be a valuable piece of knowledge for researchers, students, practitioners and scholars across the globe.

List of Contributors

Tomy Acsente
National Institute for Lasers, Plasma and Radiation Physics, Bucharest, Romania

Lavinia Gabriela Carpen, Bogdan Bita and Gheorghe Dinescu
National Institute for Lasers, Plasma and Radiation Physics, Bucharest, Romania
Faculty of Physics, University of Bucharest, Bucharest, Romania

Elena Matei and Raluca Negrea
National Institute of Materials Physics, Bucharest, Romania

Elodie Bernard and Christian Grisolia
CEA, IRFM, Saint Paul lez Durance, France

Junsheng Zeng
College of Engineering, Peking University, Beijing, China

Heng Li
School of Earth Resources, China University of Geosciences, Wuhan, China

Akio Sanpei
Kyoto Institute of Technology, Kyoto, Japan

Sandip H. Gharat
Department of Chemical Engineering, Shroff S.R. Rotary Institute of Chemical Technology, Ankleshwar, Gujarat, India
Department of Chemical Engineering, Indian Institute of Technology Bombay, Mumbai, India

Jorge Enrique García-Farieta and Rigoberto Ángel Casas Miranda
Universidad Nacional de Colombia, Bogotá, Colombia

Yasuhito Matsubayashi
Advanced Coating Technology Research Center, National Institute of Advanced Industrial Science and Technology, Tsukuba, Ibaraki, Japan

Noritaka Sakakibara, Tsuyohito Ito and Kazuo Terashima
Department of Advanced Materials Science, Graduate School of Frontier Sciences, The University of Tokyo, Chiba, Japan

Hiroo Totsuji
Okayama University (Professor Emeritus), Okayama, Japan
Saginomiya, Nakanoku, Tokyo, Japan

Yoshifumi Saitou
School of Engineering, Utsunomiya University, Utsunomiya, Tochigi, Japan

Osamu Ishihara
Chubu University, Kasugai, Aichi, Japan

Ravinder Goyal and R.P. Sharma
Centre for Energy Studies, Indian Institute of Technology, Delhi, India

Kazuo Takahashi
Faculty of Electrical Engineering and Electronics, Kyoto Institute of Technology, Kyoto, Japan

Viktor Gerasimenko
Institute of Mathematics of the NAS of Ukraine, Kyiv, Ukraine

Sukhmander Singh
Plasma Waves and Electric Propulsion Laboratory, Department of Physics, Central University of Rajasthan, Ajmer, Kishangarh, India

Muhammad Asif Shakoori and Maogang He
Key laboratory of Thermal Fluid Science and Engineering, Ministry of Education (MOE), School of Energy and Power Engineering, Xi'an Jiaotong University, Xi'an, China

Aamir Shahzad
Molecular Modeling and Simulation Laboratory, Department of Physics, Government College University Faisalabad (GCUF), Faisalabad, Pakistan

Misbah Khan
Department of Refrigeration and Cryogenics Engineering, School of Energy and Power Engineering, Xi'an Jiaotong University, Xi'an, China

Yasuaki Hayashi
Kyoto Institute of Technology, Sakyo-ku, Kyoto, Japan

Aamir Shahzad, Madiha Naheed, Aadil Mahboob, Muhammad Kashif and Alina Manzoor
Department of Physics, Molecular Modeling and Simulation Laboratory, Government College University Faisalabad (GCUF), Faisalabad, Pakistan

H.E. Maogang
Key Laboratory of Thermal Fluid Science and Engineering, Ministry of education (MOE), School of Energy and Power Engineering, Xi'an Jiaotong University, Xi'an, China

Index

R

S

T

V

Printed in the USA
CPSIA information can be obtained
at www.ICGtesting.com
JSHW052311231023
50683JS00006BA/63

9 781647 253844